Selected Titles in This Series

- 54 **Casper Goffman, Togo Nishiura, and Daniel Waterman,** Homeomorphisms in analysis, 1997
- 53 **Andreas Kriegl and Peter W. Michor,** The convenient setting of global analysis, 1997
- 52 **V. A. Kozlov, V. G. Maz′ya, and J. Rossmann,** Elliptic boundary value problems in domains with point singularities, 1997
- 51 **Jan Malý and William P. Ziemer,** Fine regularity of solutions of elliptic partial differential equations, 1997
- 50 **Jon Aaronson,** An introduction to infinite ergodic theory, 1997
- 49 **R. E. Showalter,** Monotone operators in Banach space and nonlinear partial differential equations, 1997
- 48 **Paul-Jean Cahen and Jean-Luc Chabert,** Integer-valued polynomials, 1997
- 47 **A. D. Elmendorf, I. Kriz, M. A. Mandell, and J. P. May (with an appendix by M. Cole),** Rings, modules, and algebras in stable homotopy theory, 1997
- 46 **Stephen Lipscomb,** Symmetric inverse semigroups, 1996
- 45 **George M. Bergman and Adam O. Hausknecht,** Cogroups and co-rings in categories of associative rings, 1996
- 44 **J. Amorós, M. Burger, K. Corlette, D. Kotschick, and D. Toledo,** Fundamental groups of compact Kähler manifolds, 1996
- 43 **James E. Humphreys,** Conjugacy classes in semisimple algebraic groups, 1995
- 42 **Ralph Freese, Jaroslav Ježek, and J. B. Nation,** Free lattices, 1995
- 41 **Hal L. Smith,** Monotone dynamical systems: an introduction to the theory of competitive and cooperative systems, 1995
- 40.2 **Daniel Gorenstein, Richard Lyons, and Ronald Solomon,** The classification of the finite simple groups, number 2, 1995
- 40.1 **Daniel Gorenstein, Richard Lyons, and Ronald Solomon,** The classification of the finite simple groups, number 1, 1994
- 39 **Sigurdur Helgason,** Geometric analysis on symmetric spaces, 1994
- 38 **Guy David and Stephen Semmes,** Analysis of and on uniformly rectifiable sets, 1993
- 37 **Leonard Lewin, Editor,** Structural properties of polylogarithms, 1991
- 36 **John B. Conway,** The theory of subnormal operators, 1991
- 35 **Shreeram S. Abhyankar,** Algebraic geometry for scientists and engineers, 1990
- 34 **Victor Isakov,** Inverse source problems, 1990
- 33 **Vladimir G. Berkovich,** Spectral theory and analytic geometry over non-Archimedean fields, 1990
- 32 **Howard Jacobowitz,** An introduction to CR structures, 1990
- 31 **Paul J. Sally, Jr. and David A. Vogan, Jr., Editors,** Representation theory and harmonic analysis on semisimple Lie groups, 1989
- 30 **Thomas W. Cusick and Mary E. Flahive,** The Markoff and Lagrange spectra, 1989
- 29 **Alan L. T. Paterson,** Amenability, 1988
- 28 **Richard Beals, Percy Deift, and Carlos Tomei,** Direct and inverse scattering on the line, 1988
- 27 **Nathan J. Fine,** Basic hypergeometric series and applications, 1988
- 26 **Hari Bercovici,** Operator theory and arithmetic in H^∞, 1988
- 25 **Jack K. Hale,** Asymptotic behavior of dissipative systems, 1988
- 24 **Lance W. Small, Editor,** Noetherian rings and their applications, 1987
- 23 **E. H. Rothe,** Introduction to various aspects of degree theory in Banach spaces, 1986
- 22 **Michael E. Taylor,** Noncommutative harmonic analysis, 1986
- 21 **Albert Baernstein, David Drasin, Peter Duren, and Albert Marden, Editors,** The Bieberbach conjecture: Proceedings of the symposium on the occasion of the proof, 1986
- 20 **Kenneth R. Goodearl,** Partially ordered abelian groups with interpolation, 1986

(*Continued in the back of this publication*)

Mathematical
Surveys
and
Monographs

Volume 54

Homeomorphisms in Analysis

Casper Goffman
Togo Nishiura
Daniel Waterman

American Mathematical Society

1991 *Mathematics Subject Classification.* Primary 26–02, 28–02, 42–02, 54–02; Secondary 30–02, 40–02, 49–02.

ABSTRACT. The interplay of two of the main branches of Mathematics, topology and real analysis, is presented. In the past, homeomorphisms have appeared in books on analysis in incidental or casual fashion. The presentation in this book looks deeply at the effect that homeomorphisms have, from the point of view of analysis, on Lebesgue measurability, Baire classes of functions, the derivative function, C^n and C^∞ functions, the Blumberg theorem, bounded variation in the sense of Cesari, and various theorems on Fourier series and generalized variation of functions. Unified developments of many results discovered over the past 50 years are given in book form for the first time. Among them are the Maximoff theorem on the characterization of the derivative function, various results on improvements of convergence or preservation of convergence of Fourier series by change of variables, and approximations of one-to-one mappings in higher dimensions by homeomorphisms. Included here is the von Neumann theorem on homeomorphisms of Lebesgue measure.

Library of Congress Cataloging-in-Publication Data

Goffman, Casper, 1913–
Homeomorphisms in analysis / Casper Goffman, Togo Nishiura, Daniel Waterman.
p. cm. — (Mathematical surveys and monographs, ISSN 0076-5376 ; v. 54)
Includes bibliographical references (p. –) and index.
ISBN 0-8218-0614-9 (alk. paper)
1. Homeomorphisms. 2. Mathematical analysis. I. Nishiura, Togo, 1931– . II. Waterman, Daniel. III. Title. IV. Series: Mathematical surveys and monographs ; no. 54.
TA355.T47 1998
620.3—dc21 97-25854
CIP

∞ The paper used in this book is acid-free and falls within the guidelines
established to ensure permanence and durability.
Visit the AMS homepage at URL: http://www.ams.org/

10 9 8 7 6 5 4 3 2 1 02 01 00 99 98 97

ERRATUM TO "HOMEOMORPHISMS IN ANALYSIS"

BY CASPER GOFFMAN, TOGO NISHIURA, AND DANIEL WATERMAN

An omission appears in a display on page 113. The display on line 8 from the bottom of the page should appear as follows

$$|f(x)| = \lim_{n\to\infty} |\sigma_n(x; f)| \leq \sup_n \|S_n(f)\|_\infty \leq \|f\|_{\mathrm{U}}.$$

Contents

Preface

As the title of the book indicates, there is an interesting role played by homeomorphisms in analysis. The maturing of topology in the beginning of the 20th Century changed the outlook of analysis in many ways. Earlier, the notions of curve and surface were intuitively obvious to mathematicians. In differential geometry, a curve was always discussed in terms of a specific representation with high orders of differentiability and, consequently, notions of length of curves and area of surfaces had very nice integral formulas. And in the study of Fourier series, the functions were conveniently assumed to be of bounded variation for the purposes of convergence.

But, with the advent of the then modern analysis and topology, numerous questions came to the fore. Should curves and surfaces have high order of differentiability as part of their definitions? Is there a reasonable theory of Fourier series for functions that are not of bounded variation? These questions originate from the study of functionals, both nonlinear and linear. A functional is a function whose domain is a collection of functions. So, in particular, length, differentiation, Fourier coefficients and series (which are sources of many of the problems that are treated in this book) are examples of nonlinear and linear functionals. Modern analysis forced natural expansions of the classical domains of these functionals to larger classes of functions. For example, the domain of the length functional is expanded to include functions for which such things as the integral formula for length do not apply. For such functions, the question becomes the existence of nicely differentiable representations of rectifiable curves. That is, are there homeomorphisms of the parameter domain that result in differentiable representations? Here, arc length representation turned out to be very useful. This method of change in representation by means of homeomorphisms was formalized by Lebesgue in his paper [**92**].

Topology deals with continuous functions. But functions in analysis need not be continuous. One might believe that homeomorphic changes of the domains of discontinuous functions would not have much interest. It will be shown that such beliefs are unfounded. The theme of the effects on the analytical behavior of a function f as well as the consequent functional behavior that result from self-homeomorphisms of the domain of the function f will be developed. A second theme is the approximation, with the aid of measures, by homeomorphisms of not necessarily continuous, one-to-one transformations. Clearly, it is the combination of analysis and topology that makes this theme possible. These two themes are presented in three parts.

The one dimensional case

Chapter 1 eases the reader into the subject by considering the simplest situation of subsets of $\mathbb{R}^1$, that is, characteristic functions of sets. The reader is introduced to the definition of Lebesgue's equivalence of functions that first appeared in [92]. The major theorem in the context of subsets of $\mathbb{R}^1$ is Gorman's theorem.

In Chapter 2, functions in the Baire class 1 are investigated. From elementary topology, the Baire class 1 is invariant under self-homeomorphisms of the domains of the functions in the class. Consequently, the questions investigated here are more related to analytical facts. Recall that a measurable function f on an interval I is known to be almost everywhere equal a function in the Baire class 2 but not necessarily in the Baire class 1. It is shown that if f is measurable and assumes only finitely many values then there is a homeomorphism h such that $f \circ h$ is equal almost everywhere to a function in the Baire class 1. However, if f assumes infinitely many values, this need not be true. But on the positive side, it is shown that if f is such that $f \circ h$ is measurable for every homeomorphism h, then there is an h such that $f \circ h$ is equal almost everywhere to a function in the Baire class 1.

Attention is turned next to the derivative operator. Chapter 3 deals with differentiable functions, that is, differentiable everywhere. For a function f defined on an interval I, necessary and sufficient conditions are given under which there is a homeomorphism h such that $f \circ h$ is respectively (i) everywhere differentiable with bounded derivative, (ii) continuously differentiable, (iii) n times continuously differentiable, and (iv) differentiable infinitely often. For the first of these, the condition turns out to be deceptively simple, namely, that the function be continuous and have finite total variation. The second one uses a measure theoretic necessary and sufficient condition, which turns out to be equivalent to a Hausdorff packing condition which is associated with the rarefaction index. This Hausdorff packing condition also plays a role in (iii) and (iv).

The fourth chapter concerns the image of the derivative operator, that is, the derivative function. Every derivative is known to be in the Baire class 1 and to have the Darboux property (the intermediate value property). A proof is given of the fact (Maximoff's theorem) that for functions f in the Baire class 1 having the Darboux property there are self-homeomorphisms h such that $f \circ h$ is a derivative. The use of interval functions will be shown to be natural in the development.

Mappings and measures on $\mathbb{R}^n$

The second part is concerned with one-to-one transformations defined on intervals in n-space. Observe that one-to-one, order-preserving transformations of a closed interval in $\mathbb{R}^1$ onto itself are automatically homeomorphisms, a not very interesting situation from an analysis viewpoint.

Chapter 5 is an investigation of functions f in the domain of the nonparametric surface area functional. The nonparametric surface area of a real-valued function f is the area of its graph where, of course, area means n-dimensional area. It is well known that a definition of area is very elusive when one drops the high differentiability requirement on the function f. As Lebesgue observed, it is the useful lower semicontinuity property of this functional that is important in many parts of analysis, especially in the direct method of the calculus of variations. By a lower semicontinuous functional F on a function space $\mathcal{F}$ endowed with a metric, we mean the usual notion of lower semicontinuity in metric spaces. By choosing the appropriate mode of convergence in the function space which constitutes the domain of the area functional, one introduces natural classes of functions. Here, the functions are the linearly continuous ones. A function of n variables is linearly continuous in the narrowest sense if it is continuous as a function of each of its variables separately (or, partially continuous). Of course, since the emphasis is on analysis, this partial continuity requirement must hold almost everywhere in the orthogonal coordinate spaces. The linearly continuous functions that are studied are those whose nonparametric surface areas are finite. It is shown that linear continuity of f is not preserved under changes of representation by means of arbitrary self-homeomorphisms of the domain of f. But the class of linearly continuous functions whose nonparametric areas are finite is preserved under bi-Lipschitzian self-homeomorphisms. This is achieved by a characterization of this class in terms of approximations by approximately continuous functions, the class of functions determined by the density topology on $\mathbb{R}^n$. The fact that the class of approximately continuous functions is invariant under bi-Lipschitzian homeomorphisms will be given here.

The theme of Chapter 6 is approximations of one-to-one transformations by homeomorphisms whenever $n \geq 2$. First, counter to all intuition, it is shown that each one-to-one, onto transformation (continuous or not) together with its inverse can be simultaneously approximated by a homeomorphism on sets of arbitrarily large relative outer Lebesgue measure. Under suitably mild assumptions on the transformations, there is a natural metric associated with these approximations. Indeed, the approximation may be taken to be an interpolation. Necessary and sufficient conditions are given for a one-to-one, onto transformation to be in the closure under this metric of the collection of all homeomorphisms. It then results that every one-to-one transformation T of an open n-cube of $\mathbb{R}^n$ onto itself for which both T and T^{-1} are measurable can be approximated by a homeomorphism in this metric. Finally, one-to-one bimeasurable transformations T of a closed n-cube I_n onto an open m-cube I_m, where $1 \leq n < m$, are shown to be approximated arbitrarily closely by homeomorphisms (of course, homeomorphisms are not onto when $n < m$).

The next chapter is a presentation of a not so well known theorem of J. von Neumann that characterizes those measures μ on a closed n-cube that are formed from Lebesgue measure λ upon composition with a self-homeomorphism h of the n-cube (that is, the measure $\mu = h_\# \lambda$ defined by $h_\# \lambda(E) = \lambda(h^{-1}[E])$ for Borel sets E). The theorem proved here is a sharpening of von Neumann's theorem into a deformation version.

The second part should not end without a discussion of the case $n = 1$. The final chapter, Chapter 8, addresses the connection between Blumberg's theorem when applied to one-to-one functions of a closed interval I in $\mathbb{R}^1$ onto itself and self-homeomorphisms of dense subsets of I. Blumberg's theorem says that each real-valued function f defined on I has a dense subset D of I such that the restricted function $f|D$ is continuous. Conditions on f are given under which $f|D$ is a homeomorphism of D onto $f[D]$ with D and $f[D]$ dense in I.

Fourier series

The third part is concerned with the effect on the Fourier series of a function by a change of variable, i.e., the composition of a function f with a self-homeomorphism of $[-\pi, \pi]$.

Chapter 9 deals with the question of the extent to which the behavior of the Fourier series of a function can be improved by such a composition. The basic result here is that of Pál and Bohr which shows that, for each continuous real-valued function f, there is a change of variable h such that the Fourier series of $f \circ h$ converges uniformly. This theorem was proven by complex-variable methods, but a real-variable proof was supplied by Kahane and Katznelson, who also demonstrated this result for complex-valued functions f. The Pál-Bohr theorem was generalized by Jurkat and Waterman who showed the stronger result that, for each real-valued function f, there is a change of variable h such that the function conjugate to $f \circ h$ is of bounded variation. In the same vein, Lusin raised the question: For each real-valued function f, is there a change of variable h such that the Fourier series of $f \circ h$ converges absolutely? Olevskii has answered this in the negative.

A property is said to be preserved under change of variable if $f \circ h$ has that property for every change of variable h. Chapter 10 deals with the problem of characterizing the classes of functions in which everywhere or uniform convergence of Fourier series is preserved. The class of continuous functions for the first property of everywhere convergence was characterized by Goffman and Waterman; and, the class for the second property of uniform convergence was solved by Baernstein and Waterman. The characterizing condition of Goffman and Waterman is certainly a property that is preserved under change of variables. For the class of regulated functions, the condition for preserving the property of everywhere convergence was shown by Pierce and Waterman to be the condition of Goffman and Waterman. Even more, for integrable functions, a necessary and sufficient condition for preserving the property of everywhere convergence is that the function be equal, except on a set of absolute measure zero, to a function satisfying the condition of Goffman and Waterman.

Chapter 11 is concerned with the preservation of the rate of growth of the Fourier coefficients of a function under change of variable. Hadamard introduced the functions of bounded deviation, the class of functions whose restriction to every interval have Fourier coefficients with order $O(\frac{1}{n})$. This class of functions has the functions of bounded variation as a proper subset. It may be shown that functions that remain in this class under every change of variable are those that are equal to functions of bounded variation except on a set of absolute measure zero. For general rates of growth the analogous results are valid with the ordinary variation replaced by Chanturiya variation.

Finally, to simplify the exposition, other discussions have been placed in the Appendix. And, in the remarks section of some of the chapters, there are brief discussions of the historical background of some of the theorems and other relevant matters.

The second author thanks Dickinson College, especially Professors Craig Miller and David Reed and the Department of Mathematics and Computer Science, for providing help and resources during the final stage of the manuscript preparation.

Part 1

The One Dimensional Case

CHAPTER 1

Subsets of $\mathbb{R}$

In this chapter, the first of four devoted to self-homeomorphisms of $\mathbb{R}$, the simplest of functions are studied, namely, the characteristic functions of subsets of $\mathbb{R}$. Of course, this is the study of the changes in the analytic properties of sets under homeomorphisms of $\mathbb{R}$. By further restricting the homeomorphisms to be the identity outside of some open interval, the discussion will in effect be about order-preserving self-homeomorphisms of some open interval (a, b). The notion of Lebesgue equivalence that has been mentioned in the Preface is introduced and important classes of sets are discussed in the context of this equivalence relation. The main result is Gorman's theorem (Theorem 1.3).

1.1. Equivalence classes

The two equivalences in analysis that interest us will be defined next. One involves topologies and the other involves measures. The definitions will be given in a general setting even though we shall be dealing mainly with the space $\mathbb{R}^n$ in the book.

1.1.1. Lebesgue equivalence. We generalize the definition given by Lebesgue [**92**]. For a topological space X, the group of all self-homeomorphisms of X onto itself will be denoted by $\mathcal{H}(X)$.

DEFINITION 1.1. Let X be a topological space and let Y be a set. If f and g are functions from X into Y, then f is said to be **Lebesgue equivalent to** g if there is a homeomorphism h in $\mathcal{H}(X)$ such that $f \circ h = g$. This relation will be denoted as $f \sim g$. When f and g are characteristic functions of sets S and T, the notation $S \sim T$ will be used.

Clearly, Lebesgue equivalence is an equivalence relation on the class $\mathcal{F}(X, Y)$ of all functions from X into Y. When Y is also a topological space, it is obvious that the class of all continuous functions is invariant under Lebesgue equivalence. In our applications, the space X is often the usual Euclidean space $\mathbb{R}^n$ and Y is $\mathbb{R}$. In this context, it is easy to see that the usual Baire classes of functions and the Borel classes of functions are also invariant under Lebesgue equivalence.

It is known that every Borel set is an analytic (Suslin) set. Analytic subsets of $\mathbb{R}^n$ have been characterized as those sets that are the continuous images of the set of irrational numbers with the usual topology. Clearly, analytic sets are invariant under Lebesgue equivalence. Also, analytic sets are Lebesgue measurable.

1.1.2. Equal almost everywhere. Let X be a measure space with measure μ. The collection of all sets with μ-measure equal to 0 will be denoted by $\mathcal{N}(X, \mu)$.

DEFINITION 1.2. Let (X, μ) be a measure space and let Y be a set. If f and g are functions from X into Y, then f is said to be **equal to g μ-almost everywhere** if $\{ x : f(x) \neq g(x) \} \in \mathcal{N}(X, \mu)$. This relation will be written as $f = g$ μ-ae.

We shall denote the Lebesgue measure on $\mathbb{R}^n$ by λ_n. The subscript will be dropped whenever there is only one dimension n under discussion. It is well-known that Lebesgue measurable functions are equal λ-almost everywhere to a function of Baire class 2. Chapter 2 will deal with a combination of this fact and Lebesgue equivalence.

1.2. Lebesgue equivalence of sets

For the remainder of the chapter we shall restrict ourselves to an open interval I of $\mathbb{R}$. Unless stated otherwise, homeomorphisms are understood to preserve the order of $\mathbb{R}$. We give some simple examples of properties that are invariant and that are not invariant under Lebesgue equivalence.

1.2.1. Countable sets. If S is a countable and infinite set and T is Lebesgue equivalent to S, then T is also countable and infinite. However, not all countable and infinite sets are Lebesgue equivalent to a fixed one, for if S is dense in I and T is not dense in I then S and T are not Lebesgue equivalent.

1.2.2. Nowhere dense sets. If T is Lebesgue equivalent to a nowhere dense set S, then T is nowhere dense. However, not every pair of nowhere dense sets are Lebesgue equivalent. In particular, a nowhere dense set may be finite, infinite and countable, or uncountable.

It follows that sets of first category are invariant under Lebesgue equivalence.

1.2.3. Porous sets. Let us first define the notion of porous sets in $\mathbb{R}$. The length of an interval J in $\mathbb{R}$ will be denoted by $|J|$. Let $S \subset \mathbb{R}$ and $x \in S$. For an interval J centered at x, let $\alpha(J)$ be the supremum of $|L|$ for all intervals L for which $L \subset J$ and $L \cap S = \emptyset$. Then S is said to be **porous at x** if $\limsup_{|J| \to 0} \frac{\alpha(J)}{|J|} > 0$. And, S is **porous** if it is porous at every x in S.

We give an example of a porous set which is Lebesgue equivalent to a nonporous set. Let S consist of the point 0 and the open intervals $I_n = \left(\frac{1}{2^n}, \frac{1}{2^n} + \frac{1}{2^{n+1}}\right)$, $n = 1, 2, \ldots$, and their reflections to the left of 0; and, let T consist of the point 0 and the intervals $J_n = \left(\frac{1}{n}, \frac{1}{n} + \frac{1}{n^2}\right)$ and their reflections to the left of 0. Then S is porous at 0 and T is not porous at 0. Obviously, S and T are Lebesgue equivalent.

1.2.4. Borel and analytic sets. As noted earlier, it is quite obvious that the class of Borel sets of $\mathbb{R}$ is invariant under Lebesgue equivalence. The analytic sets of $\mathbb{R}$ are those sets which are the continuous images of the space of irrational numbers. Consequently, the class of analytic sets of $\mathbb{R}$ is also invariant under Lebesgue equivalence.

1.2.5. Lebesgue measurable sets. It is well-known that the Cantor ternary set can be changed into a set of positive Lebesgue measure by means of a homeomorphism of $\mathbb{R}$. Consequently, some subset of the Cantor ternary set is changed by this homeomorphism into a non-Lebesgue measurable set. And, the inverse of this homeomorphism will change a non-Lebesgue measurable set into a measurable one. Hence the collection of Lebesgue measurable sets is not invariant under Lebesgue equivalence.

1.3. Density topology

We now consider the density topology on the real line. (See the Appendix.) A subset S of the real line is said to be **density-open** if it is measurable and if the density of S is 1 at each point x of S. The collection of all such sets is a topology called the **density topology** on $\mathbb{R}$.

1.3.1. Density-open sets. The above example of a nonporous set whose image under a homeomorphism is porous also serves as an example of an open set in the density topology whose image under a homeomorphism is not in this topology. Thus the density topology is not invariant under Lebesgue equivalence.

Perhaps the definitive result in this connection is the one by Gorman [**73**]. The statement of the theorem will require us to define the notion of a bilateral condensation set, c-set for short. A subset S of $\mathbb{R}$ is said to be a **c-set** if for each x in S and for each positive number ε the point x is a condensation point of the sets $(x-\varepsilon, x)\cap S$ and $(x, x+\varepsilon)\cap S$. It follows immediately from this definition that c-sets are invariant under Lebesgue equivalence. Clearly, the union of a collection of c-sets is again a c-set. Also, it is an easy exercise to construct two c-sets whose intersection is not a c-set. So the collection of c-sets do not form a topology unlike the collection of density-open sets.

THEOREM 1.3 (GORMAN). *If S is an F_σ c-set that is contained in $I=(0,1)$, then there is a homeomorphism h of I onto itself such that $h[S]$ is an F_σ, density-open set. Moreover, if S is dense in $(0,1)$, then h may be chosen so that $\lambda(h[S])=1$.*

Since the proof contains some interesting techniques, we shall give it in detail. Observe first that if $h\colon \mathbb{R}\to\mathbb{R}$ is a homeomorphism and if x is an interior point of a subset E of $\mathbb{R}$, then $h(x)$ is a point of density 1 of $h[E]$. Consequently, in Gorman's theorem, the set S may be assumed further to be of the first category since this case yields the general one. The proof of the modified theorem will rely on two lemmas which will be developed next.

DEFINITION 1.4. An increasing sequence G_i, $i=0,1,2,\dots$, of nowhere dense, perfect subsets of $\mathbb{R}$ is said to be a **Gorman sequence** if there is a bounded interval (a,b) such that

(1) $G_i\subset(a,b)$ for every i,
(2) G_i meets every component of $(a,b)\setminus G_{i-1}$ for each positive integer i,
(3) for each positive integer i, the end points of each component J of $(a,b)\setminus G_{i-1}$ are limit points of $G_i\cap J$ except for a and b.

A set G is said to have a **Gorman decomposition** if there exists a Gorman sequence G_i, $i=0,1,2,\dots$, such that $G=\bigcup_{i=0}^{\infty}G_i$.

PROPOSITION 1.5. *A subset G of $\mathbb{R}$ has a Gorman decomposition if and only if G is a nonempty, bounded, F_σ c-set of the first category.*

The proof of the proposition is straightforward because every uncountable Borel set contains a topological copy of the Cantor ternary set.

We now take advantage of the fact that the natural order relation on $\mathbb{R}$ induces an order relation on each subset G of $\mathbb{R}$. It is well-known that there corresponds to each pair of nonempty, bounded, nowhere dense, perfect sets G and G' an order-preserving, one-to-one, onto mapping $h\colon G\to G'$. Moreover, this mapping is a topological homeomorphism that has an extension to a self-homeomorphism of $\mathbb{R}$.

With the aid of these observations, it is a simple matter to show that two Gorman sequences G_i, $i = 0, 1, 2, \ldots$, and G'_i, $i = 0, 1, 2, \ldots$, lead naturally to an order-preserving, one-to-one mapping h of $G = \bigcup_{i=0}^{\infty} G_i$ onto $G' = \bigcup_{i=0}^{\infty} G'_i$ with the property that $h[G_i] = G'_i$. From the definition of Gorman sequences, it follows that $h\colon G \to G'$ is a homeomorphism of the topological spaces G and G' endowed with the order topologies, which are also the same as the relative Euclidean topologies on G and G'. But examples will show that every such homeomorphism constructed from these sequences need not be extendable to the intervals $[a, b]$ and $[a', b']$ that contain G and G' respectively. Indeed, the topological structures of $[a, b] \setminus G$ and $[a', b'] \setminus G'$ can be quite different. In view of this, the next two lemmas are quite nice.

LEMMA 1.6. *Let G and G' each possess Gorman decompositions. If G and G' are respectively dense subsets of $[a, b]$ and $[a', b']$, then there is a homeomorphism $h\colon [a, b] \to [a', b']$ such that $h[G] = G'$.*

PROOF. It has been already observed that there exists an order-preserving homeomorphism $h\colon G \to G'$ induced by Gorman sequences for G and G'. When G is dense in $[a, b]$ and G' is dense in $[a', b']$, the homeomorphism will have a unique extension to a homeomorphism $h\colon [a, b] \to [a', b']$.

LEMMA 1.7. *Suppose that both G and G' possess Gorman decompositions. If G and G' are nowhere dense subsets of (a, b) and (a', b') respectively, and if $\{ a, b \} \subset \overline{G}$ and $\{ a', b' \} \subset \overline{G'}$, then there is a homeomorphism $h\colon [a, b] \to [a', b']$ such that $h[G] = G'$.*

PROOF. Since G and G' are nowhere dense c-sets, the respective collections of the nondegenerate components of $[a, b] \setminus G$ and of $[a', b'] \setminus G'$ consist of infinitely many mutually disjoint closed intervals and the union of these collections are dense in their respective intervals $[a, b]$ and $[a', b']$. Let $\{ J_i : i = 1, 2, \ldots \}$ and $\{ J'_i : i = 1, 2, \ldots \}$ be well-orderings of these collections. Let G_i, $i = 0, 1, 2, \ldots$, be a Gorman decomposition for G. Observe that each J_{i_1} is contained in a unique component of $[a, b] \setminus G_i$. And, if J_{i_1} and J_{i_2} are distinct, then there is an i_3 such that J_{i_1} and J_{i_2} are contained in distinct components of $[a, b] \setminus G_{i_3}$. Noting that subsequences of Gorman decompositions of G are again Gorman decompositions of G, we may assume further that the Gorman decomposition of G has the property that, for each i, if $i_1 < i_2 < i$, then J_{i_1} and J_{i_2} are contained in distinct components of $[a, b] \setminus G_i$. Let G'_i, $i = 0, 1, 2, \ldots$, be an analogous Gorman decomposition of G'.

Let us now construct a sequence h_i, $i = 0, 1, 2, \ldots$, of order-preserving homeomorphisms of $[a, b]$ onto $[a', b']$ that converges pointwise on the set $G \cup \bigcup_{j=1}^{\infty} J_j$ to a function h that is one-to-one and onto the set $G' \cup \bigcup_{k=1}^{\infty} J'_k$ and satisfies $h[G] = G'$. The construction will go through 2-phase cycles.

Let h_0 be an order-preserving homeomorphism such that $h_0[G_0] = G'_0$. For the inductive construction, we let $j_0 = 0$, $k_0 = 0$, $i_0 = j_0 + k_0 + 0$, $I_0 = \emptyset$, $H_0 = G_{i_0}$ and $H'_0 = G'_{i_0}$.

We shall now construct h_1. With

$$j_1 = \min \{ i : i \neq j_0 \},$$

let I'_1 be the component of $[a', b'] \setminus G'_{i_0}$ that contains $h_0[J_{j_1}]$. Define

$$k_1 = \min \{ i : J'_i \subset I'_1 \quad \text{and} \quad i \neq k_0 \}.$$

Then, with

$$i_1 = (j_0 + j_1) + (k_0 + k_1) + 4,$$
$$H_1 = G_{i_1} \cup \bigcup_{m \le 1} J_{j_m} \quad \text{and} \quad H'_1 = G'_{i_1} \cup \bigcup_{m \le 1} J'_{k_m},$$

construct a homeomorphism $h_1 \colon [a,b] \to [a',b']$ that agrees with h_0 on H_0 and that satisfies $h_1[H_1] = H'_1$. Thus the first of the sequence of 2-phase cycles is now constructed.

To construct h_2, let

$$k_2 = \min\{\, i : i \neq k_0,\ i \neq k_1 \,\}.$$

Note that $k_2 < i_1$. Define I_2 to be the component of $[a,b] \setminus G_{i_1}$ that contains $h_1^{-1}[J'_{k_2}]$, and let

$$j_2 = \min\{\, i : J_i \subset I_2 \quad \text{and} \quad i \neq j_0,\ i \neq j_1 \,\}.$$

Note that $j_2 < i_1$. With

$$i_2 = (j_0 + j_1 + j_2) + (k_0 + k_1 + k_2) + 6,$$
$$H_2 = G_{i_2} \cup \bigcup_{m \le 2} J_{j_m} \quad \text{and} \quad H'_2 = G'_{i_2} \cup \bigcup_{m \le 2} J'_{k_m},$$

select a homeomorphism $h_2 \colon [a,b] \to [a',b']$ that agrees with h_1 on H_1 and satisfies $h_2[H_2] = H'_2$.

For $i > 1$, the construction of h_{2i-1} is similar to that of h_1 and the construction of h_{2i} is similar to that of h_2. The details of the constructions will be left to the reader.

From the construction, the pointwise limit of the sequence h_i, $i = 0, 1, 2, \ldots$, exists on $G \cup \bigcup_{i=1}^{\infty} J_i$ as a strictly increasing function that maps onto $G' \cup \bigcup_{i=1}^{\infty} J'_i$. Since these two sets are dense in the intervals $[a,b]$ and $[a',b']$ respectively, the sequence converges to a homeomorphism h of $[a,b]$ onto $[a',b']$ with the property that $h[G] = G'$. The lemma is proved.

We are now ready to give a proof of Gorman's theorem.

Proof of Gorman's theorem. Let S be an F_σ c-set of the first category that is contained in $I = (0,1)$. Let $\{\, [a_i,b_i] : i = 1, 2, \ldots \,\}$ be a well ordering of the closures of the nondegenerate components of the interior of the closure of S. The difficulty that one encounters is the possibility that $[a_i,b_i] \cap S$ may not be a c-set. That is, it may happen that a_i is a member of S. When this happens, a_i will be a left condensation point of S. The symmetric situation will occur when b_i is a member of S. With these observations in mind, we can construct inside of S a first category, F_σ c-set G_0 such that G_0 is dense in S and

$$G_0 \setminus \bigcup_{i=1}^{\infty} (a_i,b_i) = S \setminus \bigcup_{i=1}^{\infty} (a_i,b_i).$$

For each positive integer i, we let G_i be the F_σ c-set of the first category given by

$$G_i = S \cap (a_i,b_i).$$

Clearly, G_i is dense in (a_i,b_i) for each positive integer i. Select a nowhere dense, F_σ, density-open set T_0 contained in I. Such a set is easily constructed from a nowhere dense perfect set of positive measure. We may assume that each of 0 and 1 is a limit point of T_0 if and only if it is also a limit point of S. By Lemma 1.7, there is a homeomorphism $h_0 \colon (0,1) \to (0,1)$ such that $h_0[G_0] = T_0$. Next consider a positive integer i. With $L_i = h_0\bigl[(a_i,b_i)\bigr]$, let T_i be a density-open, F_σ c-set of the first category contained in L_i that is dense in L_i and has $\lambda(T_i) = \lambda(L_i)$. By Lemma 1.6, there exists a homeomorphism $h_i\colon (a_i, b_i) \to L_i$ such that $h_i[G_i] = T_i$. To complete the proof, replace each homeomorphism $h_0|(a_i, b_i)$ with the homeomorphism h_i to produce a strictly increasing function $h\colon (0,1) \to (0,1)$ that maps S onto the measurable set

$$T = \left(T_0 \setminus \bigcup_{i=1}^{\infty} L_i\right) \cup \bigcup_{i=1}^{\infty} T_i.$$

Since h is onto $(0,1)$, it is a homeomorphism. Clearly T is a density-open set since each set $T_0 \cap L_i$ has been replaced by the set T_i which has the same measure as L_i. This concludes the proof.

1.3.2. Applications. Many applications of the density topology have been found in analysis and in topology. This topology is nonnormal and completely regular (see the Appendix). We shall denote the density topology by $\mathcal{T}_d$. The density-interior of a measurable set is precisely the collection of points of the set that are density points of the set. Consequently, the dispersion points of a measurable set are members of the complement of the density-closure of the density-interior of the set.

We will give two applications in analysis. For the first one, we investigate the preservation of density points of measurable sets under one-to-one mappings.

The definition of Lebesgue equivalence uses the usual topology $\mathcal{T}$ on the interval (a,b) to determine the continuity of the homeomorphism $h\colon (a,b) \to (a,b)$. Gorman's theorem shows that density-open sets are not invariant under Lebesgue equivalence. An obvious remark is that if the density topology $\mathcal{T}_d$ was to replace the topology $\mathcal{T}$ in determining the continuity of the maps h and h^{-1}, then the invariance of density-open sets will occur. Such homeomorphisms have been studied by Ciesielski [**34**]. (For a related reference, see [**35**].) Clearly such homeomorphisms must preserve density points of measurable sets.

In the earlier paper [**18**] by Bruckner concerning homeomorphisms that preserve density points, there is the following theorem.

THEOREM 1.8 (BRUCKNER). *If $h\colon \mathbb{R} \to \mathbb{R}$ is a homeomorphism with the property that for each positive number ε there exists a positive number δ such that $\lambda(h[A]) < \varepsilon\lambda(h[I])$ whenever A is a measurable set contained in an interval I with $\lambda(A) < \delta\lambda(I)$, then h preserves points of density of each measurable set.*

PROOF. With the aid of the Vitali covering theorem, the condition on h yields $\lambda(h[Z]) = 0$ whenever Z is a G_δ set of measure 0. Hence $h[A]$ is measurable for every measurable set A. Suppose that x is a point of density of a measurable set S. Let $\varepsilon > 0$ and select a corresponding δ. Let γ be a positive number so that if I is an interval containing x with $\lambda(I) < \gamma$ then $\lambda(I \setminus S) < \delta\lambda(I)$. Finally, with the aid of the continuity of h^{-1}, select a positive number η such that if J is a interval containing $h(x)$ with $\lambda(J) < \eta$ then $\lambda(h^{-1}[J]) < \gamma$. As $h^{-1}[J]$ is an interval that contains x, we have $\lambda(h^{-1}[J] \setminus S) < \delta\lambda(h^{-1}[J])$ for each such J, and consequently

$$\lambda(J \setminus h[S]) = \lambda\big(h\big[h^{-1}[J] \setminus S\big]\big) < \varepsilon\lambda\big(h\big[h^{-1}[J]\big]\big) = \varepsilon\lambda(J).$$

And, $\lambda(J \cap h[S]) > (1-\varepsilon)\lambda(J)$ follows. That is, $h(x)$ is a point of density of $h[S]$.

Clearly, as we shall see, bi-Lipschitzian maps possess the property of the above theorem (see page 189 for the definition of bi-Lipschitzian). Indeed, the function h need be bi-Lipschitzian only on a Euclidean neighborhood of the measurable set S. Suppose that $h\colon (a,b) \to (c,d)$ is a bi-Lipschitzian map and let M be the larger of the Lipschitz constants $\text{Lip}(h)$ and $\text{Lip}(h^{-1})$. Then for every measurable set A contained in (a, b) and for every measurable set B contained in (c, d) we have $\lambda(h[A]) \le M\lambda(A)$ and $\lambda(h^{-1}[B]) \le M\lambda(B)$. Consequently, $\delta = M^{-2}\varepsilon$ will work.

The above Bruckner's theorem uses the property that $h^{-1}[J]$ is an interval whenever J is. This property is not true in the higher dimensional spaces $\mathbb{R}^n$. Nonetheless, we shall see in Chapter 5 that the result about bi-Lipschitzian homeomorphisms is still correct in the higher dimensional case (see also Buczolich's theorem, Theorem A.9 of the Appendix).

For the second example, observe that countable sets are closed in the density topology. This fact was exploited to yield a quick proof of the existence of an everywhere differentiable function that is not monotone on any interval (see [**59**]). Early proofs of the existence of such functions were quite complicated. A simpler but still elaborate proof was given by Katznelson and Stromberg [**87**]. Finally, a very elementary proof that exploited the Baire category theorem was given by Weil [**139**].

We give here the proof found in [**59**] to show the utility of the complete regularity of the topology $\mathcal{T}_d$. Let $\{ x_i : i = 1, 2, \dots \}$ and $\{ y_i : i = 1, 2, \dots \}$ be disjoint subsets of $\mathbb{R}$ that are dense in the usual topology. For each i let $\varphi_i \colon \mathbb{R} \to [0, 1]$ and $\psi_i \colon \mathbb{R} \to [0, 1]$ be approximately continuous functions such that

$$\varphi_i(x_i) = 1, \quad \varphi_i(y_j) = 0 \quad \text{for} \quad j = 1, 2, \dots ,$$

and

$$\psi_i(y_i) = 1, \quad \psi_i(x_j) = 0 \quad \text{for} \quad j = 1, 2, \dots .$$

Clearly the function

$$g = \sum_{i=1}^{\infty} \frac{1}{2^i} (\varphi_i - \psi_i)$$

is bounded and approximately continuous, whence a bounded derivative. Its antiderivative is the required function f. Note that a Urysohn theorem is not available because the density topology is not normal.

An application of the density topology in general topology began with an interesting counterexample to generalizations of a theorem of Blumberg [**12**] concerning arbitrary real-valued functions of a real variable. Blumberg's theorem states that there is, for each real-valued function f, a dense subset D of $\mathbb{R}$ for which the restriction of f to D is continuous. It is natural to want to generalize the domain of the functions in this theorem to other classes of spaces. The class of metric spaces for which the Blumberg theorem held was characterized by Bradford and Goffman [**15**]. Here, metrizable Baire spaces played a prominent role. (Baire spaces are defined on page 187 of the Appendix.) It was shown by White [**140**] that, under the continuum hypothesis, the topological space $\mathbb{R}$ with the density topology $\mathcal{T}_d$ fails to have Blumberg's property. Clearly, $\mathcal{T}_d$ is a Baire space. Being a completely regular space, it is also a uniform space (see for example, [**80**, page 9] or [**88**, page 188]). Thus it was shown that the class of uniform Baire spaces fails to have the Blumberg theorem. (See also [**141**].) (A full development along the line of this paragraph will be included in Chapter 8, the chapter in which connections between homeomorphisms and Blumberg's theorem are investigated.)

It was recognized by Tall [**124**] that the density topology and other modifications of it have properties that are well suited for problems in set-theoretic topology. (See also White [**142**].)

It is clear that the notion of bitopology (that is, a space with two topological structures) is key to the study of the density topology. Of course, the interest is in the case where one topology is finer than the other as in the case of the density and Euclidean topologies. Such bitopologies have been studied under the name of fine topologies. The readers who want a detailed reference on this subject are directed to the book [**95**] by Lukeš, Malý and Zajíček. In this vein, connections between measure and category can be found also in the book [**106**] by Oxtoby.

1.4. The Zahorski classes

The classes of subsets of $\mathbb{R}$ which appear in the statement of Gorman's theorem are among those listed below that were studied by Zahorski [**146**]. (See Bruckner [**21**] or [**20**, pages 85–97].) These classes are related to derivatives of functions.

1.4.1. The classes. Let E be a nonempty F_σ subset of $\mathbb{R}$. We say that E belongs to the class

M_0 if every point of E is a bilateral accumulation point of E;

M_1 if every point of E is a bilateral condensation point of E;

M_2 if each one-sided neighborhood of each point of E intersects E in a set of positive measure;

M_3 if for each point x of E

$$\lim_{k\to\infty} \frac{\lambda(I_k)}{\operatorname{dist}(x, I_k)} = 0$$

whenever I_k, $k = 1, 2, \ldots$, is a sequence of closed intervals converging to x with $x \notin I_k$ and $\lambda(I_k \cap E) = 0$ for each k;

M_4 if there exists a sequence K_k, $k = 1, 2, \ldots$, of closed sets with $E = \bigcup_{k=1}^{\infty} K_k$ such that for each k there exists a positive number η_k with the property that for each x in K_k and for each positive number c there exists a positive number $\varepsilon(x, c)$ for which

$$h\,h_1 > 0, \quad \frac{h}{h_1} < c, \quad |h + h_1| < \varepsilon(x, c) \qquad \text{imply}$$
$$\frac{\lambda\big(E \cap (x + h, x + h + h_1)\big)}{|h_1|} > \eta_k\,;$$

M_5 if every point of E is a point of density of E.

It is known that the successive classes are smaller. The classes M_1 and M_5 are the ones in Gorman's theorem. It is clear that the classes M_0 and M_1 are invariant under Lebesgue equivalence. Gorman's theorem shows that the remaining classes are not invariant under Lebesgue equivalence.

1.4.2. Derivatives. We shall establish the existence of homeomorphisms that possess very special properties. The proof of the lemma is an immediate consequence of the density topology on $\mathbb{R}$.

LEMMA 1.9. *Let Z be a G_δ subset of $\mathbb{R}$ with $\lambda(Z) = 0$. Then there exists a homeomorphism $f\colon \mathbb{R} \to \mathbb{R}$ such that f is differentiable and such that its derivative is a bounded, nonnegative, approximately continuous function with $(f')^{-1}[0] = Z$.*

PROOF. As Z is a G_δ set with $\lambda(Z) = 0$, the set Z is a zero-set in the density topology (see page 195 for zero-set). There is an approximately continuous function $\varphi\colon \mathbb{R} \to [0,1]$ such that $Z = \varphi^{-1}[0]$. Let f be any antiderivative of φ. It is clear that f is a homeomorphism because the set Z contains no intervals.

The set $\mathbb{R} \setminus Z$ in the above lemma is a set of the Zahorski type M_5. Zahorski has proved stronger results for sets of the type M_4 [**146**, Theorem 8].

1.4.3. Homeomorphisms. Gorman's theorem deals with two sets, the set S and its complement. We shall extend the theorem to a finite collection of disjoint sets.

An analysis of the proof of Gorman's theorem will show the need of the following new 'closure' operator. For each set E, denote by $\mathrm{P}(E)$ the set of points x of $\mathbb{R}$ with the property that every neighborhood of x that meets E contains a nonempty perfect set that is contained in E. It is clear that $\mathrm{P}(E)$ is a perfect set and that the boundary $\partial\,\mathrm{P}(E)$ of $\mathrm{P}(E)$ is nowhere dense in $\mathbb{R}$.

PROPOSITION 1.10. *The following statements hold.*

(1) *If* $E \subset F$, *then* $\mathrm{P}(E) \subset \mathrm{P}(F)$.

(2) *If* E *and* F *are Borel sets, then*

$$\mathrm{P}(E \cup F) = \mathrm{P}(E) \cup \mathrm{P}(F),$$

whence

$$\mathrm{P}(E) \setminus \mathrm{P}(F) \subset \mathrm{P}(E \setminus F).$$

PROOF. Statement (1) is obvious.

For statement (2), let H be a nonempty perfect set contained in $E \cup F$. Then $E \cap H$ and $F \cap H$ are Borel sets. The uncountable one contains a nonempty perfect set.

Note that the closure operator P is not a true closure operator in the sense of topology since each uncountable Borel set can be decomposed into two disjoint totally imperfect sets. Clearly this operator is aimed at the members of the class M_1.

There is an analogous closure operator for the class M_2. For each set E, denote by $\mathrm{P}_\lambda(E)$ the set of points x of $\mathbb{R}$ with the property that every neighborhood of x that meets E contains a nonempty perfect set with positive Lebesgue measure that is contained in E.

PROPOSITION 1.11. *The following statements hold.*

(1) *If* $E \subset F$, *then* $\mathrm{P}_\lambda(E) \subset \mathrm{P}_\lambda(F)$.

(2) *If* E *and* F *are Lebesgue measurable sets, then*

$$\mathrm{P}_\lambda(E \cup F) = \mathrm{P}_\lambda(E) \cup \mathrm{P}_\lambda(F),$$

whence

$$\mathrm{P}_\lambda(E) \setminus \mathrm{P}_\lambda(F) \subset \mathrm{P}_\lambda(E \setminus F).$$

The proof of this proposition is similar to that of the previous proposition.

NOTATION 1.12. For each set E we shall employ the following notation.

(i) $\mathrm{G}(E) = \mathrm{P}(E) \setminus \partial\,\mathrm{P}(E)$ and $\mathrm{G}_\lambda(E) = \mathrm{P}_\lambda(E) \setminus \partial\,\mathrm{P}_\lambda(E)$.

(ii) $\mathrm{C}(E) = E \setminus \mathrm{P}(E)$ and $\mathrm{C}_\lambda(E) = E \setminus \mathrm{P}_\lambda(E)$.

(iii) $\mathrm{B}(E) = E \cap \partial\,\mathrm{P}(E)$ and $\mathrm{B}_\lambda(E) = E \cap \partial\,\mathrm{P}_\lambda(E)$.

(iv) $\mathrm{I}(E) = E \cap \mathrm{G}(E)$ and $\mathrm{I}_\lambda(E) = E \cap \mathrm{G}_\lambda(E)$.

Clearly, the sets $\mathrm{G}(E)$ and $\mathrm{G}_\lambda(E)$ are open and the sets $\mathrm{I}(E)$ and $\mathrm{I}_\lambda(E)$ are respectively dense subsets of them. Moreover, $\mathrm{G}(E) = \emptyset$ if and only if $\mathrm{I}(E) = \emptyset$, and analogously for $\mathrm{G}_\lambda(E)$ and $\mathrm{I}_\lambda(E)$.

PROPOSITION 1.13. *Let E be a Lebesgue measurable set. Then*

(1) $\mathrm{P}_\lambda(E) \subset \mathrm{P}(E)$, *whence* $\mathrm{C}(E)$ *and* $\mathrm{C}_\lambda(E)$ *both have Lebesgue measure* 0,
(2) *there is subset F of* $\mathrm{I}(E)$ *such that F is in Zahorski's class M_1 and is dense in* $\mathrm{G}(E)$,
(3) *there is subset F of* $\mathrm{I}_\lambda(E)$ *such that F is in Zahorski's class M_2 and is dense in* $\mathrm{G}_\lambda(E)$,
(4) $\mathrm{B}(E)$ *and* $\mathrm{B}_\lambda(E)$ *are nowhere dense, Lebesgue measurable sets.*

PROOF. Statement (1) is trivial since $\mathrm{C}_\lambda(E)$ is a measurable set that contains no perfect set of positive Lebesgue measure.

To prove (2), suppose that $\mathrm{I}(E) \neq \emptyset$ and let $\{ U_j \}_{j=1}^{\infty}$ be a basis of open sets of the open subspace $\mathrm{G}(E)$ of $\mathbb{R}$. Each set $U_j \cap E$ contains a nonempty perfect set F_j with positive measure. It follows that

$$F = \bigcup_{j=1}^{\infty} F_j$$

is an F_σ set that is dense in $\mathrm{I}(E)$. Because $\mathrm{G}(E)$ is an open set, the set F is in M_1.

The proof of statement (3) is similar to that of (2).

Statement (4) is obvious because $\partial\,\mathrm{P}(E)$ and $\partial\,\mathrm{P}_\lambda(E)$ are nowhere dense.

We can now give an extension of Gorman's theorem.

LEMMA 1.14. *Let $E_1, E_2, \ldots, E_n$ be finitely many mutually disjoint Lebesgue measurable sets such that*

$$(a,b) = \bigcup_{i=1}^{n} E_i.$$

Then there is a collection of mutually disjoint open sets $H_1, H_2, \ldots, H_n$ such that

$$H_i \subset \mathrm{G}_\lambda(E_i) \quad \textit{for each } i, \quad \textit{and} \quad H = \bigcup_{i=1}^{n} H_i \textit{ is dense in } (a,b).$$

Hence there exists a homeomorphism $h\colon (a,b) \to (a,b)$ such that

$$\lambda(h[H]) = b-a, \quad \textit{and} \quad \lambda(h[H_i]) = \lambda(h[\mathrm{I}_\lambda(E_i) \cap H_i]) \quad \textit{for each} \quad i.$$

PROOF. Let

$$H_1 = \mathrm{G}_\lambda(E_1), \quad \text{and} \quad H_i = \mathrm{G}_\lambda(E_i) \setminus \bigcup_{j<i} \mathrm{P}_\lambda(E_j) \quad \text{for} \quad i>1.$$

We infer from Proposition 1.11 that

$$\begin{aligned}[a,b] \setminus \bigcup_{j=1}^{n} \partial\,\mathrm{P}_\lambda(E_j) &= \Big(\bigcup_{i=1}^{n} \mathrm{P}_\lambda(E_i) \setminus \bigcup_{j<i} \mathrm{P}_\lambda(E_j)\Big) \setminus \bigcup_{j=1}^{n} \partial\,\mathrm{P}_\lambda(E_j) \\ &= \Big(\bigcup_{i=1}^{n} H_i \setminus \bigcup_{j<i} \mathrm{P}_\lambda(E_j)\Big) \setminus \bigcup_{j=1}^{n} \partial\,\mathrm{P}_\lambda(E_j) \\ &\subset \bigcup_{i=1}^{n} H_i.\end{aligned}$$

Hence the first statement follows. The second statement follows from Proposition 1.13 and Gorman's theorem applied to each component of H.

A second extension which involves the class M_1 will be given in the next chapter (see Lemma 2.13).

CHAPTER 2

Baire Class 1

It is a trivial fact that the class of continuous functions is invariant under Lebesgue equivalence. Even more, each Baire class of real-valued functions is invariant under Lebesgue equivalence. The next obvious collection to consider is the class of Lebesgue measurable functions. Observe that the Cantor ternary set is Lebesgue equivalent to a set of positive Lebesgue measure. Consequently the class of Lebesgue measurable functions is not invariant as one can easily show by considering the characteristic functions of subsets of the Cantor ternary set. But, the class of Borel measurable real-valued functions is invariant.

There is a second equivalence relation on the collection of Lebesgue measurable functions, that which is determined by equality λ-almost everywhere. It is well-known that a measurable function is equal λ-almost everywhere to a function in the Baire class 2. Of course there is a Baire class 2 function that is not equal λ-almost everywhere to any Baire class 1 function, for example the characteristic function of a G_δ set with the property that each nonempty open set meets both E and its complement in a set of positive Lebesgue measure (see the characterization of functions of the first Baire class found in Section A.1.3 of the Appendix). Thus the question that comes to the fore is whether it is possible to find a function that is Lebesgue equivalent to a given measurable function which is also equal λ-almost everywhere to a function in the Baire class 1. A little reflection will show that the existence of a Baire class 2 function equal λ-almost everywhere to a measurable function does not help since sets of Lebesgue measure 0 are not invariant under Lebesgue equivalence.

The problem leads naturally to functions which will be called absolutely measurable with respect to Lebesgue measure (see Section 2.2 for the definition). As it will turn out, these functions have the property that there is always a Lebesgue equivalent function that is equal λ-almost everywhere to a Baire class 1 function. Also, it will be shown that any Lebesgue measurable function whose image consists of finitely many points will have the last property as well. We shall see later that such functions need not be absolutely measurable. Surprisingly, there is a Lebesgue measurable function whose image is a countably infinite set with the property that each of its Lebesgue equivalent function is not λ-almost everywhere equal to a function in the Baire class 1 (see Section 2.3). Clearly, this example cannot be an absolutely measurable function with respect to Lebesgue measure.

2.1. Characterization

The following key characterization first appeared in [**22**]. It is worthy to note that no assumption of Lebesgue measurability is made on the function f in the lemma.

LEMMA 2.1. *Let f be a function from $(0,1)$ into $\mathbb{R}$. Then there exists a homeomorphism h of $(0,1)$ onto itself such that $f \circ h$ is equal λ-almost everywhere to a function in the Baire class 1 if and only if there exist a set E in Zahorski's class M_1 and a function g in the Baire class 1 such that*

$$E \text{ is dense in } (0,1) \quad \text{and} \quad f|E = g|E.$$

PROOF. Suppose that there is a homeomorphism h such that $f \circ h$ is equal λ-almost everywhere to a function φ in the Baire class 1. Let G be a G_δ set with Lebesgue measure 0 such that $f \circ h$ and φ agree on $(0,1) \setminus G$. Clearly, $E = h\big[(0,1) \setminus G\big]$ is a member of Zahorski's class M_1 that is dense in $(0,1)$ and $g = \varphi \circ h^{-1}$ is a Baire class 1 function such that $f|E = g|E$.

Conversely, suppose that E is member of Zahorski's class M_1 that is dense in $(0,1)$ and g is a Baire class 1 function such that $f|E = g|E$. Then by Gorman's theorem there is a homeomorphism h such that $\lambda(h^{-1}[E]) = 1$. Clearly, $g \circ h$ is in the Baire class 1 and $g \circ h$ is equal λ-almost everywhere to $f \circ h$.

2.1.1. Lattice properties. It is well-known that the composition of a continuous function and a function in the Baire class 1 is a function in the Baire class 1. Consequently, the above characterization yields the following proposition.

PROPOSITION 2.2. *If f has the property that it is Lebesgue equivalent to a function that is equal λ-almost everywhere to a function in the Baire class 1 and if φ and ψ are functions that are respectively continuous and in the Baire class 1, then $\varphi \circ f$ and $f + \psi$ are each Lebesgue equivalent to a function that is equal λ-almost everywhere to a function in the Baire class 1.*

Consequently, $|f|$, f^+ and f^- are each Lebesgue equivalent to a function that is equal λ-almost everywhere to a function in the Baire class 1 whenever f is.

It will be established in the last section of the chapter that there exists a Lebesgue measurable function f such that f^+ and f^- are each Lebesgue equivalent to a function that is equal λ-almost everywhere to a function in the Baire class 1 and yet f is not Lebesgue equivalent to any function that is equal λ-almost everywhere to a function in the Baire class 1. Consequently, one cannot extend the above lattice proposition to a vector lattice one.

2.1.2. Simple functions. As usual, a **simple function** is a measurable function whose image is a finite set. The characterization theorem will yield the following [**74**].

THEOREM 2.3. *Each simple function $f\colon (0,1) \to \mathbb{R}$ is Lebesgue equivalent to a function that is equal λ-almost everywhere to a function in the Baire class 1.*

PROOF. Let $f\big[(0,1)\big]$ be indexed so as to satisfy $a_1 < a_2 < \cdots < a_n$. There is no loss in assuming $0 < a_1$. Recall the closure operator P_λ from the last chapter. With $E_i = f^{-1}[a_i]$ in Lemma 1.14, we have a collection of mutually disjoint open sets H_i such that $H = \bigcup_{i=1}^n H_i$ is dense in $(0,1)$ and such that $\mathrm{I}_\lambda(E_i) \cap H_i$ contains a set F_i in Zahorski's class M_1 that is dense in H_i for each i with $H_i \neq \emptyset$. Let

$$F = \bigcup_{i=1}^n F_i$$

and g be the lower semicontinuous function given by

$$g(x) = \begin{cases} a_i & \text{if } x \in H_i,\ 1 \le i \le n, \\ 0 & \text{if } x \in (0,1) \setminus H. \end{cases}$$

Since $\mathrm{I}_\lambda(E_i) \subset E_i$ for each i, we have $f|F = g|F$. As F is dense in $(0,1)$, the theorem is proved.

Observe that the collection of all simple functions is a vector lattice.

2.2. Absolutely measurable functions

A function $f\colon \mathbb{R} \to \mathbb{R}$ is said to be **λ-absolutely measurable** if $f \circ h$ is Lebesgue measurable for every homeomorphism h of $\mathbb{R}$ onto itself. Clearly each Borel measurable function is λ-absolutely measurable. In Section 1.2.4 of Chapter 1, we observed that analytic sets are invariant under Lebesgue equivalence. Hence there are non-Borel measurable functions that are λ-absolutely measurable since characteristic functions of analytic sets are invariant under Lebesgue equivalence. It has already been pointed out in Section 1.2.5 that the collection of Lebesgue measurable sets is not invariant under Lebesgue equivalence. Hence there are simple functions that are not λ-absolutely measurable.

2.2.1. Elementary properties. It is immediate from the definition that the sum and product of λ-absolutely measurable functions are also λ-absolutely measurable. We shall denote the class of λ-absolutely measurable functions by $\mathrm{ab}\mathcal{F}$. Observe that, for each Borel measurable function φ, the composition $\varphi \circ f$ is λ-absolutely measurable whenever f is. Hence the ring of functions $\mathrm{ab}\mathcal{F}$ is also a vector lattice. Moreover, if f_n, $n = 1, 2, \ldots$, is a sequence of functions in $\mathrm{ab}\mathcal{F}$ that converges pointwise to f, then f is also in $\mathrm{ab}\mathcal{F}$.

If f is a λ-absolutely measurable function and B is a Borel set, then $h\big[f^{-1}[B]\big]$ is measurable for every homeomorphism h of $\mathbb{R}$ onto itself. Hence the collection

$$\{\, f^{-1}[\{\, y : y > 0 \,\}] : f \in \mathrm{ab}\mathcal{F} \,\}$$

forms a σ-algebra that is invariant under all homeomorphisms of $\mathbb{R}$ onto itself. The members of this σ-algebra are called **λ-absolutely measurable sets**. Clearly, it is the largest class of Lebesgue measurable sets that is invariant under Lebesgue equivalence.

A set Z is said to have **λ-absolute measure 0** if $\lambda(h[Z]) = 0$ for every homeomorphism h of $\mathbb{R}$ onto itself. Clearly, countable sets are examples of such sets. For the existence of uncountable ones, the reader is referred to the discussion in Oxtoby [**106**, pages 81 and 99], for example.

PROPOSITION 2.4. *If Z_i has λ-absolute measure 0 for $i = 1, 2, \ldots$, then the union $\bigcup_{i=1}^{\infty} Z_i$ has λ-absolute measure 0.*

If P is a nonempty perfect set and if Z has λ-absolute measure 0, then there is a countable collection of nonempty perfect sets P_i, $i = 1, 2, \ldots$, such that $P_i \subset P$ for each i and such that $\bigcup_{i=1}^{\infty} P_i$ is dense in $P \setminus Z$. Consequently, $P \setminus Z \neq \emptyset$.

PROOF. The proof of the first statement is left to the reader.

For the second statement, let h be a homeomorphism such that $\lambda(U \cap h[P]) > 0$ whenever U is an open set with $U \cap h[P] \neq \emptyset$. Since $\lambda(h[Z]) = 0$, there is a G_δ set G containing $h[Z]$ with $\lambda(G) = 0$. One can now find the requisite perfect sets P_i by using the nonempty F_σ set $h[P] \setminus G$.

PROPOSITION 2.5. *If E is a set in Zahorski's class M_1 and if Z has λ-absolute measure 0, then there is a subset F of $E \setminus Z$ that is in M_1 and is dense in E.*

PROOF. We have already seen that E contains nonempty, perfect subsets E_i, $i = 1, 2, \dots$, such that $E_0 = \bigcup_{i=1}^{\infty} E_i$ is dense in E. The proof is easily completed by an application of the second statement of Proposition 2.4 to each E_i.

The next proposition uses the notion of Baire space. The definition of these spaces is given in Section 8.2 of Chapter 8. For now, we can briefly say that such a space has a valid Baire category theorem.

PROPOSITION 2.6. *Let U be a nonempty open set. If Z has λ-absolute measure* 0*, then $U \setminus Z$ is a Baire space.*

PROOF. Denote the space $U \setminus Z$ by X. Suppose that F_i, $i = 1, 2, \dots$, is a collection of sets F_i that are closed and nowhere dense in the space X with

$$X = \bigcup_{i=1}^{\infty} F_i.$$

Denote the closure in $\mathbb{R}$ of F_i by H_i. Then H_i is nowhere dense and hence

$$Y = U \setminus \bigcup_{i=1}^{\infty} H_i$$

is an uncountable G_δ set such that $X \cap Y = \emptyset$. Let P be a nonempty perfect subset of Y. Then $P \subset U \setminus X \subset Z$, in contradiction with Proposition 2.4. Hence $U \setminus Z$ is a Baire space.

Recall the closure operator P defined on page 11 in Section 1.4.3.

PROPOSITION 2.7. *If E is a λ-absolutely measurable set, then $E \setminus \mathrm{P}(E)$ has λ-absolute measure* 0.

PROOF. We shall use the notation of Section 1.4.3. By the definition of the closure operator P, we have that the set $\mathrm{C}(E) = E \setminus \mathrm{P}(E)$ contains no nonempty perfect set. As $\mathrm{C}(E)$ is a λ-absolutely measurable set, it follows that $\lambda\big(h[\mathrm{C}(E)]\big) = 0$ for each homeomorphism of $\mathbb{R}$ onto itself.

The following is an important consequence of the propositions.

LEMMA 2.8. *Let A be a λ-absolutely measurable set, E be a dense set in the Zahorski class M_1 and Z be a set with λ-absolute measure* 0*. If $A \subset E$, then there is a subset F of E such that $F \in M_1$, F is dense, $F \cap Z = \emptyset$, and*

$$\chi_A \,|F = \chi_{\mathrm{P}(A)} \,|F.$$

PROOF. Statement (2) of Proposition 1.13 yields a subset A_0 of A that is dense in $\mathrm{P}(A) \setminus \partial\,\mathrm{P}(A)$ and is in M_1 whenever $\mathrm{P}(A) \setminus \partial\,\mathrm{P}(A)$ is not empty. Clearly, $E_0 = A_0 \cup \big(E \setminus \mathrm{P}(A)\big)$ is dense in E and is in M_1. By Proposition 2.7, we have $A \setminus \mathrm{P}(A)$ has λ-absolute measure 0 and hence $Z \cup \big(A \setminus \mathrm{P}(A)\big)$ has λ-absolute measure 0. And, by Proposition 2.5, there is a subset F of $E_0 \setminus \big(Z \cup (A \setminus \mathrm{P}(A))\big)$ such that F is in M_1 and F is dense in E_0, whence dense in E. Obviously,

$$(F \cap A) \setminus \mathrm{P}(A) = \emptyset \quad \text{and} \quad F \cap \mathrm{P}(A) \subset A_0 \subset A.$$

Since $F \cap Z = \emptyset$, the lemma is proved.

PROPOSITION 2.9. *If $f \colon \mathbb{R} \to \mathbb{R}$ is λ-absolutely measurable and if P is a nonempty perfect set, then there is a nonempty perfect set Q such that $Q \subset P$ and $f|Q$ is continuous.*

PROOF. There is a homeomorphism h such that $h^{-1}[P]$ has positive Lebesgue measure. By Lusin's theorem there is a nonempty perfect subset R of $h^{-1}[P]$ for which the restriction $(f \circ h)|R$ is continuous. Let Q be $h[R]$.

2.2.2. The main theorem. We shall prove the following theorem of Bruckner, Davies and Goffman [**22**].

THEOREM 2.10 (BRUCKNER-DAVIES-GOFFMAN). *If $f\colon (0,1) \to \mathbb{R}$ is λ-absolutely measurable, then there is a homeomorphism h of $(0,1)$ onto itself and a function g in the Baire class 1 such that $f \circ h = g$ λ-almost everywhere.*

The proof will rely on three lemmas. The first lemma provides a sufficient condition under which a λ-absolutely measurable function will satisfy the conclusion of the main theorem. The other two lemmas permit us to show that this sufficient condition is met by every λ-absolutely measurable function.

LEMMA 2.11. *Suppose that f is a λ-absolutely measurable function. If there exists a function φ in the Baire class 1 with the property that for each positive number ε there is a dense set F in Zahorski's class M_1 such that $|f(x) - \varphi(x)| < \varepsilon$ for every x in F, then f is Lebesgue equivalent to a function that is equal λ-almost everywhere to a function in the Baire class 1.*

PROOF. Suppose that such a function φ exists. Let (a_k, b_k), $k = 1, 2, \ldots$, be a basis for the topology. Let us inductively construct a sequence E_k, $k = 1, 2, \ldots$, of mutually disjoint, nonempty, nowhere dense, perfect sets such that

$$E_k \subset (a_k, b_k) \setminus \bigcup_{i<k} E_i$$

and such that

$$\psi_k = (f - \varphi)\,\chi_{E_k} \quad \text{is in the Baire class 1 with} \quad |\psi_k| < \tfrac{1}{2^{k-1}}.$$

For $k = 1$, from the property possessed by φ, we have a set F_1 in the class M_1 that is dense and satisfies

$$\big|(f - \varphi)|F_1\big| < 1.$$

There is a nonempty, nowhere dense, perfect set P_1 such that

$$P_1 \subset F_1 \cap (a_1, b_1).$$

By Proposition 2.9, there is a nonempty perfect set E_1 contained in P_1 such that $(f - \varphi)|E_1$ is continuous. We infer from the Tietze extension theorem that

$$\psi_1 = (f - \varphi)\,\chi_{E_1}$$

is in the Baire class 1. Certainly, $|\psi_1| < 1$.

Observe that $(a_2, b_2) \setminus E_1$ is dense in (a_2, b_2). Hence the inductive step should be obvious to the reader.

Clearly, $E = \bigcup_{k=1}^{\infty} E_k$ is a dense set that is in M_1. Also $\psi = \sum_{k=1}^{\infty} \psi_k$ is in the Baire class 1 because the series is uniformly convergent. Finally, Lemma 2.1 finishes the proof since $f|E = (\varphi + \psi)|E$.

LEMMA 2.12. *Suppose that $f_1\colon (0,1) \to \mathbb{R}$ is a λ-absolutely measurable function and suppose that E_1 is a set in the Zahorski's class M_1 and $g_1\colon (0,1) \to \mathbb{R}$ is in the Baire class 1 such that E_1 is dense in $(0,1)$ and $f_1|E_1 = g_1|E_1$.*

If $f_2\colon (0,1) \to \mathbb{R}$ is a λ-absolutely measurable function such that the image of $f_2 - f_1$ is contained in $\{0,1\}$, then there exists a set E_2 in Zahorski's class M_1 and a function $g_2\colon (0,1) \to \mathbb{R}$ in the Baire class 1 such that

(1) *E_2 is a dense subset of E_1,*
(2) *$g_2|E_2 = f_2|E_2$,*

(3) $(g_2 - g_1)\big[(0,1)\big] \subset \{0,1\}$.

PROOF. Let A be the λ-absolutely measurable set $E_1 \cap \{\, x : f_1(x) = f_2(x)\,\}$. Then it is clear that $f_2 = f_1 + 1 - \chi_{_A}$ holds on E_1. By Lemma 2.8, there is a set E_2 in Zahorski's class M_1 that satisfies condition (1) and $\chi_{_A}|E_2 = \chi_{_{P(A)}}|E_2$. By defining g_2 to be $g_1 + 1 - \chi_{_{P(A)}}$, we have that f_2 equals g_2 on E_2, whence (2) holds. Condition (3) follows trivially.

The final lemma is the countable analogue for the Zahorski's class M_1, in the context of λ-absolutely measurable sets, of Lemma 1.14 in Chapter 1. Recalling Notation 1.12 on page 11, we have $\mathrm{C}(A) = A \setminus \mathrm{P}(A)$, and $\mathrm{G}(A) = \mathrm{P}(A) \setminus \partial\,\mathrm{P}(A)$.

LEMMA 2.13. *Let A_i, $i = 1, 2, \ldots$, be a sequence of mutually disjoint sets that are λ-absolutely measurable. Then*

$$Z = \textstyle\bigcup_{i=1}^{\infty} \mathrm{C}(A_i).$$

has λ-absolute measure 0. *Moreover, if*

$$(0,1) = \textstyle\bigcup_{i=1}^{\infty} A_i,$$

then

$$G = \big(\textstyle\bigcup_{i=1}^{\infty} (\mathrm{G}(A_i) \setminus \bigcup_{j<i} \mathrm{P}(A_j))\big) \setminus Z$$

is a dense subset of $(0,1)$.

PROOF. From Proposition 2.7 we have that each $\mathrm{C}(A_i)$ has λ-absolute measure 0. The set Z clearly has λ-absolute measure 0 by Proposition 2.4. Only the denseness statement of G needs proof. Let $(0,1) = \bigcup_{i=1}^{\infty} A_i$ and let (a,b) be a subinterval of $(0,1)$. By Lemma 2.6, the subspace $(a,b) \setminus Z$ is a Baire space. Since $A_i \subset \mathrm{C}(A_i) \cup \mathrm{P}(A_i)$ for each i, we have

$$(0,1) \setminus Z = \big(\textstyle\bigcup_{i=1}^{\infty} \mathrm{P}(A_i)\big) \setminus Z.$$

The sets

$$F_i = \big((a,b) \setminus Z\big) \cap \big(\mathrm{P}(A_i) \setminus \textstyle\bigcup_{j<i} \mathrm{P}(A_j)\big)$$

being F_σ in the topology of the subspace $(a,b) \setminus Z$, there is an index i such that F_i has a nonempty interior relative to the space $(a,b) \setminus Z$. Now, from

$$\mathrm{G}(A_i) \setminus \big(\mathrm{P}(A_i) \setminus \textstyle\bigcup_{j<i} \mathrm{P}(A_j)\big) = \big(\mathrm{P}(A_i) \setminus \big(\mathrm{P}(A_i) \setminus \bigcup_{j<i} \mathrm{P}(A_j)\big)\big) \setminus \partial\,\mathrm{P}(A_i),$$

we have that

$$\big((a,b) \setminus Z\big) \cap \big(\mathrm{G}(A_i) \setminus \big(\mathrm{P}(A_i) \setminus \textstyle\bigcup_{j<i} \mathrm{P}(A_j)\big)\big)$$

is not empty. Hence, $(a,b) \cap G \neq \emptyset$ and the denseness of G is established.

It is now immediate that a λ-absolutely measurable function f with a countable image set has a Lebesgue equivalent function that is equal λ-almost everywhere to a function in the Baire class 1. Indeed, if $A_i = f^{-1}[y_i]$ for each y_i in the image of f, then we infer from statement (2) of Proposition 1.13 and Proposition 2.4 that the λ-absolutely measurable set $A_i \cap \mathrm{G}(A_i) \setminus Z$ contains a set E_i in Zahorski's class M_1 that is dense in the open set $\mathrm{G}(A_i) \setminus \bigcup_{j<i} \mathrm{P}(A_j)$ whenever it is not empty. Then $E = \bigcup_{i=1}^{\infty} E_i$ is in Zahorski's class M_1 and is dense in $(0,1)$. With the open set $\mathrm{G}(A_i) \setminus \bigcup_{j<i} \mathrm{P}(A_j)$ denoted by H_i, the function $\varphi = \sum_{i=1}^{\infty} y_i\, \chi_{_{H_i}}$ is the difference of 2 lower semicontinuous functions. Lemma 2.1, the characterization lemma, completes the proof.

We are now ready for the proof of the main theorem.

PROOF OF MAIN THEOREM. For each positive integer k let f_k be the function that takes values only from $\frac{m}{2^k}$, $m = 0, \pm 1, \pm 2, \ldots$, and satisfies

$$f_k(x) \leq f(x) < f_k(x) + \tfrac{1}{2^k} \quad \text{for every} \quad x.$$

Clearly, for every k, the function f_k is λ-absolutely measurable,

$$f_k \leq f_{k+1} \quad \text{and} \quad 2^{k+1}(f_{k+1} - f_k)\big[(0,1)\big] \subset \{0,1\}.$$

We shall inductively construct a sequence φ_k, $k = 1, 2, \ldots$, of functions in the Baire class 1 and a sequence E_k, $k = 1, 2, \ldots$, of sets in Zahorski's class M_1 satisfying

(1) E_k is dense in $(0,1)$,
(2) $E_k \supset E_{k+1}$,
(3) $f_k|E_k = \varphi_k|E_k$
(4) $0 \leq \varphi_{k+1}(x) - \varphi_k(x) \leq \frac{1}{2^{k+1}}$ for every x.

We already know that there is a set E_1 and there is a function φ_1 in the Baire class 1 such that (1) and (3) are satisfied. Assuming that E_k and φ_k have been constructed, we can use Lemma 2.12 to get E_{k+1} and φ_{k+1}. The construction is now completed.

From (4) we have φ_k, $k = 1, 2, \ldots$, is a Cauchy sequence of functions in the Baire class 1. Hence it converges uniformly to a Baire class 1 function φ. Let us show that this function φ satisfies the property in the sufficient condition of Lemma 2.11. To this end, we have for each k

$$\begin{aligned} \big|(f - \varphi)|E_k\big| &\leq \big|(f - f_k)|E_k\big| + \big|(f_k - \varphi)|E_k\big| \\ &= \big|(f - f_k)|E_k\big| + \big|(\varphi_k - \varphi)|E_k\big| \\ &\leq \tfrac{1}{2^k} + \tfrac{1}{2^k} = \tfrac{1}{2^{k-1}}. \end{aligned}$$

Consequently, Lemma 2.11 completes the proof.

2.3. Example

The example is a Lebesgue measurable function f such that $f \circ h$ is not equal λ-almost everywhere to any function in the Baire class 1 for every homeomorphism h of $\mathbb{R}$ onto itself. In fact, one can also require that $f[\mathbb{R}]$ be a countably infinite set. The example is essentially the same as the one given by Gorman [**74**].

2.3.1. The construction. Let P_k, $k = 1, 2, \ldots$, be a sequence of mutually disjoint, nonempty, nowhere dense, perfect sets such that $P = \bigcup_{k=1}^{\infty} P_k$ is dense in $\mathbb{R}$. Then $G = \mathbb{R} \setminus P$ is a dense, totally disconnected, G_δ set. It is not difficult to construct one with $\lambda(G) = 0$. The set G can be written as the disjoint union of two totally imperfect sets G_1 and G_2 (see Section A.1.2 of the Appendix). Then the function

$$f = -1\,\chi_{G_1} - 2\,\chi_{G_2} + \sum_{k=1}^{\infty} k\,\chi_{P_k}$$

is an integer-valued function that is equal λ-almost everywhere to a function in the Baire class 2. Suppose that there is a function that is Lebesgue equivalent to f and that is equal λ-almost everywhere to a function in the Baire class 1. Then, by Lemma 2.1, there is a set E in Zahorski's class M_1 and there is a function g in the

Baire class 1 such that E is dense and $f|E = g|E$. Suppose that $E \cap G$ is uncountable. Then there is a nonempty perfect set F contained in $E \cap G$. Observe that $g|F$ is in the Baire class 1. Since G_1 and G_2 are totally imperfect, the function $f|F$ is not a function in the Baire class 1 (see Section A.1.3 of the Appendix). As $g|F = f|F$, we have that $g|F$ is not in the Baire class 1, a contradiction. Consequently, $E \cap G$ must be a countable set, whence $E \setminus G$ is a dense set. Let U be any nonempty open set. Since G is dense, there is a point x in $G \cap U$, whence $f(x) \leq -1$. There also is a point x' in $U \cap (E \setminus G)$ because $E \setminus G$ is dense, whence $f(x') \geq 1$. We now conclude that g is not in the Baire class 1, a contradiction. We have verified that the constructed function is indeed an example.

2.3.2. Remarks. We make some observations on the example f constructed above.

Note that f is integer-valued, and that f^+ and f^- are respectively λ-absolutely measurable and simple. Consequently, the collection of functions with the property that they are Lebesgue equivalent to a function that is λ-almost everywhere equal to a function in the Baire class 1 does not form a vector space.

Since f also has $f^{-1}[0] = \emptyset$, the function $\frac{1}{f}$ is real-valued and is measurable. It is easy to see that $\frac{1}{f \vee 1}$ is upper semicontinuous and is equal to $\frac{1}{f}$ λ-almost everywhere. Hence $\frac{1}{f}$ is Lebesgue equivalent to a function that is λ-almost everywhere equal to a function in the Baire class 1. Observe that the image of $\frac{1}{f}$ is a bounded countable set.

Next consider the composed function $\arctan \circ f$. Clearly the image of this function is a bounded, countable set. Yet it is not Lebesgue equivalent to a function that is λ-almost everywhere equal to some function in the Baire class 1.

Finally, the delicate proof of the main theorem concerning λ-absolutely measurable functions has been taken from the paper [**22**] by Bruckner, Davies and Goffman. The lattice of λ-absolutely measurable functions is closed under uniform convergence, and the collection of all such functions that have countable image sets is a dense linear sublattice. Theorem 2.3 concerning the linear lattice of simple functions is due to Gorman [**74**]. In the same paper, it is shown that the uniform closure of this linear lattice has a member f with the property that f is not Lebesgue equivalent to a function that is λ-almost everywhere equal to some function in the Baire class 1.

The notion of λ-absolute measure 0 presented here appears to depend on the Lebesgue measure λ. Analogous notions are certainly available for each σ-finite, complete, nonatomic Borel regular measure μ. In the book [**106**] by Oxtoby, the stronger notion of absolute measure 0 (where $\mu(E) = 0$ is to hold for every σ-finite, complete, nonatomic Borel measure μ) is investigated by using Baire category sets of the first class. In Part 3 of the book we shall revisit λ-absolute measurability and λ-absolute measure 0 sets in a very natural setting. Further properties of λ-absolutely measurable functions that are related to the analytic notions of essential supremum and Lebesgue integrability are developed in Section 11.1 of Chapter 11 under the name of absolutely measurable functions. The development leads to what we call absolutely essentially bounded functions.

CHAPTER 3

Differentiability Classes

This chapter concerns the equivalence classes under Lebesgue equivalence of continuously n-times differentiable functions defined on $[0,1]$, $n = 1, 2, \ldots, \infty$. These functions are obviously continuous and have finite total variation. As usual, CBV will denote the class of all continuous functions whose total variations are finite. It is easily seen that the class CBV is invariant under Lebesgue equivalence. The chapter begins with a characterization of the collection of functions which are Lebesgue equivalent to some function in CBV. The main theorem here is the one due to Bruckner and Goffman [**23**] given in the first section. With the aid of the notions of the set of varying monotonicity of a function (see Definition 3.7) and of packing indices, the subclasses of CBV that correspond to the Lebesgue equivalence classes of the classes $\mathrm{C}^n[0,1]$ will be characterized.

3.1. Continuous functions of bounded variation

We shall begin with a characterization of the Lebesgue equivalence classes of continuous functions with finite total variations. Here, a surprising connection with functions with bounded derivatives is made (Theorem 3.1 below).

3.1.1. Bounded derivative. Functions of bounded variation play a major role in the calculus of variations since they arise naturally in the study of curves in $\mathbb{R}^m$ with finite Jordan length. It is well-known that such functions are differentiable λ-almost everywhere and that the total variation of the function can be recovered from the derivative if and only if the function is absolutely continuous. This recovery is made even when the derivative is not defined everywhere and may even be unbounded. In contrast to these classical facts is the following theorem of Bruckner and Goffman [**23**]. In passing, we remark that the proof of the theorem is easily modified to yield a corresponding theorem for continuous rectifiable curves in $\mathbb{R}^m$.

THEOREM 3.1 (BRUCKNER-GOFFMAN). *Let $f\colon [0,1] \to \mathbb{R}$. Then a necessary and sufficient condition that f be Lebesgue equivalent to a function with bounded derivative is that f be continuous and have finite total variation.*

The proof will use the following simple lemma. For a real-valued function f of bounded variation defined on $[a,b]$ we define its **variation function** v_f to be

$$v_f(x) = \begin{cases} 0 & \text{for } x = a, \\ \mathrm{V}(f,[a,x]) & \text{for } a < x \leq b. \end{cases}$$

Of course, $\mathrm{V}(f,[a,b])$ is the total variation of f on the interval $[a,b]$, and a function of bounded variation on $[a,b]$ is one for which the total variation is finite on $[a,b]$.

LEMMA 3.2. *If $f\colon [0,1] \to \mathbb{R}$ is continuous and of bounded variation, then there exists a homeomorphism h of $[0,1]$ onto itself such that $f \circ h$ and $v_f \circ h$ are Lipschitzian functions.*

PROOF. Let $\varphi\colon [0,1] \to [\,0, 1 + \mathrm{V}(f, [0,1])\,]$ be defined by

$$\varphi(x) = x + v_f(x) \quad \text{for} \quad 0 \le x \le 1.$$

Clearly, φ is a homeomorphism of $[0,1]$ onto $\big[0, 1 + \mathrm{V}(f,[0,1])\big]$ and

$$|f(x) - f(x')| \le |v_f(x) - v_f(x')| \le |\varphi(x) - \varphi(x')|$$

whenever x, $x' \in [0,1]$. Define

$$h(t) = \varphi^{-1}\big(t\,(1 + \mathrm{V}(f,[0,1])\big) \quad \text{for} \quad t \in [0,1].$$

Then, for t, $t' \in [0,1]$,

$$\begin{aligned} |f \circ h(t) - f \circ h(t')| &\le \big|v_f\big(h(t)\big) - v_f\big(h(t')\big)\big| \\ &\le \big|\varphi\big(h(t)\big) - \varphi\big(h(t')\big)\big| \\ &\le \big(1 + \mathrm{V}(f,[0,1])\big)\,|t - t'|. \end{aligned}$$

PROOF OF THEOREM 3.1. The necessity of the condition is immediate. Indeed, if there is a homeomorphism h such that $f \circ h$ has a bounded derivative, then $f \circ h$ is continuous and has finite total variation. As $f = (f \circ h) \circ h^{-1}$, we have that f is continuous and has the same total variation.

Turning to the sufficiency of the condition, we may assume further that f is Lipschitzian by the last lemma. Then f is differentiable λ-almost everywhere. For this exceptional set Z of Lebesgue measure 0, let G be a G_δ set such that $Z \subset G$ and $\lambda(G) = 0$. We assert that there is a differentiable homeomorphism h such that

$$\frac{dh}{dx}(x) = 0 \quad \text{exactly on the set} \quad h^{-1}[G]$$

and

$$\frac{dh}{dx} \quad \text{is bounded.}$$

(The existence of such a differentiable homeomorphism will be given at the end of this section as Zahorski's theorem.) Clearly,

$$\frac{(f \circ h)(t) - (f \circ h)(x)}{t - x} = \frac{f\big(h(t)\big) - f\big(h(x)\big)}{h(t) - h(x)}\,\frac{h(t) - h(x)}{t - x}.$$

For $h(x) \in Z$, since f is Lipschitzian, the right-hand side always has a limit of 0 as t tends to x. Moreover, the same identity yields, for every x,

$$\left|\frac{d}{dx}\,(f \circ h)(x)\right| \le \mathrm{Lip}(f)\,\mathrm{Lip}(h).$$

Notice that the last lemma permits the assertion that the same homeomorphism will also apply to the function v_f.

3.1.2. The Cantor function. The Cantor function f presents a very interesting example. It is a continuous nondecreasing function from $[0,1]$ onto $[0,1]$ such that its derivative at each point outside of the Cantor set K is 0. The above theorem of Bruckner and Goffman asserts that there is a homeomorphism h such that $f \circ h$ has a bounded derivative. The derivative of $f \circ h$ at each point of the complement of $h^{-1}[K]$ is equal to 0. We assert that $h^{-1}[K]$ is a set of positive Lebesgue measure. Indeed, if its Lebesgue measure is 0, then we would have $\lambda(f[K]) = 0$ because $f \circ h$ is Lipschitzian. But the Cantor function has the property that $\lambda(f[K]) = 1$. Therefore the derivative of $f \circ h$ cannot be equal to 0 λ-almost everywhere. Functions like the Cantor functions have been investigated by Bruckner and Leonard in [**24**].

3.1.3. Differentiable homeomorphisms. In the proof of Theorem 3.1, we asserted the existence of certain differentiable homeomorphisms. This assertion is due to Zahorski [**146**]. The proof given below is modeled after the one given by him in [**145**].

We shall begin with two elementary facts about the density topology.

PROPOSITION 3.3. *Let F be a nonempty compact subset of $[0,1]$ and G be a G_δ set of Lebesgue measure 0 with $F \cap G = \emptyset$. Then, for each positive number η and each compact set K with $0 \notin K$, there exists an approximately continuous function $\psi \colon \mathbb{R} \to [0,1]$ such that*

$$F = \psi^{-1}[0] \quad \textit{and} \quad G = \psi^{-1}[1],$$

and

$$\frac{1}{h} \int_x^{x+h} \psi(t)\, dt < \eta \quad \textit{whenever} \quad h \in K \textit{ and } x \in F.$$

If in addition G is compact, then ψ may be chosen to be continuous.

PROOF. Since F and G are disjoint density topology zero-sets, there is an approximately continuous function $\psi_0 \colon \mathbb{R} \to [0,1]$ such that

$$F = {\psi_0}^{-1}[0] \quad \text{and} \quad G = {\psi_0}^{-1}[1].$$

For each positive integer k, the function $\Psi_k \colon F \times K \to [0,1]$ given by the formula

$$\Psi_k(x,h) = \frac{1}{h} \int_x^{x+h} \big(\psi_0(t)\big)^k \, dt$$

is continuous. Clearly

$$\Psi_k \geq \Psi_{k+1} \geq 0.$$

Since the sequence ${\psi_0}^k$, $k = 1, 2. \dots,$ decreases to 0 λ-almost everywhere, we have that $\Psi_k(x,h)$ converges to 0 for each (x,h) in $F \times K$. Hence the sequence Ψ_k converges uniformly to 0 on $F \times K$. Observe that ψ_0 can be chosen to be continuous whenever G is compact. The proposition now follows.

PROPOSITION 3.4. *Let F be a closed set and let Z_k, $k = 1, 2, \dots,$ be a sequence of density topology zero-sets and δ_k, $k = 1, 2, \dots,$ be a sequence of positive numbers such that $F \subset Z_k$ and Z_k is contained in the δ_k-neighborhood of F. If $\lim_{k \to \infty} \delta_k = 0$, then*

$$Z = \textstyle\bigcup_{k=1}^{\infty} Z_k$$

is a density topology zero-set. If in addition each Z_k is closed, then so is Z.

PROOF. Let F_k be the closure of the δ_k-neighborhood of F. Then $Z \cup F_k$ is a density topology zero-set. Let $\psi_k\colon \mathbb{R} \to [0,1]$ be an approximately continuous function with $Z \cup F_k = {\psi_k}^{-1}[0]$. Then $\psi = \sum_{k=1}^{\infty} \frac{1}{2^k} \psi_k$ is an approximately continuous function such that $Z = \psi^{-1}[0]$. For the last statement, use the usual continuity.

The following lemma is similar to a statement in [**145**].

LEMMA 3.5. *Let F be a nonempty compact set, G be a G_δ set of Lebesgue measure 0 with $F \cap G = \emptyset$, and H be a density topology zero-set with $G \cap H = \emptyset$. Then, for each positive number ε, there exists an approximately continuous function $\varphi\colon \mathbb{R} \to [0,1]$ such that*

$$F \cup H \subset \varphi^{-1}[0] \quad \textit{and} \quad G = \varphi^{-1}[1],$$

and, for each positive integer i,

$$\frac{1}{h}\int_x^{x+h} \varphi(t)\,dt < \frac{1}{2^{i+1}}\,\varepsilon \quad \textit{whenever} \quad 0 < |h| \le \frac{1}{2^{i-1}} \quad \textit{and} \quad x \in F.$$

If in addition G and H are compact, then φ may be chosen to be continuous.

PROOF. For each pair of positive integers i and m let $\psi_{i,m}$ be the function given by Proposition 3.3 that satisfies

$$\frac{1}{h}\int_x^{x+h} \psi_{i,m}(t)\,dt < \frac{1}{2^{i+m+2}}\,\varepsilon \qquad \text{whenever} \quad \frac{1}{2^{i+m}} \le |h| \le \frac{1}{2^{i+m-1}} \quad \text{and} \quad x \in F.$$

The set

$$H_{i,m} = \{\, x : \psi_{i,m}(x) \le \tfrac{1}{2} \,\}$$

is a density topology zero-set such that

$$\frac{1}{h}\frac{1}{2}\int_x^{x+h} \bigl(1 - \chi_{H_{i,m}}(t)\bigr)\,dt \le \frac{1}{h}\int_x^{x+h} \psi_{i,m}(t)\,dt < \frac{1}{2^{i+m+2}}\,\varepsilon \qquad \text{whenever} \quad \frac{1}{2^{i+m}} \le |h| \le \frac{1}{2^{i+m-1}} \quad \text{and} \quad x \in F.$$

Let $Z_{i,m}$ be the intersection of $H_{i,m}$ and the closure of the $\frac{1}{2^{i+m-1}}$-neighborhood of F. By Proposition 3.4, we have that

$$Z = \bigcup_{i=1}^{\infty} \bigcup_{m=1}^{\infty} Z_{i,m}$$

is a density topology zero-set that contains F and is disjoint with G. Next let h be such that $0 < |h| \le \frac{1}{2^{i-1}}$. Then there is an m_0 such that

$$\frac{1}{2^{i+m_0}} < |h| \le \frac{1}{2^{i+m_0-1}}\,,$$

whence

$$\frac{1}{h}\int_x^{x+h} \chi_Z(t)\,dt \ge \frac{1}{h}\int_x^{x+h} \chi_{Z_{i,m_0}}(t)\,dt > 1 - \frac{1}{2^{i+m_0+1}}\,\varepsilon$$

because $(x, x+h) \cap Z_{i,m_0} = (x, x+h) \cap H_{i,m_0}$ whenever $x \in F$.

Now let $\varphi\colon \mathbb{R} \to [0,1]$ be an approximately continuous function such that

$$Z \cup H = \varphi^{-1}[0] \quad \text{and} \quad G = \varphi^{-1}[1].$$

If $0 < |h| \le \frac{1}{2^{i-1}}$ and $x \in F$, then

$$\frac{1}{h} \int_x^{x+h} \varphi(t)\, dt \le \frac{1}{h} \int_x^{x+h} \big(1 - \chi_z(t)\big)\, dt < \frac{1}{2^{i+1}}\, \varepsilon.$$

Thereby the function φ has been constructed. Finally observe that the sets $H_{i,m}$ can be chosen to be compact whenever H is compact, whence $Z \cup H$ will also be compact. So, the function φ can be selected to be continuous in the event that both H and G are compact.

LEMMA 3.6. *Let G be a G_δ set such that $G \subset [0,1]$ and $\lambda(G) = 0$. Then there exists an approximately continuous, extended real-valued function φ defined on $\mathbb{R}$ such that*

$$\varphi \ge 0, \qquad \int_{-\infty}^{+\infty} \varphi(t)\, dx < \infty,$$

$$\varphi(x) = +\infty \quad \textit{if and only if} \quad x \in G,$$

and

$$\lim_{h \to 0} \frac{1}{h} \int_x^{x+h} \varphi(t)\, dt = \varphi(x) \quad \textit{for every} \quad x\,.$$

If in addition the set G is compact, then φ can chosen to be continuous as well. Consequently, for a suitably chosen constant C, the function

$$H(x) = C \int_0^x (\varphi(t) + 1)\, dt \quad \textit{for} \quad 0 \le x \le 1,$$

is a homeomorphism of $[0,1]$ onto itself such that its derivative, in the extended real-valued sense, exists everywhere and is $C(\varphi + 1)$.

PROOF. As G is a bounded G_δ set with $\lambda(G) = 0$ it is the intersection of a nested sequence U_n, $n = 1, 2, \dots$, of bounded open sets with $\lambda(U_n) < \frac{1}{2^n}$. Let F_n denote $[-1,2] \setminus U_n$.

We shall construct a decreasing sequence $\varphi_n\colon \mathbb{R} \to [0,1]$, $n = 1, 2, \dots$, of approximately continuous functions and an increasing sequence Z_n, $n = 1, 2, \dots$, of sets such that, for each n and for each i,

(1) F_n is contained in the density topology interior of ${\varphi_n}^{-1}[0]$,
(2) $Z_n = {\varphi_n}^{-1}\big[[0, \frac{1}{2^n}]\big]$ and $G = {\varphi_n}^{-1}[1]$,
(3) if $x \in Z_n$ and $Z_n \subset {\varphi_n}^{-1}[0]$, then

$$\sum_{j=1}^{n} \varphi_j(x) \le n - 1 + \frac{1}{2^n},$$

(4) if $x \in F_n$ and $0 < |h| \le \frac{1}{2^{i-1}}$, then

$$\frac{1}{h} \int_x^{x+h} \varphi_n(t)\, dt < \frac{1}{2^{2n+i+1}}.$$

For $n = 1$ let φ_1 be the function given by Lemma 3.5 in which $F = F_1$, $H = \emptyset$ and $\varepsilon = \frac{1}{4}$ have been used. Since G is contained in $[0,1]$ we may assume that $\varphi_1(x) = 0$ whenever $x \notin [-1,2]$. Define Z_1 to be ${\varphi_1}^{-1}\big[[0,\frac{1}{2}]\big]$. Clearly, φ_1 satisfies the requirements.

For $n > 1$, assume that φ_k and Z_k have been constructed for each k with $k \le n-1$. Again, let φ_n be the function given by Lemma 3.5 in which $F = F_n$, $H = Z_{n-1}$ and $\varepsilon = \frac{1}{2^{2n}}$ have been used. And again we may assume that $\varphi_n(x) = 0$ whenever $x \notin [-1,2]$. Define Z_n to be ${\varphi_n}^{-1}\big[[0,\frac{1}{2^n}]\big]$. It is easy to verify conditions (1)–(4). The induction is completed.

Observe that F_n is contained in the set $W_n = {\varphi_n}^{-1}\big[[0,\frac{1}{2^n})\big]$ which is open in the density topology. Now define

$$\varphi = \sum_{n=1}^{\infty} \varphi_n .$$

Obviously, support$(\varphi) \subset [-1,2]$. We shall use Φ_n to denote $\sum_{k=1}^{n} \varphi_k$. Observe that Φ_n is approximately continuous. Clearly,

$$\varphi(u) = +\infty \quad \text{whenever} \quad u \in G,$$

and

$$\varphi(u) = \Phi_n(u) \le n - 1 + \tfrac{1}{2^n} \quad \text{whenever} \quad u \in Z_n .$$

Let us prove the approximate continuity of φ. Let α be a real number and suppose x is such that $\varphi(x) > \alpha$. As Φ_n increases pointwise to φ, there is an n such that $\Phi_n(x) > \alpha$, whence

$$x \in {\Phi_n}^{-1}\big[(\alpha,+\infty]\big] \subset \varphi^{-1}\big[(\alpha,+\infty]\big].$$

Thereby, we have shown that $\varphi^{-1}\big[(\alpha,+\infty]\big]$ is open in the density topology. Next, let β be a real number with $0 < \beta$ and suppose x is such that $\varphi(x) < \beta$. Then there is an integer n such that $x \in F_n$. As $F_n \subset W_n \subset Z_n$ and $\Phi_n(u) = \varphi(u)$ whenever $u \in Z_n$, we have

$$x \in {\Phi_n}^{-1}\big[[0,\beta)\big] \cap W_n \subset \varphi^{-1}\big[[0,\beta)\big].$$

Thus we have shown that $\varphi^{-1}\big[[0,\beta)\big]$ is open in the density topology. Hence φ is approximately continuous.

Since $F_n \subset Z_n$ and $\lambda([-1,2] \setminus F_n) = \lambda(U_n) < \frac{1}{2^n}$, we infer from (2) above that the integral of φ is finite. Hence only the limit statement remains to be proved. If $x \in G$, then

$$n = \lim_{h\to 0} \frac{1}{h} \int_x^{x+h} \Phi_n(t)\,dt \le \liminf_{h\to 0} \frac{1}{h} \int_x^{x+h} \varphi(t)\,dt \quad \text{for each} \quad n\,.$$

If $x \in [-1,2] \setminus G$, then there is an n such that $x \in F_n$. By (3), we have $\varphi(z) = \Phi_n(z) < n$ for every z in Z_n. Let ψ be the bounded, approximately continuous function $\varphi \wedge n$. Suppose $0 < |h| \le \frac{1}{2^{i-1}}$. We shall give the proof for the case $0 < h$, leaving the reader to supply the simple modifications for the case $0 > h$.

Again by (3),

$$\int_x^{x+h} (\varphi - \psi)(t)\,dt \leq \sum_{k=1}^{\infty} k\,\lambda\big((x,x+h)\cap\{\,u : \varphi(u) \geq n+k\,\}\big)$$
$$\leq \sum_{k=1}^{\infty} k\,\lambda\big((x,x+h)\setminus Z_{n+k}\big).$$

To compute a bound for the last line, we use the identity

$$(x,x+h)\setminus Z_{n+k} = (x,x+h)\cap\{\,u : \varphi_{n+k}(u) > \tfrac{1}{2^{n+k}}\,\}$$

to further compute, with the aid of (4),

$$\frac{1}{2^{n+k}}\big(\lambda\big((x,x+h)\setminus Z_{n+k}\big)\big) \leq \int_x^{x+h} \varphi_{n+k}(t)\,dt < \frac{1}{2^{2(n+k)+i+1}}\,h.$$

Consequently,

$$\left|\frac{1}{h}\int_x^{x+h} \varphi(t)\,dt - \frac{1}{h}\int_x^{x+h} \psi(t)\,dt\right| \leq \sum_{k=1}^{\infty} k\,\frac{1}{2^{n+k+i+1}} \leq \frac{1}{2^i}\sum_{k=1}^{\infty}\frac{k}{2^k}.$$

The limit statement is proved by letting h tend to 0 and then letting i tend to ∞. As support$(\varphi) \subset [-1,2]$ we have that the remaining case of $x \notin [-1,2]$ is trivial. The lemma now follows.

We now give the theorem of Zahorski [**146, 145**].

THEOREM (ZAHORSKI). *Let G be a G_δ subset of $[0,1]$ with $\lambda(G) = 0$. Then there exists a differentiable homeomorphism H of $[0,1]$ onto itself such that H' is bounded and $H'(x) = 0$ precisely on the set $H^{-1}[G]$. If in addition G is compact, then H may be chosen so that H' is continuous as well.*

PROOF. Let H be the inverse function of the homeomorphism provided by Lemma 3.6.

3.2. Continuously differentiable functions

The Cantor function example indicates that there is a set on which obstructions to continuous differentiability can occur. We shall present a necessary and sufficient condition on this obstruction set under which a CBV function will be Lebesgue equivalent of a continuously differentiable function. (See [**23**].)

3.2.1. Set of varying monotonicity. Motivated by the Cantor function, we make the following definition.

DEFINITION 3.7. Let $f\colon [a,b] \to \mathbb{R}$. A point x in $[a,b]$ is called a **point of varying monotonicity** of f if there is no neighborhood in $\mathbb{R}$ of x on which f is either strictly monotonic or strictly constant. The set of all such points will be denoted by K_f and called the **set of varying monotonicity**. Note that both a and b are points of K_f since f is undefined in the whole of any neighborhood in $\mathbb{R}$ of these points and hence is not strictly monotonic nor constant on the neighborhood.

An important observation is that

$$|f(d) - f(c)| = \mathrm{V}(f,[c,d]) \quad\text{whenever}\quad [c,d] \subset [a,b]\setminus K_f.$$

LEMMA 3.8. *Let $f\colon [a,b] \to \mathbb{R}$ be continuous. Then the following statements hold.*

(1) *K_f is a compact set.*
(2) *If $h\colon [a,b] \to [a,b]$ is a homeomorphism, then*

$$K_{f\circ h} = h^{-1}[K_f].$$

(3) *If $f[K_f]$ is nowhere dense, then K_f is nowhere dense.*

PROOF. Statements (1) and (2) are easily proved.

To prove (3), suppose that K_f is dense in some interval (c,d). Since K_f is closed, (c,d) is a subset of K_f. As $f[K_f]$ is nowhere dense and f is continuous, we have that $f\big[(c,d)\big]$ is a singleton set. We have arrived at the contradiction $(c,d) \cap K_f = \emptyset$.

Observe that if $f\colon (c,d) \to \mathbb{R}$ is a continuous, bounded, strictly increasing function, then there is a homeomorphism h of (c,d) onto itself such that $f \circ h$ is linear. Consequently, we have the following simple lemma.

LEMMA 3.9. *Let $f\colon [a,b] \to \mathbb{R}$ be a continuous function with finite total variation. Then there is a homeomorphism h of $[a,b]$ onto itself such that*

(1) *for each component (c,d) of $(a,b) \setminus K_f$,*

$$\operatorname{Lip}\big((f\circ h)|h^{-1}[[c,d]]\big) \le \frac{\mathrm{V}\big(f|[c,d],[c,d]\big)}{d-c},$$

(2) *$h[K_f] = K_f$,*
(3) *$f\circ h$ is continuously differentiable on the open set $[a,b]\setminus K_f$,*
(4) *$\operatorname{Lip}(f\circ h) \le \operatorname{Lip}(f)$.*

PROOF. Statements (1), (2) and (3) are immediate from the observation.

The proof of (4) is rather straightforward if one considers the various cases of $a \le x < x' \le b$. We only indicate the proof of the case where x and x' are in different components of $[a,b]\setminus K_f$. Let (c,d) and (c',d') be the components of $[a,b]\setminus K_f$ that contain x and x', respectively. Then we have

$$|(f\circ h)(x) - f(d)| \le \operatorname{Lip}(f)\,|x-d|,$$
$$|f(d) - f(c')| \le \operatorname{Lip}(f)\,|d-c'|$$

and

$$|f(c') - (f\circ h)(x')| \le \operatorname{Lip}(f)\,|c'-x'|.$$

Hence,

$$|(f\circ h)(x) - (f\circ h)(x')| \le \operatorname{Lip}(f)\,|x-x'|.$$

We have the following invariance lemma.

LEMMA 3.10. *Let $f\colon [a,b] \to \mathbb{R}$ be a continuous function with finite total variation. Then, for each homeomorphism h of $[a,b]$ onto itself,*

$$v_{f\circ h}[K_{f\circ h}] = v_f[K_f].$$

That is, the set $E = v_f[K_f]$ is invariant under Lebesgue equivalence.

PROOF. We know that $K_{f\circ h} = h^{-1}[K_f]$. Also, we have $v_{f\circ h}(x) = v_f\big(h(x)\big)$. The lemma follows.

3.2.2. The $\mathrm{C}^1[0,1]$ case. The characterization theorem by Bruckner and Goffman is

THEOREM 3.11 (BRUCKNER-GOFFMAN). *Let $f\colon [0,1] \to \mathbb{R}$. Necessary and sufficient conditions for the existence of a homeomorphism h of $[0,1]$ onto itself such that $f \circ h$ is continuously differentiable are that f be continuous, have finite total variation and satisfy $\lambda\big(f[K_f]\big) = 0$.*

The following proof is simpler than that in [**23**].

PROOF OF NECESSITY. That f be continuous and have finite total variation follows from Theorem 3.1. Suppose that h is a homeomorphism with $g = f \circ h$ being continuously differentiable. Hence g is Lipschitzian. By Lemma 3.8, we have $h^{-1}[K_f] = K_g$. The continuous differentiability of g yields $K_g \setminus \{0,1\} \subset g'^{-1}[0]$. Consequently, $\lambda\big(g[K_g]\big) = 0$ (see Section A.2.2). Observe that $f[K_f] = g[K_g]$.

PROOF OF SUFFICIENCY. In view of Lemmas 3.9 and 3.8, we may assume that $\lambda(K_f) = 0$ and that f is continuously differentiable on $[0,1] \setminus K_f$. The function

$$\varphi(x) = x + v_f(x), \quad x \in [0,1]$$

is a homeomorphism of $[0,1]$ onto $[0,B]$, where $B = 1 + \mathrm{V}(f,[0,1])$, with the properties that

$$f \circ \varphi^{-1} \quad \text{is Lipschitzian}$$

and

$$f \circ \varphi^{-1} \quad \text{is continuously differentiable on} \quad [0,B] \setminus K_{f\circ\varphi^{-1}}\,.$$

We assert that $\lambda\big(\varphi[K_f]\big) = 0$. Indeed, let $\varepsilon > 0$. There exists a positive number η such that the Lebesgue measure of the η-neighborhood U of K_f has $\lambda(U) < \varepsilon$. From Proposition A.11 of the Appendix, there is a positive number δ such that

$$\sum_{i=1}^{n} \mathrm{V}(f,[a_i,b_i]) < \varepsilon \tag{3.1}$$

whenever $[a_i,b_i]$, $i = 1,\ldots,n$, is a nonoverlapping collection of intervals such that $f\big[[a_i,b_i]\big]$ is contained in the δ-neighborhood of $f[K_f]$. As f is continuous, there is a collection $[a_i,b_i]$, $i = 1,\ldots,n$, of nonoverlapping intervals contained in U such that $\bigcup_{i=1}^{n}[a_i,b_i] \supset K_f$ and $f\big[[a_i,b_i]\big]$ is contained in the η-neighborhood of $f[K_f]$ for each i. For each i,

$$\varphi(b_i) - \varphi(a_i) = b_i - a_i + v_f(b_i) - v_f(a_i).$$

Now we have that the collection $\big[\varphi(a_i),\varphi(b_i)\big]$, $i = 1,\ldots,n$, covers $\varphi[K_f]$ and

$$\lambda(\varphi[K_f]) \le \sum_{i=1}^{n}(b_i - a_i) + \sum_{i=1}^{n} \mathrm{V}(f,[a_i,b_i]) < 2\,\varepsilon.$$

By the theorem of Zahorski from Section 3.1.3, there is a continuously differentiable homeomorphism ψ from $[0,1]$ onto $[0,B]$ such that $\psi'(t) = 0$ precisely when $t \in \psi^{-1}\big[\varphi[K_f]\big]$. The continuous differentiability on $[0,1] \setminus \psi^{-1}\big[\varphi[K_f]\big]$ of $f \circ \varphi^{-1} \circ \psi$ is clear. Let us show

$$(f \circ \varphi^{-1} \circ \psi)'(t) = \psi'(t) = 0 \quad \text{whenever} \quad t \in \psi^{-1}\big[\varphi[K_f]\big]. \tag{3.2}$$

We have, for any s and u with $s \neq u$, that

$$\begin{aligned}(3.3)\quad &\frac{(f\circ\varphi^{-1}\circ\psi)(s)-(f\circ\varphi^{-1}\circ\psi)(u)}{s-u}\\ &\qquad=\frac{(f\circ\varphi^{-1})\big(\psi(s)\big)-(f\circ\varphi^{-1})\big(\psi(u)\big)}{\psi(s)-\psi(u)}\;\frac{\psi(s)-\psi(u)}{s-u}\end{aligned}$$

and that $f\circ\varphi^{-1}$ is Lipschitzian, whence formula (3.2) follows. From this identity follows also the continuity of $(f\circ\varphi^{-1}\circ\psi)'$ at any t in $\psi^{-1}\big[\varphi[K_f]\big]$. Indeed, let $\varepsilon > 0$. Then there is a neighborhood U of t such that

$$0 \le \operatorname{Lip}(f\circ\varphi^{-1})\,\psi'(u) < \varepsilon \quad\text{whenever}\quad u \in U.$$

From (3.3) above, we infer that

$$|(f\circ\varphi^{-1}\circ\psi)'(u)| \le \varepsilon \quad\text{whenever}\quad u \in U.$$

The theorem is completely proved.

3.2.3. Observations. In anticipation of the next section, we make some observations on the last sufficiency proof. The condition associated with the inequality (3.1) is very important. In essence, it asserts that the naive 1-dimensional packing of the set $v_f[K_f]$ is 0, that is, $\mathsf{npH}_1(v_f[K_f]) = 0$. (See Section A.7 of the Appendix.) Also, key to the proof is the fact that $v_f[K_f]$ has Hausdorff 1-dimensional measure equal to 0, that is, $\mathsf{H}_1(v_f[K_f]) = 0$. The condition $\mathsf{npH}_{\frac{1}{n}}(v_f[K_f]) = 0$ will appear in the next section on functions in $\mathrm{C}^n[0,1]$. Observe that, in the sufficiency proof, we could have chosen the set K_f to be any compact set of Lebesgue measure equal to 0. Indeed, it could be chosen to be a set of Hausdorff dimension equal to 0, that is, $\dim_{\mathsf{H}} K_f = 0$.

3.3. The class $\mathrm{C}^n[0,1]$

The characterization of the class $\mathrm{C}^n[0,1]$ for $n > 1$ turns out to be slightly different from that for $n = 1$. As was observed at the end of the last section, the naive Hausdorff packing will play a major role; in particular, $\mathsf{npH}_{\frac{1}{n}}(v_f[K_f]) = 0$ is the condition that is involved. For $n > 1$, this condition automatically assures that $\lambda(v_f[K_f]) = 0$ since

$$\lambda(E) = \mathsf{H}_1(E) \le \mathsf{H}_{\frac{1}{n}}(E) \le \mathsf{npH}_{\frac{1}{n}}(E)$$

when E is a compact subset of $\mathbb{R}$.

3.3.1. Characterization theorem. We shall discover that the proof of the characterization theorem is essentially the same as that of the characterization in the case $\mathrm{C}^1[0,1]$. The necessary condition is a consequence of the Fundamental Theorem of Calculus. And the sufficiency is a consequence of a straightforward modification of the function that is produced in the $\mathrm{C}^1[0,1]$ case. We state the theorem.

THEOREM 3.12. *Let n be a positive integer and f be a function from $[0,1]$ into $\mathbb{R}$. In order for f to have a homeomorphism h of $[0,1]$ onto itself such that $f\circ h$ be in $\mathrm{C}^n[0,1]$ it is necessary and sufficient that f be continuous, have finite total variation and have $\mathsf{npH}_{\frac{1}{n}}\big(v_f[K_f]\big) = 0$.*

The proof given here is similar to that given by Laczkovich and Preiss in [**91**] of their characterization theorem. The theorem of Laczkovich and Preiss does not use the variation function v_f; rather it uses another condition called the $\frac{1}{n}$-variation of f on K_f, denoted by $V_{\frac{1}{n}}(f, K_f)$, which will be defined later. Their theorem is stated next.

THEOREM 3.13 (LACZKOVICH-PREISS). *If $n > 1$, then $f\colon [0,1] \to \mathbb{R}$ has a homeomorphism h of $[0,1]$ onto itself such that $f \circ h$ is in $C^n[0,1]$ if and only if f is continuous and has $V_{\frac{1}{n}}(f, K_f) < +\infty$.*

The condition $\mathsf{npH}_{\frac{1}{n}}(v_f[K_f]) = 0$ is geometric in nature and the condition $V_{\frac{1}{n}}(f, K_f) < +\infty$ is function theoretic in nature. We shall prove the equivalence of the two theorems at the end of the section.

3.3.2. Necessary conditions. The necessity in Theorem 3.12 for $n = 1$ has been proved in the previous section (see the observations made there). Hence we shall assume that $n > 1$.

Let us begin with two elementary facts about the total variation of a function in $C^n[0,1]$. The first one is trivial.

PROPOSITION 3.14. *For $n > 1$, let $f \in C^n[0,1]$ and let η and γ be such that $0 < \eta < \gamma < 1$. If $[c,d]$ is a subset of $[0,1]$ such that $\eta \le d - c < \gamma$, then*

$$\eta^n \, V(f,[c,d]) \le (d-c)^n \sup\{\, V(f,[x,x+\gamma]) : x \in [0,1-\gamma]\,\}.$$

PROPOSITION 3.15. *For $n > 1$, let $f \in C^n[0,1]$. If $[c,d]$ is a subset of $[0,1]$ such that the set of points in $[c,d]$ at which the value of f' is 0 has at least $n-1$ members, then*

$$V(f,[c,d]) \le \|f^{(n)}\|_\infty \, (d-c)^n.$$

PROOF. By repeatedly applying Rolle's theorem, we may select points x_j, $j = 1, \ldots, n-1$, such that $f^{(j)}(x_j) = 0$ for each j. From

$$|f^{(n-1)}(x) - f^{(n-1)}(x')| \le \|f^{(n)}\|_\infty \, (d-c), \quad c \le x \le x' \le d,$$

we infer that

$$|f^{(n-1)}(x)| \le \|f^{(n)}\|_\infty \, (d-c) \quad \text{whenever} \quad x \in [c,d].$$

Repeating the same argument, we have

$$|f^{(1)}(x)| \le \|f^{(n)}\|_\infty \, (d-c)^{n-1} \quad \text{whenever} \quad x \in [c,d].$$

Consequently,

$$V(f,[c,d]) \le \|f^{(n)}\|_\infty \, (d-c)^n$$

and the proposition follows.

We are now ready to prove that $\mathsf{npH}_{\frac{1}{n}}(v_f[K_f]) = 0$ for functions f in $C^n[0,1]$ when $n > 1$. As $f \in C^1[0,1]$ we have $\lambda(v_f[K_f]) = 0$. Hence Proposition A.22 of the Appendix applies. So showing $\mathsf{npH}_{\frac{1}{n}}(v_f[K_f]) = 0$ is equivalent to showing

$$\Gamma_{\frac{1}{n}}(v_f[K_f]) < \infty, \tag{3.4}$$

where $\Gamma_\alpha(E)$ is defined on page 200. Since v_f is an increasing function, we may consider finite partitions of $[0,1]$ formed by points of K_f because $\Gamma_{\frac{1}{n}}(v_f[K_f])$ does not exceed the bound that results from using these finite partitions. Recall that 0 and 1 are members of K_f. Since f has a continuous derivative, each point of $(0,1)\cap K_f$ is a point at which the value of f' is 0.

Let $0 = x_0 < x_2 < \cdots < x_k = 1$ be a partition of $[0,1]$ with $x_i \in K_f$ for each i. If $\operatorname{card}\big((0,1)\cap K_f\big) < n+2$, then we infer from the first proposition that

$$\big(v_f(x_i) - v_f(x_{i-1})\big)^{\frac{1}{n}} \le \tfrac{1}{\eta}\ \mathrm{V}(f,[0,1])^{\frac{1}{n}}\,(x_i - x_{i-1}),$$

where η is the minimum of the lengths of the finitely many components of $(0,1)\backslash K_f$. Hence

$$\sum_{i=1}^{k}\big(v_f(x_i) - v_f(x_{i-1})\big)^{\frac{1}{n}} \le \tfrac{1}{\eta}\ \mathrm{V}(f,[0,1])^{\frac{1}{n}},$$

whence the inequality (3.4) holds for $\operatorname{card}\big((0,1)\cap K_f\big) < n+2$.

Suppose next that $\operatorname{card}\big((0,1)\cap K_f\big) \ge n+2$. It may happen that the number k of subintervals of the partition is smaller than $n+3$. Fortunately, we have

$$(a+b)^{\frac{1}{n}} \le a^{\frac{1}{n}} + b^{\frac{1}{n}} \quad \text{whenever} \quad 0 \le a \quad \text{and} \quad 0 \le b.$$

So we may refine the partition so as to achieve one with $k \ge n+3$ because we are searching for an upper bound. Thus we consider partitions with

$$0 = x_0 < \cdots < x_n < \cdots < x_{n+3} < \cdots < x_{k-1} < x_k = 1$$

and

$$f'(x_i) = 0 \quad \text{for} \quad 0 < i < k.$$

It may happen that $f'(x_0) \ne 0$ or $f'(x_k) \ne 0$. We shall consider the worst case in which both x_0 and x_k (that is, 0 and 1) are isolated points of K_f. In this case let

$$M = \max\big\{\,(x_1 - x_0)^{-1},\ (x_k - x_{k-1})^{-1}\,\big\}.$$

Then, by the first of the 2 propositions,

$$\begin{aligned}\big(v_f(x_1) - v_f(x_0)\big)^{\frac{1}{n}} &+ \big(v_f(x_k) - v_f(x_{k-1})\big)^{\frac{1}{n}} \\ &\le M\,\big(\mathrm{V}(f,[0,1])\big)^{\frac{1}{n}}\big((x_1 - x_0) + (x_k - x_{k-1})\big)\end{aligned} \tag{3.5}$$

for every partition under consideration.

Let $1 \le j \le k-n-1$ and consider the intervals $[x_j, x_{j+n}]$. These intervals have at least $n-1$ points at which f' has the value 0. So, by the second of the above propositions, we have

$$\begin{aligned}\big(v_f(x_{j+m}) - v_f(x_{j+m-1})\big)^{\frac{1}{n}} &\le \big(v_f(x_{j+n}) - v_f(x_j)\big)^{\frac{1}{n}} \\ &\le \big(\|f^{(n)}\|_\infty\big)^{\frac{1}{n}}\,(x_{j+n} - x_j)\end{aligned}$$

for $1 \le m \le n$. Observe that, for each i with $2 \le i \le k-1$, the interval $[x_i, x_{i-1}]$ is contained in at least one and at most n of the intervals $[x_{j+n}, x_j]$,

$j = 1, \dots, k-n-1$. So,

$$\begin{aligned}\sum_{i=2}^{k-1}\big(v_f(x_i) - v_f(x_{i-1})\big)^{\frac{1}{n}} &\le \sum_{j=1}^{k-n-1}\big(\|f^{(n)}\|_\infty\big)^{\frac{1}{n}}(x_{j+n} - x_j)\\ &= \big(\|f^{(n)}\|_\infty\big)^{\frac{1}{n}}\sum_{j=1}^{k-n-1}\sum_{m=1}^{n}(x_{j+m} - x_{j+m-1})\\ &\le n\,\big(\|f^{(n)}\|_\infty\big)^{\frac{1}{n}}(x_{k-1} - x_1).\end{aligned}$$

This together with the inequality (3.5) will yield the required inequality (3.4).
Thereby the necessity is proved.

3.3.3. Sufficient conditions. As the sufficiency has already been shown for the case $n = 1$, we shall assume $n > 1$. We need a preliminary lemma. The lemma will use a standard function $w\colon [0,1] \to [0,1]$, given below, in the class $C^\infty[0,1]$ that strictly increases from 0 to 1.

$$w(x) = c\int_0^x \exp\big(-t^{-2} - (1-t)^{-2}\big)\,dt, \quad x \in [0,1],$$

where the positive constant is chosen so that $w(1) = 1$. The function w satisfies

$$w^{(k)}(0) = w^{(k)}(1) = 0, \quad k = 1, 2, \dots .$$

Now let $f\colon [0,1] \to \mathbb{R}$ be a continuous function. We shall modify f on $(0,1)\setminus K_f$ into a function g_f so that g_f 'looks like' w on each component of $(0,1)\setminus K_f$. It will be evident from the construction that f and g_f are Lebesgue equivalent. Let (a_j, b_j), $j \in J$, be an enumeration of the components of $(0,1)\setminus K_f$. Then define

$$g_f(x) = \begin{cases} f(x) & \text{if } x \in K_f,\\ f(a_j) + \big(f(b_j) - f(a_j)\big)\,w\big(\frac{x-a_j}{b_j-a_j}\big) & \text{if } x \in (a_j, b_j),\ j \in J.\end{cases}$$

LEMMA 3.16. *Let $n > 1$ and let $f\colon [0,1] \to \mathbb{R}$ be a continuous function that satisfies the the following condition on K_f.*

For each positive number ε there is a positive number η such that

$$|f(x) - f(x')| \le \varepsilon\,|x - x'|^n \quad \textit{whenever} \quad x, x' \in K_f \quad \textit{and} \quad |x - x'| < \eta.$$

Then $g_f \in C^n[0,1]$, and f and g_f are Lebesgue equivalent.

PROOF. It is clear that ${g_f}^{(k)}(x) = 0$ for $k > 0$ whenever x is an isolated point of K_f. Also, for each j in J with $b_j - a_j < \eta$, we have

$$|{g_f}^{(k)}(t)| \le \frac{|f(b_j) - f(a_j)|}{(b_j - a_j)^k}\,\|w^{(k)}\|_\infty \le \varepsilon\,(b_j - a_j)^{n-k}\,\|w^{(k)}\|_\infty$$

whenever $t \in (a_j, b_j)$ and $1 \le k \le n$.

Let us prove that g_f has a continuous derivative. First, since $n > 1$, it is immediate that $\lambda_1\big(f[K_f]\big) = 0$. So, by Lemma 3.8, K_f is nowhere dense. If $x = a_j$ for some j in J, then g_f has continuous derivatives of every positive order from the right at x with the value of 0. So, assume that $x \in K_f$ with $x \ne a_j$ for every j. Let t be such that $0 < t - x < \eta$ and $t \notin K_f$. Then there exists a j in J such that

$$x < a_j \le t \le b_j.$$

We may assume further that $b_j - a_j < \eta$ holds. Then

$$|g_f(t) - g_f(x)| \le |g_f(t) - g_f(a_j)| + |f(a_j) - f(x)|$$
$$\le \varepsilon\,(b_j - a_j)^n\,\|w\|_\infty + \varepsilon\,|a_j - x|^n,$$

whence g_f is continuous from the right at x. To establish the differentiability from the right at x, we have

$$|g_f(t) - g_f(x)| \le \varepsilon\,(b_j - a_j)^{n-1}\,\|w^{(1)}\|_\infty\,|t - a_j| + \varepsilon\,|a_j - x|^n$$
$$\le \varepsilon\left((b_j - a_j)^{n-1}\,\|w^{(1)}\|_\infty + |a_j - x|^{n-1}\right)|t - x|.$$

Hence the derivative from the right at x of g_f exists with value 0. Since the argument from the left is the same, the derivative ${g_f}^{(1)}$ exists. It is also clear that this derivative is continuous. The proof of the existence and continuity of ${g_f}^{(k)}$ follows the same pattern for $2 \le k \le n$. The remainder of the proof is left to the reader.

Let us now prove the sufficiency. Let $E = v_f[K_f]$. We have that

$$\lim_{\delta\to 0} \mathsf{npH}^{\delta}_{\frac{1}{k}}(E) = \mathsf{npH}_{\frac{1}{k}}(E) = 0 \quad \text{for} \quad 2 \le k \le n.$$

For each positive integer m and for each k with $2 \le k \le n$ we select a positive number $\delta_{k,m}$ such that

$$\mathsf{npH}^{\delta_{k,m}}_{\frac{1}{k}}(E) < \frac{1}{m\,2^{k+m}}\,.$$

We infer from Proposition A.23 of the Appendix that there is a continuous, nondecreasing function $F_{k,m}\colon [0, v_f(1)] \to \left[0, \frac{1}{m\,2^{k+m}}\right]$ such that

$$|x' - x|^{\frac{1}{k}} \le F_{k,m}(x') - F_{k,m}(x) \le \mathsf{npH}^{\delta_{k,m}}_{\frac{1}{k}}\left([x, x'] \cap E\right) \tag{3.6}$$

whenever $0 < x' - x < \delta_{k,m}$ and $x, x' \in E$.

As

$$F = \sum_{k=2}^{n} \sum_{m=1}^{\infty} m\,F_{k,m}$$

is a uniformly convergent series of nondecreasing, continuous functions, we have that F is also a nondecreasing, continuous function. Consequently the function φ defined as

$$\varphi(x) = x + v_f(x) + \left(F \circ v_f\right)(x), \quad x \in [0, 1],$$

satisfies the inequalities

$$|f(x) - f(x')| \le |v_f(x) - v_f(x')| \le |\varphi(x) - \varphi(x')|$$

and

$$m\left|\left(F_{n,m} \circ v_f\right)(x) - \left(F_{n,m} \circ v_f\right)(x')\right|$$
$$\le \left|\left(F \circ v_f\right)(x) - \left(F \circ v_f\right)(x')\right| \le \left|\varphi(x) - \varphi(x')\right|.$$

That is,

$$\left|(f \circ \varphi^{-1})(t) - (f \circ \varphi^{-1})(t')\right| \le \left|t - t'\right|$$

and

$$m\left|\left(F_{n,m} \circ v_f \circ \varphi^{-1}\right)(t) - \left(F_{n,m} \circ v_f \circ \varphi^{-1}\right)(t')\right| \le \left|t - t'\right|.$$

If t and t' are points of $\varphi[K_f]$ with $t > t'$ and

$$\left|(v_f \circ \varphi^{-1})(t) - (v_f \circ \varphi^{-1})(t')\right| < \delta_{n,m},$$

then we have from the inequality (3.6) that

$$\begin{aligned}\left|(f \circ \varphi^{-1})(t) - (f \circ \varphi^{-1})(t)\right|^{\frac{1}{n}} &\le \left|(v_f \circ \varphi^{-1})(t) - (v_f \circ \varphi^{-1})(t)\right|^{\frac{1}{n}} \\ &\le \left(F_{n,m} \circ v_f \circ \varphi^{-1}\right)(t') - \left(F_{n,m} \circ v_f \circ \varphi^{-1}\right)(t) \le \frac{1}{m}|t-t'|.\end{aligned}$$

Now we make a linear change of scale ψ from $[0,1]$ to $[0, \varphi(1)]$. Since

$$K_{f\circ\varphi^{-1}\circ\psi} = \left(\varphi^{-1} \circ \psi\right)^{-1}[K_f] = \psi^{-1}\left[\varphi[K_f]\right],$$

we infer from the continuity of $v_f \circ \varphi^{-1} \circ \psi$ that $f \circ \varphi^{-1} \circ \psi$ satisfies the property of Lemma 3.16. The sufficiency now follows from that lemma.

3.3.4. $C^\infty[0,1]$ and examples. The proofs in the preceding section have been designed to make the proof of the $C^\infty[0,1]$ case immediately obvious. Indeed, the sufficiency is proved by just passing to the infinite series.

THEOREM 3.17. *In order for a function $f\colon [0,1] \to \mathbb{R}$ to have a homeomorphism h of $[0,1]$ onto itself such that $f \circ h$ be in $C^\infty[0,1]$ it is necessary and sufficient that f be continuous, have finite total variation and have $\mathsf{npH}_{\frac{1}{n}}(v_f[K_f]) = 0$ for every positive integer n.*

Let us exhibit for each positive integer n a continuous function f with finite total variation such that $\mathsf{npH}_\alpha(v_f[K_f]) = +\infty$ for $0 \le \alpha < \frac{1}{n}$ and $\mathsf{npH}_\beta(v_f[K_f]) = 0$ for $\frac{1}{n} \le \beta$. For each positive integer k, define

$$x_1 = 0 \quad \text{and} \quad x_k = \frac{1}{\log_2 k} \quad \text{for} \quad k \ge 2.$$

Then construct $f\colon [0,1] \to \mathbb{R}$ by first defining

$$f(x_k) = 0 \quad \text{and} \quad f\left(\tfrac{x_{k+1}+x_{k+2}}{2}\right) = (k+1)^{-n}\left(\log_2(k+1)\right)^{-2n-2} \quad \text{for} \quad k \ge 1,$$

and then extending linearly to the rest of $[0,1]$. Obviously,

$$K_f = \{\, x_k : k = 1, 2, \dots \} \cup \{\, \tfrac{x_k + x_{k+1}}{2} : k = 2, 3, \dots \}.$$

It is clear that the function f is continuous and that its total variation

$$\mathrm{V}(f, [0,1]) = 2\sum_{k=2}^{\infty} k^{-n}\left(\log_2 k\right)^{-2n-2}$$

is finite. The set $v_f[K_f]$ consists of $v_f(0) = 0$ together with the points

$$v_f(x_k) = 2\sum_{j=k+1}^{\infty} j^{-n}\left(\log_2 j\right)^{-2n-2},$$

$$v_f\left(\tfrac{x_k+x_{k+1}}{2}\right) = v_f(x_{k+1}) + k^{-n}\left(\log_2 k\right)^{-2n-2}$$

for $k \ge 2$. According to Proposition A.22 of the Appendix, we can determine the value of $\mathsf{npH}_\alpha(v_f[K_f])$ by computing the convergence or divergence of

$$2\sum_{k=1}^{\infty}\left(k^{-n}\left(\log_2 k\right)^{-2n-2}\right)^{\alpha}.$$

It is easily shown that the series converges for $\alpha \geq \frac{1}{n}$ and diverges for $\alpha < \frac{1}{n}$.

By virtue of the characterization theorem, we are able to conclude that there is a homeomorphism h such that $f \circ h$ is in $\mathrm{C}^n[0,1]$ and that no homeomorphism h will place $f \circ h$ into $\mathrm{C}^{n+1}[0,1]$. In closing, we observe that the set $v_f[K_f]$ is countable. This shows that the rarefaction index $\Lambda(v_f[K_f])$ (see Section A.7.2 of the Appendix) is more appropriate as a sufficient condition than the Hausdorff packing dimension $\mathrm{Dim}(v_f[K_f])$ as defined by Taylor and Tricot in [**125**] since the the latter number has the value 0.

3.3.5. Proof of equivalence. To prove the equivalence of the characterization theorem by Laczkovich and Preiss and the one that was just proved, we will need their definition of the α-variation of a function on a set.

Let $f\colon [0,1] \to \mathbb{R}$ be a function and K be a nonempty subset of $[0,1]$. For a positive real number α, the **α-variation of f on K** is defined to be

$$\mathrm{V}_\alpha(f,K) = \sup \sum_{i=1}^{m} |f(b_i) - f(a_i)|^\alpha,$$

where the supremum is taken over all collections $\{\,[a_i,b_i] : i = 1,2,\ldots,m\}$ of nonoverlapping intervals $[a_i,b_i]$ such that $\partial\,[a_i,b_i] \subset K$. Clearly, if $0 < \alpha \leq \beta \leq 1$, then

$$(2\,\|f\|_\infty)^{\beta-\alpha}\,\mathrm{V}_\alpha(f,K) \geq \mathrm{V}_\beta(f,K). \tag{3.7}$$

As a variant of this form of total variation, they consider a positive number δ and define

$$\mathrm{V}_\alpha^\delta(f,K) = \sup \sum_{i=1}^{m} |f(b_i) - f(a_i)|^\alpha,$$

where the supremum is taken over all collections $\{\,[a_i,b_i] : i = 1,2,\ldots,m\}$ of nonoverlapping intervals $[a_i,b_i]$ such that $\partial\,[a_i,b_i] \subset K$ and $b_i - a_i \leq \delta$.

And finally, they define

$$\mathrm{SV}_\alpha(f,K) = \lim_{\delta \to 0} \mathrm{V}_\alpha^\delta(f,K).$$

There is an extensive discussion of these notions in their paper [**91**] with references to other earlier related concepts. They do not, as we do, restrict α to satisfy $0 < \alpha \leq 1$.

In our presentation we shall always assume that $f\colon [0,1] \to \mathbb{R}$ is continuous. Let us give some elementary properties of these variations. As f is uniformly continuous there is a positive number δ such that $|f(x) - f(x')| < 1$ whenever $|x - x'| < \delta$. Consequently, if $0 < \alpha < \beta < 1$, then

$$\mathrm{V}_\alpha^\delta(f,K) \geq \mathrm{V}_\beta^\delta(f,K) \geq \mathrm{V}_1^\delta(f,K),$$

whence

$$\mathrm{SV}_\alpha(f,K) \geq \mathrm{SV}_\beta(f,K_f) \geq \mathrm{SV}_1(f,K_f).$$

Because $\partial\,[0,1] \subset K_f$, we have from the definition of the set K_f that

$$\mathrm{V}_1(f,K_f) = \mathrm{V}(f,[0,1]). \tag{3.8}$$

The following is a simple consequence of the definitions.

PROPOSITION 3.18. *Suppose that* $f\colon [0,1] \to \mathbb{R}$ *is a continuous function and* α *satisfies* $0 < \alpha \le 1$. *If* $\mathrm{SV}_\alpha(f, K_f) = 0$, *then* $\lambda_1\big(f[K_f]\big) = 0$, *whence* f *is Lebesgue equivalent to a function in* $\mathrm{C}^1[0,1]$ *and* $\lambda_1\big(v_f[K_f]\big) = 0$.

PROOF. Let ε be a positive number. We know that $\mathrm{SV}_1(f, K_f) = 0$. Hence there exists a positive number δ such that $\mathrm{V}_1^\delta(f, K_f) < \varepsilon$. Let $[a_i, b_i]$, $i = 1, 2, \dots, m$, be a finite, mutually disjoint collection that covers K_f with $\partial[a_i, b_i] \subset K_f$ and $b_i - a_i < \delta$ for each i. (Here, we permit $a_i = b_i$ if necessary.) As K_f is compact, we have for each i that $\operatorname{diam} f\big[[a_i, b_i] \cap K_f\big] \le |f(a_i') - f(b_i')|$ for some a_i' and b_i' in $[a_i, b_i] \cap K_f$. Consequently, $f\big[[a_i, b_i] \cap K_f\big]$, $i = 1, 2, \dots, m$, is a cover $f[K_f]$ with $\sum_{i=1}^m \lambda_1\big(f\big[[a_i, b_i] \cap K_f\big]\big) < \varepsilon$. Thereby, we have shown $\lambda_1\big(f[K_f]\big) = 0$.

To complete the proof, we shall show, for $0 < \alpha \le 1$, that $\mathrm{V}_\alpha^\delta(f, K) < +\infty$ if and only if $\mathrm{V}_\alpha(f, K) < +\infty$, where K is a compact subset of $[0,1]$. One direction is obvious. Let $\mathcal{C}$ be the collection $[c_j, d_j]$, $j = 1, 2, \dots, k$, consisting of the closures of the components $[0,1] \setminus K$ whose diameters exceed $\frac{\delta}{2}$. Let $[a, b]$ be a closed interval with $\partial[a, b] \subset K$. Then, for each j, we have either $[c_j, d_j] \subset [a, b]$ or $[c_j, d_j] \cap [a, b] = \emptyset$. Hence $[a, b]$ can be partitioned into intervals from the collection $\mathcal{C}$ and intervals whose diameters do not exceed δ. As $\alpha \le 1$, we have $\mathrm{V}_\alpha(f, K) \le \mathrm{V}_\alpha^\delta(f, K) + 2\,k\,\|f\|_\infty$.

The proposition follows from (3.7) and (3.8).

PROPOSITION 3.19. *Suppose that* $f\colon [0,1] \to \mathbb{R}$ *is a continuous function and* α *satisfies* $0 < \alpha \le 1$. *If* $\mathrm{SV}_\alpha(f, K_f) = 0$, *then* $\mathsf{npH}_\alpha\big(v_f[K_f]\big) = 0$.

PROOF. The last proposition gives us that f has finite total variation. Let δ be such that $\mathrm{V}_\alpha^\delta(f, K_f) < 1$. As $\lambda_1(K_f) = 0$ we may apply Proposition A.22 of the Appendix. Let $\mathcal{P} = \{\, x_i : i = 0, 1, 2, \dots, k \,\}$ be a partition of $[0,1]$ whose mesh is less than δ. For each i, let $K_i = K_f \cap [x_{i-1}, x_i]$ and $E_i = v_f[K_i]$. Observe that each member of E_i is a lower bound of E_j whenever $i < j$ and E_i and E_j are not empty, whence the distance between them does not exceed $\mathrm{V}(f, [0,1])$. Moreover, $v_f[K_f] = \bigcup_{i=1}^k E_i$. Using the notation of Proposition A.22, we have $\Gamma_\alpha(E_i) \le \mathrm{V}_\alpha^\delta(f, K_i)$. A straightforward computation will yield

$$\Gamma_\alpha\big(v_f[K_f]\big) \le k\,\big(1 + (\mathrm{V}(f, [0,1]))^\alpha\big) < +\infty.$$

By Proposition A.22, we have $\mathsf{npH}_\alpha\big(v_f[K_f]\big) = 0$.

PROPOSITION 3.20. *Suppose that* $f\colon [0,1] \to \mathbb{R}$ *is a continuous function with finite total variation and* α *satisfies* $0 < \alpha \le 1$. *Then*

$$\mathrm{V}_\alpha^\delta(f, K_f) \le \mathsf{npH}_\alpha^\eta\big(v_f[K_f]\big)$$

whenever δ *is such that* $\sup\{\, v_f(b) - v_f(a) : b - a \le \delta \,\} \le \eta$.

PROOF. Let $\{\, [a_i, b_i] : i = 1, 2, \dots, m \,\}$ be a collection of nonoverlapping intervals $[a_i, b_i]$ with $\partial\,[a_i, b_i] \subset K_f$ and $b_i - a_i \le \delta$. Then

$$\sum_{i=1}^m |f(b_i) - f(a_i)|^\alpha \le \sum_{i=1}^m |v_f(b_i) - v_f(a_i)|^\alpha \le \mathsf{npH}_\alpha^\eta\big(v_f[K_f]\big)$$

and the proposition follows.

We now have the following theorem, which has the Laczkovich-Preiss characterization theorem as a corollary.

THEOREM 3.21. *Let $f: [0,1] \to \mathbb{R}$ be a continuous function and let $0 < \alpha \leq 1$. Then the following conditions are equivalent.*

(1) $\mathrm{V}(f,[0,1]) < +\infty$ *and* $\mathsf{npH}_\alpha(v_f[K_f]) = 0$.
(2) $\mathrm{V}_\alpha(f, K_f) < +\infty$.
(3) $\mathrm{SV}_\alpha(f, K_f) = 0$.

PROOF. Condition (1) implies (2) because of Proposition 3.20. Indeed, by selecting $\eta = \mathrm{V}(f,[0,1])$ and $\delta = 1$, we have by Proposition A.22 of the Appendix that $\mathsf{npH}^\eta_\alpha(v_f[K_f]) < +\infty$. And, we infer from the definition of the set K_f and the same Proposition A.22 that condition (2) implies condition (1).

That condition (3) implies (1) is Proposition 3.19. Finally, that condition (1) implies (3) follows from Proposition 3.20 and the definition of $\mathrm{SV}_\alpha(f, K_f) = 0$.

3.4. Remarks

3.4.1. Packing characterizations. The Theorems 3.12 and 3.17 are new to the literature. And the preliminary propositions (Propositions 3.19 and 3.20) to the equivalence theorem, Theorem 3.21, are also new.

3.4.2. Almost everywhere differentiability. In the spirit of Chapter 2, it is possible to combine the Lebesgue equivalence with differentiability λ-almost everywhere. We state without proof a theorem due to Bruckner [**19**] along this line.

THEOREM 3.22 (BRUCKNER). *Let f be a function defined on $[0,1]$. A necessary and sufficient condition for there to exist a homeomorphism h of $[0,1]$ onto itself such that $f \circ h$ is differentiable λ-almost everywhere is that f be continuous on a dense subset of $[0,1]$.*

3.4.3. Representations. Many questions on representations surfaced with the arrival of modern analysis and topology. Questions such as whether the unit square was a curve. That is, can one use a parameter domain that consists of a unit interval? As we now know, this led to the notion of space filling curves of topology. Another question concerned the existence of nicely differentiable representations of a rectifiable curve (that is, continuous curves for which only the finiteness of the length is assumed). In particular, are there homeomorphisms of the parameter domain that result in differentiable representations for rectifiable curves? We have seen in this chapter that the answer is yes in a very pleasing way (see Theorem 3.1). Such questions were a natural part of the direct method of the calculus of variations in the early part of the 20th Century due to the introduction of compactness in function spaces. In this context there is another equivalence called Fréchet equivalence (see Cesari [**29**]). Two curves $f: [0,1] \to \mathbb{R}^n$ and $g: [0,1] \to \mathbb{R}^n$ are said to be **Fréchet equivalent** if for each positive number ε there is a self-homeomorphism of $[0,1]$ such that $\|f - g\circ h\|_\infty < \varepsilon$. Fréchet equivalence has many nice properties. Among them is the fact that there is a representation in which intervals of constancy do not exist for parametric curves of finite, positive length. In the parlance of topology, there exists a 'light representation'. Clearly, Lebesgue equivalence does not yield this property.

For continuous surfaces, the question of representation is much more difficult. The topological problems become very nonintuitive and homeomorphic changes in the domain of a 2-dimensional surface are not enough. As in the curve case, only the finiteness of the surface area is available in the direct method of the calculus of variations. It turns out that Fréchet equivalence of surfaces is more useful in the 2-dimensional case due to the Moore Cactoid Theorem and the monotone-light factorizations of topology. (See, for example, Youngs [**144**] and Cesari [**28, 29, 30, 31**].) But for higher dimensions, even this equivalence is not good enough to get reasonable representations from the point of view of analysis (see Breckenridge and Nishiura [**16**] for more details).

Thus it appears that Lebesgue equivalence is not appropriate for some kinds of analysis even when continuity is assumed. But, as we have seen in the first three chapters of the book, Lebesgue equivalence in $\mathbb{R}$ does have content. Our final chapter on $\mathbb{R}$ (Chapter 4, on the derivative function) will add further proof of the content of Lebesgue equivalence. Self-homeomorphisms in $\mathbb{R}$ will be investigated again in context of Fourier series in the last part of the book.

CHAPTER 4

The Derivative Function

In the last chapter we investigated the class of functions whose derivatives are bounded functions. Here, the focus will be the derivative function itself. This Lebesgue equivalence problem is more complicated than that in Chapter 3. The main theorem is Maximoff's theorem [**97**].

THEOREM (MAXIMOFF). *In order that a function* $f\colon (0,1) \to \mathbb{R}$ *be Lebesgue equivalent to a derivative function it is necessary and sufficient that* f *be in the Baire class* 1 *and satisfy the intermediate value property.*

The Zahorski classes of functions will play a substantial role in the analysis. We shall model our presentation of the proof of the Maximoff theorem after that of Preiss [**110**]. And, using the results of [**64**], we shall show that approximate derivative functions also have the same characterization.

4.1. Properties of derivatives

It is well-known that derivative functions are in the Baire class 1 and it was observed by Darboux that derivative functions satisfy the intermediate value property. It is now conventional to refer to functions that satisfy the intermediate value property as **Darboux** functions. We shall denote the class of all Darboux functions in the Baire class 1 by $\mathcal{DB}_1$.

4.1.1. The Zahorski classes $\mathcal{M}_0$ and $\mathcal{M}_1$. In Chapter 1 we introduced the classes M_0 and M_1 of subsets of $\mathbb{R}$ defined by Zahorski. Zahorski used these classes to define his classes $\mathcal{M}_0$ and $\mathcal{M}_1$ of real-valued functions as follows.

For $i = 1, 2$, a function $f\colon \mathbb{R} \to \mathbb{R}$ is in the class $\mathcal{M}_i$ if, for each real number α,

$$f^{-1}\big[\{y : y > \alpha\}\big] \in M_i \cup \{\emptyset\} \quad \text{and} \quad f^{-1}\big[\{y : y < \alpha\}\big] \in M_i \cup \{\emptyset\}.$$

We shall give a proof of Zahorski's assertion that the classes $\mathcal{M}_0$, $\mathcal{M}_1$ and $\mathcal{DB}_1$ are the same. Let us begin with characterizations of the classes M_0 and M_1. We infer from [**110**] the following easily proved proposition.

PROPOSITION 4.1. *A nonempty* F_σ *set* E *of* $\mathbb{R}$ *is in the class* M_0 (M_1) *if and only if, for each closed interval* I, *the set* $I \cap E$ *is infinite* (*uncountable*) *whenever* $I \cap E$ *is not empty.*

A second characterization of the class M_0 that is also easily proved is the following.

PROPOSITION 4.2. *A nonempty* F_σ *set* E *of* $\mathbb{R}$ *is in the class* M_0 *if and only if each component of* $\mathbb{R} \setminus E$ *is a closed set.*

We are now ready to prove a proposition due to Zahorski concerning connected subsets of $\mathbb{R}$. Our proof is somewhat easier than that of Zahorski. In the proof we shall use the assertion that if I is an interval and if E and F are F_σ sets such that $I \subset E \cup F$ then $(I \cap E)^o \cup (I \cap F)^o$ is dense in I. This assertion is a simple consequence of the Baire category theorem and we shall leave the proof to the reader.

PROPOSITION 4.3. *Let I be a closed interval of $\mathbb{R}$ and let E and F be members of M_0 such that $I \subset E \cup F$ and the sets $I \cap E$ and $I \cap F$ are not empty. Then $I \cap E \cap F \neq \emptyset$.*

PROOF. We shall show that $I \cap E \cap F = \emptyset$ leads to a contradiction.

Assume that $I \cap E$ and $I \cap F$ are disjoint. Clearly, $I \cap E$ and $I \cap F$ are σ-compact and nonempty. Since E and F are members of M_0, the components of $I \cap E$ and $I \cap F$ will be compact, and there are infinitely many components of $I \cap F$ between any two distinct components of $I \cap E$. Also, there are infinitely many components of $I \cap E$ between any two distinct components of $I \cap F$.

Let E_0 and F_0 be the respective interiors of $I \cap E$ and $I \cap F$. Then the closure of $E_0 \cup F_0$ is equal to I and E_0 and F_0 are disjoint. From the preceding paragraph we have that $I \setminus (E_0 \cup F_0)$ is a nonempty, nowhere dense, perfect subset of I. (For convenience, we shall assume that the end points of I are in this set.) There is a homeomorphism φ of I onto $[0,1]$ such that $\varphi\big[I \setminus (E_0 \cup F_0)\big]$ is the usual Cantor set in $[0,1]$. The composition $f = \psi \circ \varphi$ of φ with the Cantor function ψ yields a monotone function from I onto $[0,1]$. Clearly, by Proposition 4.2, the intervals of constancy of the function f are precisely the nondegenerate components of $I \cap E$ and of $I \cap F$. By way of contradiction, let us show

$$f[I \cap E] \cap f[I \cap F] = \emptyset. \tag{4.1}$$

Suppose that there are two points $x_{_E}$ and $x_{_F}$ in $I \cap E$ and $I \cap F$ respectively such that $f(x_{_E}) = f(x_{_F})$. We may assume $x_{_E} < x_{_F}$. Because f is monotone, we have $[x_{_E}, x_{_F}] \subset I \cap E \cap F$. But, $I \cap E \cap F = \emptyset$ has been assumed. A contradiction has been reached and (4.1) now holds. Since both $f[I \cap E]$ and $f[I \cap F]$ are σ-compact, one of them has a nonempty interior, say $f[I \cap E]$. Let J_0 be a component of this interior. Since f is monotone, $f^{-1}[J_0]$ is a nondegenerate interval which, by (4.1), is contained in $I \cap E$. Hence there is a component C of $I \cap E$ that contains $f^{-1}[J_0]$. As the intervals of constancy of f are components of $I \cap E$ or of $I \cap F$, we have that $f[C]$ is a singleton set. But, from (4.1) we have $f\big[f^{-1}[J_0]\big] = J_0 \subset f[C]$. Thereby our final contradiction has occurred and the proposition is proved.

An immediate consequence of the above proposition is the following theorem due to Zahorski.

THEOREM 4.4 (ZAHORSKI). $\mathcal{M}_0 = \mathcal{M}_1 = \mathcal{DB}_1$.

PROOF. As the inclusions $\mathcal{M}_0 \supset \mathcal{M}_1 \supset \mathcal{DB}_1$ are quite clear, only the inclusion $\mathcal{DB}_1 \supset \mathcal{M}_0$ remains to be proved. Suppose that there is a function $f\colon (0,1) \to \mathbb{R}$ that is a member of $\mathcal{M}_0 \setminus \mathcal{DB}_1$. Then there is a closed interval $I = [a,b]$ contained in $(0,1)$ and a number y_0 between $f(a)$ and $f(b)$ such that $I \cap f^{-1}[y_0] = \emptyset$. We have that the sets $E = \{\, y : f(x) > y_0 \,\}$ and $F = \{\, y : f(x) < y_0 \,\}$ are members of M_0 and that $I \cap E \cap F = \emptyset$. Clearly, $I \subset E \cup F$, $I \cap E \neq \emptyset$ and $I \cap F \neq \emptyset$. The above proposition now yields $I \cap E \cap F \neq \emptyset$. That is, $\emptyset \neq \emptyset$. The theorem now follows.

It is clear that the classes M_0 and M_1 are each closed under countable unions. Consequently, the following corollary holds.

COROLLARY 4.5. *$f \in \mathcal{M}_0$ if and only if $f^{-1}[U] \in M_0 \cup \{\emptyset\}$ whenever U is an open set of $\mathbb{R}$. $f \in \mathcal{M}_1$ if and only if $f^{-1}[U] \in M_1 \cup \{\emptyset\}$ whenever U is an open set of $\mathbb{R}$.*

PROOF. If $f^{-1}[U] \in M_0 \cup \{\emptyset\}$ ($f^{-1}[U] \in M_1 \cup \{\emptyset\}$) for each open set U of $\mathbb{R}$, then it is obvious that $f \in \mathcal{M}_0$ ($f \in \mathcal{M}_1$).

Conversely, assume that $f \in \mathcal{M}_0$. For each open interval (a, b) and each closed interval I such that $I \cap f^{-1}[(a, b)] \neq \emptyset$, we must show that the set $I \cap f^{-1}[(a, b)]$ is uncountable. From the theorem we have that f is in $\mathcal{DB}_1$. So the set $f[I]$ is a connected set that intersects the open interval (a, b). It follows that $(a, b) \cap f[I]$ is either a singleton or an uncountable set. From this we infer that $I \cap f^{-1}[(a, b)]$ is uncountable.

4.1.2. Characterizations of $\mathcal{DB}_1$. The above Theorem 4.4 characterizes the class $\mathcal{DB}_1$. Let us provide two more that will prove useful. The first will require Borel measurable functions and measures in its statement.

We will not deal with arbitrary measures. Indeed, in the remainder of the chapter, we shall assume that *all measures on an open interval of $\mathbb{R}$ are nonnegative, nonatomic, finite, regular Borel measures.* Hence, for each nonnegative Borel measurable function g and each Borel set A, the integral $\int_A g\, d\mu$ exists (possibly as $+\infty$). The measure induced by such an integral will be denoted by $g\mu$. We shall say that two measures μ and ν are **equivalent** if there are Borel measurable functions g and h such that $\nu = g\mu$ and $\mu = h\nu$. A measure μ on an open interval J is called **Lebesgue-like** if $\mu(U) > 0$ for each nonempty open subset U of J. (We shall revisit the notion of Lebesgue-like measures in the context of $\mathbb{R}^n$ in Chapter 7 where the Lebesgue equivalence of measures is investigated.) In his paper [**110**], Preiss uses the name **positive** for Lebesgue-like measures on $\mathbb{R}$.

According to [**110**], for a measure μ defined on an open interval J of $\mathbb{R}$, a Borel measurable function $f\colon J \to \mathbb{R}$ is said to possess the **μ-Denjoy property** if $\mu\big(I \cap f^{-1}[U]\big) > 0$ whenever U is an open set and I is a closed interval of $\mathbb{R}$ with $I \cap f^{-1}[U] \neq \emptyset$. Note that μ is necessarily Lebesgue-like.

THEOREM 4.6 (PREISS). *In order for a function $f\colon (0,1) \to \mathbb{R}$ to be in $\mathcal{DB}_1$ it is necessary and sufficient for the function to be in the Baire class 1 and possesses the μ-Denjoy property for some measure μ on $(0,1)$.*

PROOF. Sufficiency follows immediately from Theorem 4.4.

Suppose that f is in $\mathcal{DB}_1$. Let (p, q) be an open interval with rational end points such that $f^{-1}[(p, q)] \neq \emptyset$. From Corollary 4.5 we infer that there is a countable collection $\{F_n : n = 1, 2, \dots\}$ of nonempty, nowhere dense, perfect subsets of $f^{-1}[(p, q)]$ whose union is dense in $f^{-1}[(p, q)]$. As the set F_n is homeomorphic to the Cantor set, the measure associated with the Cantor function will yield a measure μ_n whose support is F_n and satisfies $\mu_n\big((0,1)\big) = 1$. The measure $\mu_{(p,q)} = \sum_{n=1}^{\infty} \frac{1}{2^n} \mu_n$ also will satisfy $\mu_{(p,q)}\big((0,1)\big) = 1$. As there are only countably many such intervals (p, q), the necessity will follow.

The other characterization is due to Neugebauer [**101**]. In the statement of the theorem we shall need the notion of a selection. By a **selection** for a collection $\mathcal{I}$

of sets we mean a function s on $\mathcal{I}$ that satisfies $s(I) \in I$ for each I in $\mathcal{I}$. For Neugebauer's characterization theorem, the collection $\mathcal{I}$ will be the collection of all closed intervals I that are contained in $(0,1)$. For this collection $\mathcal{I}$ and a point x in $(0,1)$, the notation $I \to x$ will mean that $\lambda(I) \to 0$ where only those I in $\mathcal{I}$ with $x \in I$ are used. The characterization theorem is inspired by a theorem due to Gleyzal [**48**] which characterizes Baire class 1, real-valued functions defined on $(0,1)$. Since we shall have occasion to refer again to Gleyzal's theorem later in the chapter, this will be a good place to state and prove it.

With $\mathcal{I}$ as above, a real-valued function φ defined on $\mathcal{I}$ will be called an **interval function**. And an interval function φ is said to **converge** to $f\colon (0,1) \to \mathbb{R}$ if

$$\lim_{I \to x} \varphi(I) = f(x) \quad \text{for every} \quad x.$$

Here is Gleyzal's theorem.

THEOREM 4.7 (GLEYZAL). *Let $f\colon (0,1) \to \mathbb{R}$. Then f is in the Baire class 1 if and only if there is an interval function $\varphi\colon \mathcal{I} \to \mathbb{R}$ that converges to f.*

PROOF. Suppose that φ converges to f. We shall construct a sequence of continuous functions that converges pointwise to f. For each positive integer n let $\mathcal{I}_n$ be the collection of all closed intervals I contained in $(0,1)$ with $\lambda(I) < \frac{1}{n}$. Obviously the collection $\{\, I^o : I \in \mathcal{I}_n \,\}$ is an open cover of $(0,1)$. Let

$$\sum_{i=1}^{\infty} p_i(x) = 1 \quad \text{for} \quad x \in (0,1)$$

be a partition of unity that subordinates this cover. That is, the collection p_i, $i = 1, 2, \ldots$, satisfies the two properties

(1) each p_i is a nonnegative continuous function such that, for some I_i in $\mathcal{I}_n$,
$$U_i = \{\, x : p_i(x) > 0 \,\} \subset I_i^o,$$

(2) for each x in $(0,1)$ there is a neighborhood U of x and a positive integer k such that $U_i \cap U = \emptyset$ for $i > k$.

Let

$$f_n(x) = \sum_{i=1}^{\infty} \varphi(I_i)\, p_i(x) \quad \text{for} \quad x \in (0,1).$$

Condition (2) assures us that f_n is continuous. Moreover, conditions (1) and (2) yield

$$\begin{aligned} |f_n(x) - f(x)| &= \Bigl|\sum_{i=1}^{\infty} \bigl(\varphi(I_i) - f(x)\bigr)\, p_i(x)\Bigr| \\ &\le \sum_{p_i(x) \neq 0} |\varphi(I_i) - f(x)|\, p_i(x) \\ &\le \sup\{\, |\varphi(I) - f(x)| : x \in I \in \mathcal{I}_n \,\}. \end{aligned}$$

As $\lim_{I \to x} \varphi(I) = f(x)$, the pointwise convergence of f_n to f now follows. That is, f is in the Baire class 1.

To establish the converse, suppose that f is in the Baire class 1. Let f_n, $n = 1, 2, \ldots$, be a sequence of continuous functions on $(0,1)$ that converges pointwise to f. There is no loss in assuming that $f_1 = 0$. For each positive integer n, let the collection $\mathcal{I}_n$ of closed intervals be defined as

$$\mathcal{I}_n = \{\, I : \lambda(I) < \tfrac{1}{n} \quad \text{and} \quad \lambda\big(f_n[I]\big) < \tfrac{1}{n}\,\}.$$

Each closed interval I is a member of $\mathcal{I}_1$ because $f_1 = 0$ and $I \subset (0,1)$. For each I in $\mathcal{I}$, let

$$n(I) = \max\{\, n : I \in \mathcal{I}_n \,\}.$$

From the definition of $n(I)$, it follows that

$$\lim_{I \to x} n(I) = +\infty \quad \text{for each} \quad x.$$

For each I in $\mathcal{I}$, select a point x_I in I and let $\varphi(I) = f_{n(I)}(x_I)$. Let us show that this interval function φ converges to f. Suppose $x \in I$. Then

$$\begin{aligned} |f(x) - \varphi(I)| &\le |f(x) - f_{n(I)}(x)| + |f_{n(I)}(x) - f_{n(I)}(x_I)| \\ &\le |f(x) - f_{n(I)}(x)| + \frac{1}{n(I)}. \end{aligned}$$

It now follows that $\lim_{I \to x} \varphi(I) = f(x)$.

The proof is completed.

In passing we remark that the above proof is easily generalized to functions defined on an open subset of $\mathbb{R}^n$ if one employs n-cubes for intervals.

We turn now to Neugebauer's characterization of the class $\mathcal{DB}_1$.

THEOREM 4.8 (NEUGEBAUER). *Let f be a real-valued function on $(0,1)$. Then $f \in \mathcal{DB}_1$ when and only when for each closed interval $I = [a,b]$ contained in $(0,1)$ there is a selection x_I in (a,b) such that*

$$\lim_{I \to x} f(x_I) = f(x) \quad \textit{whenever} \quad x \in (0,1).$$

PROOF. Suppose that the convergence condition is satisfied by f. Then the interval function φ defined by $\varphi(I) = f(x_I)$ converges to f. Hence f is in the Baire class 1 by Gleyzal's theorem above. Theorem 4.4 gives us $f \in \mathcal{DB}_1$.

To establish the converse, let $f \in \mathcal{DB}_1$. From the fact that f is in the Baire class 1, there is by Gleyzal's theorem an interval function φ defined on the collection $\mathcal{I}$ that converges to f. For each I in $\mathcal{I}$ the set $f[I]$ is a connected set because f satisfies the Darboux property. As $\operatorname{dist}\big(\varphi(I), f[I]\big) < +\infty$, there is a point x_I in I^o such that

$$|\varphi(I) - f(x_I)| < \operatorname{dist}\big(\varphi(I), f[I]\big) + \lambda(I).$$

Observe that, for each x in I, we have

$$\operatorname{dist}\big(\varphi(I), f[I]\big) \le |\varphi(I) - f(x)|.$$

So, when $x \in I$,

$$\begin{aligned} |f(x) - f(x_I)| &\le |f(x) - \varphi(I)| + |\varphi(I) - f(x_I)| \\ &\le 2\,|\varphi(I) - f(x)| + \lambda(I). \end{aligned}$$

It now follows that $\lim_{I \to x} f(x_I) = f(x)$ for every x.

4.2. Characterization of the derivative

As the proof of the Maximoff theorem will use interval functions, it will be useful to have a characterization of the derivative in terms of such functions. We have already seen an interval function characterization of $\mathcal{DB}_1$.

4.2.1. Neugebauer's theorem.

The characterization appears in [**101**].

THEOREM 4.9 (NEUGEBAUER). *Let $f\colon (0,1) \to \mathbb{R}$. Then f is a derivative if and only if for each closed interval I contained in $(0,1)$ there corresponds a point x_I in I^o such that*

(1) $\lim_{I\to x} f(x_I) = f(x)$ *for every* x,

and

(2) $f(x_I) = \dfrac{f(x_J)\lambda(J) + f(x_H)\lambda(H)}{\lambda(I)}$ *whenever I, J and H are in $\mathcal{I}$ and are such that $I = J \cup H$ and $J^o \cap H^o = \emptyset$.*

PROOF. Suppose that f has a primitive F. For each interval $I = [a,b]$ there exists a point x_I in I^o such that $F(b) - F(a) = f(x_I)\,(b-a)$. The verification of condition (2) is a computation, and then (1) follows from the fact that F is a primitive for f.

For the converse, suppose that such a selection exists. Let us show the existence of a primitive F. Fix a point x_0 in $(0,1)$. Define the function $F\colon (0,1) \to \mathbb{R}$ as follows:

$$F(x) = \begin{cases} f(x_{[x_0,x]})\,\lambda([x_0,x]) & \text{if } x_0 < x < 1, \\ 0 & \text{if } x_0 = x, \\ -f(x_{[x,x_0]})\,\lambda([x,x_0]) & \text{if } 0 < x < x_0. \end{cases}$$

Condition (2) yields $F(b) - F(a) = f(x_I)\,(b-a)$ for each closed interval $I = [a,b]$ contained in $(0,1)$. Condition (1) yields that F is differentiable and that $F' = f$.

The theorem is proved.

The following is an immediate consequence of the above theorem and the definition of Lebesgue-like measures given earlier.

PROPOSITION 4.10. *Let $f\colon (0,1) \to \mathbb{R}$. Then there exists a homeomorphism $h\colon (0,1) \to (0,1)$ such that $f \circ h$ is a derivative if and only if there exists a Lebesgue-like measure ν on $(0,1)$ with the property that for each closed interval I contained in $(0,1)$ there corresponds a point x_I in I^o such that*

(1) $f(x_I) \to f(x)$ *whenever* $I \to x$,

and

(2) $f(x_I) = \dfrac{f(x_J)\nu(J) + f(x_H)\nu(H)}{\nu(I)}$ *whenever I, J and H are in $\mathcal{I}$ and are such that $I = J \cup H$ and $J^o \cap H^o = \emptyset$.*

There is no loss in assuming $\nu\big((0,1)\big) = 1$. Observe that

$$\psi(x) = \nu\big((0,x)\big) \quad \text{for} \quad x \in [0,1]$$

defines a self-homeomorphism of $(0,1)$ for which

$$\lambda([a,b]) = \nu\big(\psi^{-1}[[a,b]]\big)$$

and

$$f(x) = \big(f \circ \psi^{-1}\big)(\psi(x)).$$

The proof of the proposition is left to the reader.

4.3. Proof of Maximoff's theorem

Although our proof is based on the above selection characterizations of the class $\mathcal{DB}_1$ and the class of derivatives, it is essentially the proof of Preiss [**110**].

4.3.1. Interval functions. Let us begin with some preliminary observations.

PROPOSITION 4.11. *Let $f\colon (0,1) \to \mathbb{R}$ be in $\mathcal{DB}_1$. Then there exists a Lebesgue-like measure ν on $(0,1)$ such that f has the ν-Denjoy property and is ν-integrable, and there is a selection x_I for each closed interval I contained in $(0,1)$ such that*

(1) $x_I \in I^o$,

and

(2) $f(x_I)\,\nu(I) = \int_I f(t)\,d\nu(t)$.

PROOF. We know that f is Borel measurable and, by Theorem 4.6, that there exists a Lebesgue-like measure μ on $(0,1)$ such that f has the μ-Denjoy property. Let $\nu = \frac{1}{1+|f|}\,\mu$ (see page 43). Then one can easily verify that ν is a Lebesgue-like measure, f is ν-integrable, and f has the ν-Denjoy property. To prove the existence of the selection, we observe that, for each closed interval I, the set $f[I]$ is connected and $\inf f[I]\,\nu(I) \le \int_I f(t)\,d\nu(t) \le \sup f[I]\,\nu(I)$. If the set $f[I]$ is a singleton, the selection is trivial. If the connected set $f[I]$ is not a singleton, then the fact that f has the ν-Denjoy property yields the strict inequalities

$$\inf\, f[I] < \frac{\int_I f(t)\,d\nu(t)}{\nu(I)} < \sup\, f[I].$$

The Darboux property of f completes the proof.

Observe that the Radon-Nykodym derivative of the signed measure $f\,\nu$ with respect to ν is ν-almost everywhere equal to f. Unfortunately, the selection need not satisfy $f(x_I) \to f(x)$ as $I \to x$ for every x. That is, the Lebesgue-like measure ν may fail to have the property that

$$\lim_{I\to x} \frac{\int_I f(t)\,d\nu(t)}{\nu(I)} = f(x) \quad \text{for every} \quad x\,.$$

According to Proposition 4.10, it is precisely this last fact that must be established for some Lebesgue-like measure ν for which f has the ν-Denjoy property. The construction of such a measure ν is the point of the next section.

4.3.2. The proof. The necessity of the condition in Maximoff's theorem (that is, f be in $\mathcal{DB}_1$) is well-known. The proof of the sufficiency will require three preliminary lemmas.

We shall begin with a lemma concerning the set of points of discontinuity of a function in the first Baire class.

LEMMA 4.12. *Let $f\colon (0,1) \to \mathbb{R}$ be in the Baire class 1 and let D be its set of points of discontinuity. Then there is a sequence F_n, $n = 1,2,\ldots$, of compact nowhere dense subsets of $(0,1)$ such that*

(1) *if $m < n$, then either $F_m \supset F_n$ or $F_m \cap F_n = \emptyset$,*

and

(2) *if $x \in D$ and $m > 0$, then there exists an integer n with $m < n$ such that $x \in F_n$ and* $\operatorname{diam} f[F_n] < \frac{1}{m}$.

PROOF. Let us first observe that, for every σ-compact set M of the first category contained in $(0,1)$ and for every positive number ε, there exists a sequence M_n, $n = 1, 2, \dots$, of mutually disjoint compact sets with $M = \bigcup_{n=1}^{\infty} M_n$ and $\operatorname{diam} f[M_n] < \varepsilon$. To see this, we use the fact that $f^{-1}\big[(a,b)\big]$ is an F_σ set whenever f is in the Baire class 1 to construct a sequence H_n, $n = 1, 2, \dots$, of compact sets such that $M = \bigcup_{n=1}^{\infty} H_n$ and $\operatorname{diam} f[H_n] < \varepsilon$. As M is of the first category, we have that each H_n is nowhere dense. With this and the fact that $H_n \setminus \bigcup_{k=1}^{n-1} H_k$ is locally compact, we can construct a sequence $M_{n,j}$, $j = 1, 2, \dots$, of mutually disjoint compact sets such that $H_n \setminus \bigcup_{k=1}^{n-1} H_k = \bigcup_{j=1}^{\infty} M_{n,j}$.

Now we can complete the proof of the lemma. The set D of the points of discontinuity of the function f is an F_σ set of the first category in $(0,1)$ because f is in the Baire class 1. Using the above observation, for each positive integer m, we can write $D = \bigcup_{n=1}^{\infty} D_{m,n}$ where the summands form a mutually disjoint collection and such that $\operatorname{diam} f[D_{m,n}] < \frac{1}{m}$. Moreover, it is not difficult to have the decompositions satisfy the further property that each $D_{m,n}$ is contained in some $D_{m-1,k}$. Now well order the collection $\{\, D_{m,n} : m = 1, 2, \dots, n = 1, 2, \dots \}$.

The next two lemmas concern the construction of measures that will appear in the proof of the Maximoff theorem.

LEMMA 4.13. *Let ν be a measure on $(0,1)$ and let (a,b) and E be respectively an open interval and a Borel set contained in $(0,1)$. Denote the midpoint of the interval (a,b) by $\bar{t}$.*

If $\nu\big(E \cap (a,t)\big) > 0$ whenever $t \in (a,b]$ and if $\alpha\colon (a,b] \to \mathbb{R}$ is a monotone, positive function such that $\lim_{t\to a+} \alpha(t) = 0$, then there exists a nonnegative, Borel measurable function φ on $(0,1)$ such that

(1) $\{\, t : \varphi(t) \neq 0 \,\} \subset (a, \bar{t}\,] \cap E$,
(2) $\int_{(a,b)} \varphi(t)\, d\nu(t) \leq 2\,\alpha(b)$,
(3) $\int_{(a,s)} \varphi(t)\, d\nu(t) \geq \alpha(s)$ *for* $s \in (a,b]$.

Analogously, if $\nu\big(E \cap (t,b)\big) > 0$ whenever $t \in [a,b)$ and if $\beta\colon [a,b) \to \mathbb{R}$ is a monotone, positive function such that $\lim_{t\to b-} \beta(t) = 0$, then there exists a nonnegative, Borel measurable function ψ on $(0,1)$ such that

(4) $\{\, t : \psi(t) \neq 0 \,\} \subset [\,\bar{t}, b) \cap E$,
(5) $\int_{(a,b)} \psi(t)\, d\nu(t) \leq 2\,\beta(a)$,
(6) $\int_{(s,b)} \psi(t)\, d\nu(t) \geq \beta(s)$ *for* $s \in [a,b)$.

PROOF. Let b_n, $n = 1, 2, \dots$, be a decreasing sequence in $(a,b]$ such that $b_1 = \bar{t}$, $\lim_{n\to\infty} b_n = a$ and $\sum_{n=1}^{\infty} \alpha(b_n) \leq 2\,\alpha(b)$. With $b_0 = b$, define

$$\varphi(t) = \sum_{n=1}^{\infty} \frac{\alpha(b_{n-1})}{\nu\big(E \cap (a,b_n]\big)}\, \chi_{E\cap(a,b_n]}(t) \quad \text{for} \quad t \in (0,1).$$

As conditions (1) and (2) are obvious, let us show (3). Observe that

$$\begin{aligned} \int_{(a,s)} \varphi(t)\, d\nu(t) &\geq \frac{\alpha(b_{n-1})}{\nu\big(E \cap (a,b_n]\big)}\, \nu\big(E \cap (a,b_n]\big) \\ &= \alpha(b_{n-1}) \geq \alpha(s) \end{aligned}$$

whenever $s \in (b_n, b_{n-1}]$, whence (3) follows.

The proof of the second assertion is obvious.

LEMMA 4.14. *Let ν be a measure on $(0,1)$. Suppose that F is a nonempty, compact, nowhere dense subset of a Borel subset E of $(0,1)$ such that $\nu(I \cap E) > 0$ for every closed interval I with $I \cap F \neq \emptyset$. Then, for each positive number ε, there exists a nonnegative, Borel measurable function $\eta\colon (0,1) \to \mathbb{R}$ such that*

(1) $\{t : \eta(t) \neq 0\} \subset (E \setminus F) \cap \{t : \operatorname{dist}(t, F) < \varepsilon\}$,
(2) $\int \eta(t)\, d\nu(t) < \varepsilon$,

and

(3) *if $x \in F$, then*

$$\lim_{I \to x} \frac{\nu(I \setminus F)}{\int_I \eta(t)\, d\nu(t)} = 0.$$

PROOF. Let Γ denote the countable collection consisting of the components of $(0,1) \setminus F$. Clearly these components are open intervals. Moreover, at most one of them has 0 as an end point and similarly for 1. We shall denote by Γ_0 the subcollection of those intervals whose end points are not 0, and analogously by Γ_1 in the case of the end point 1.

Observe that

$$\sum_{(c,d)\in\Gamma} \nu\big((c,d)\big) \leq \nu\big((0,1)\big) < +\infty.$$

There exists a nondecreasing, positive, monotone function ω on $(0,+\infty)$ (see, for example, [**9**, Proposition 3.3]) such that

$$\lim_{t\to 0+} \omega(t) = 0, \tag{4.2}$$

$$\omega'(0) = \lim_{t\to 0+} \frac{\omega(t)}{t} = +\infty, \tag{4.3}$$

$$\sum_{(c,d)\in\Gamma} \omega\big(\nu\big((c,d)\big)\big) < +\infty. \tag{4.4}$$

For each (c,d) in Γ_0, let

$$\alpha_{(c,d)}(t) = \omega\big(\nu\big((c,t)\big)\big) \quad \text{for} \quad t \in (c,d]\,;$$

and, for each (c,d) in Γ_1, let

$$\beta_{(c,d)}(t) = \omega\big(\nu\big((t,d)\big)\big) \quad \text{for} \quad t \in [c,d).$$

Then, in the respective cases, by Lemma 4.13 and (4.2) there are Borel measurable functions $\varphi_{(c,d)}$ and $\psi_{(c,d)}$ defined on $(0,1)$ so that the 6 conditions of that lemma hold. Now let

$$\gamma = \sum_{(c,d)\in\Gamma_0} \varphi_{(c,d)} + \sum_{(c,d)\in\Gamma_1} \psi_{(c,d)}.$$

Then γ is a nonnegative, Borel measurable function on $(0,1)$ such that

$$\{t : \gamma(t) \neq 0\} \subset E \setminus F$$

and

$$\int \gamma(t)\, d\nu(t) \leq 4 \sum_{(c,d)\in\Gamma} \omega\big(\nu\big((c,d)\big)\big) < +\infty,$$

where the last inequality follows from (4.4).

Let $x \in F$ and let $I = [a, b]$ be a closed interval such that

$$x \in [a, b] \subset (0, 1).$$

If $a \in (c, d) \in \Gamma$, then by condition (6) in Lemma 4.13 we have

$$\begin{aligned}\int_{(c,d)\cap I} \gamma(t)\, d\nu(t) &\geq \omega\big(\nu\big((a, d)\big)\big) \\ &= \frac{\omega\big(\nu\big((a, d)\big)\big)}{\nu\big((a, d)\big)}\, \nu\big((a, d)\big) \\ &\geq \inf\left\{ \tfrac{\omega(t)}{t} : 0 < t \leq \nu\big([a, b]\big) \right\} \nu\big((a, d)\big).\end{aligned}$$

Similarly, if $b \in (c, d) \in \Gamma$, then

$$\int_{(c,d)\cap I} \gamma(t)\, d\nu(t) \geq \inf\left\{ \tfrac{\omega(t)}{t} : 0 < t \leq \nu\big([a, b]\big) \right\} \nu\big((c, b)\big).$$

If $(c, d) \subset [a, b]$ and $(c, d) \in \Gamma$, then

$$\int_{(c,d)\cap I} \gamma(t)\, d\nu(t) \geq \inf\left\{ \tfrac{\omega(t)}{t} : 0 < t \leq \nu\big([a, b]\big) \right\} \nu\big((c, d)\big).$$

Consequently,

$$\begin{aligned}\int_I \gamma(t)\, d\nu(t) &= \sum_{(c,d)\cap I \neq \emptyset} \int_{(c,d)\cap I} \gamma(t)\, d\nu(t) \\ &\geq \inf\left\{ \tfrac{\omega(t)}{t} : 0 < t \leq \nu\big([a, b]\big) \right\} \sum_{(c,d)\cap I \neq \emptyset} \nu\big((c, d) \cap I\big) \\ &\geq \inf\left\{ \tfrac{\omega(t)}{t} : 0 < t \leq \nu\big([a, b]\big) \right\} \nu(I \setminus F).\end{aligned}$$

As F is compact and nowhere dense, we have $\nu(I \setminus F) > 0$. Consequently,

$$\frac{1}{\inf\left\{ \tfrac{\omega(t)}{t} : 0 < t \leq \nu\big([a, b]\big) \right\}} \geq \frac{\nu(I \setminus F)}{\int_I \gamma(t)\, d\nu(t)} > 0.$$

With the aid of (4.3) we have

$$\lim_{I \to x} \frac{\nu(I \setminus F)}{\int_I \gamma(t)\, d\nu(t)} = 0.$$

To conclude the proof, we use the compactness of F to assert that there is a δ-neighborhood $U(\delta, F)$ of F with $\delta < \varepsilon$ such that

$$\int_{U(\delta,F)} \gamma(t)\, d\nu(t) < \varepsilon.$$

Obviously, $\eta = \gamma\, \chi_{U(\delta,F)}$ is the required Borel measurable function and the lemma is proved.

We are now ready to complete the proof of the Maximoff theorem. Let $f \in \mathcal{DB}_1$. Then there is a Lebesgue-like measure μ for which f has the μ-Denjoy property. From μ we shall construct another Lebesgue-like measure ν for which f also has the ν-Denjoy property and satisfies

$$\lim_{I \to x} \frac{\int_I f(t)\, d\nu(t)}{\nu(I)} = f(x) \quad \text{for every} \quad x, \tag{4.5}$$

which, by Propositions 4.11 and 4.10, will complete the proof. To this end, let F_n, $n = 1, 2, \dots$, be the sequence of nowhere dense, nonempty, compact sets provided by Lemma 4.12. This sequence satisfies the conditions:

(i) If $m < n$, then either $F_m \supset F_n$ or $F_m \cap F_n = \emptyset$.

(ii) If $x \in D$ and $m > 0$, then there exists an integer n with $m < n$ such that $x \in F_n$ and $\operatorname{diam} f[F_n] < \frac{1}{m}$, where D is the set of points of discontinuity of f.

As f has the μ-Denjoy property, for each n, we have

$$\mu\big(I \cap f^{-1}\big[\{\, y : \operatorname{dist}(y, f[F_n]) < \tfrac{1}{n} \,\}\big]\,\big) > 0 \qquad \text{whenever} \quad I \cap f^{-1}\big[\{\, y : \operatorname{dist}\big(y, f[F_n]\big) < \tfrac{1}{n} \,\}\big] \neq \emptyset.$$

Consequently, one easily constructs a compact set E_n such that

(iii) $F_n \subset E_n \subset f^{-1}\big[\{\, y : \operatorname{dist}(y, f[F_n]) < \frac{1}{n} \,\}\big]$,

(iv) $\mu(I \cap E_n) > 0$ whenever $I \cap F_n \neq \emptyset$.

There will be no loss in assuming $\mu\big((0,1)\big) = 1$. Our next task is to define an appropriate sequence ε_n, $n = 1, 2, \dots$, of positive numbers. Letting

$$H_n = \bigcup \{\, F_m : m < n \quad \text{and} \quad F_m \cap F_n = \emptyset \,\},$$

we define

$$\delta_n = \begin{cases} 1 & \text{if} \quad H_n = \emptyset, \\ \frac{1}{2} \operatorname{dist}(H_n, F_n) & \text{if} \quad H_n \neq \emptyset. \end{cases}$$

(Observe that H_1 is $\emptyset$.) Next, with

$$U(\gamma, F_n) = \{\, x : \operatorname{dist}(x, F_n) < \gamma \,\} \quad \text{for} \quad \gamma > 0,$$

define the positive number

$$\zeta_n = \inf \big\{\, \big(\mu(I)\big)^2 : I \in \mathcal{I},\, I \cap F_n \neq \emptyset,\, I \setminus \overline{U(\delta_n, F_n)} \neq \emptyset \,\big\} \qquad \text{whenever} \quad H_n \neq \emptyset.$$

Finally, let

$$\varepsilon_n = \begin{cases} \frac{1}{2^n} & \text{if} \quad H_n = \emptyset, \\ \frac{1}{2^n} \min\{\, \delta_n, \zeta_n \,\} & \text{if} \quad H_n \neq \emptyset. \end{cases} \tag{4.6}$$

Now that the positive numbers ε_n, $n = 1, 2, \dots$, have been determined, we can inductively construct for each n a nonnegative, Borel measurable function η_n and a Lebesgue-like measure μ_n such that

(v) $\{\, x : \eta_n(x) \neq 0 \,\} \subset (E_n \setminus F_n) \cap U(\varepsilon_n, F_n)$,

(vi) $\mu_n = (1 + \eta_n)\, \mu_{n-1}$,

(vii) $\int \eta_n(t)\, d\mu_{n-1}(t) < \varepsilon_n$,

(viii) if $x \in F_n$, then

$$\lim_{I \to x} \frac{\mu_{n-1}(I \setminus F_n)}{\int_I \eta_n(t)\, d\mu_{n-1}(t)} = 0,$$

by defining $\mu_0 = \mu$ and applying Lemma 4.14 with $\nu = \mu_{n-1}$, $F = F_n$ and $\varepsilon = \varepsilon_n$ since (iii) and (iv) hold.

From (vi) and (vii) we infer that

$$1 \leq \prod_{k=1}^{n}(1+\eta_k) \leq \prod_{k=1}^{n+1}(1+\eta_k),$$

$$\int \prod_{k=1}^{n}\big(1+\eta_k(t)\big)\, d\mu(t) < \mu\big((0,1)\big)+1.$$

Hence $\psi = \prod_{n=1}^{\infty}(1+\eta_n)$ is an extended real-valued, Borel measurable function with a finite μ-integral. We infer from (vi) that

$$\int_E \psi(t)\, d\mu(t) = \sum_{k=n}^{\infty}\int_E \eta_k(t)\, d\mu_{k-1}(t) + \int_E d\mu_{n-1}(t) \tag{4.7}$$

for every Borel set E.

Let us make our penultimate calculation. Suppose that $G\colon \mathbb{R} \to \mathbb{R}$ is a bounded continuous function and that ε is a positive number. We assert that, for each x in $(0,1)$, there is a positive number δ such that the composed function $G \circ f$ satisfies

$$\frac{\left|\int_I \big(G\circ f(t) - G\circ f(x)\big)\, \psi(t)\, d\mu(t)\right|}{\int_I \psi(t)\, d\mu(t)} < \varepsilon \qquad \text{whenever} \quad x \in I \quad \text{and} \quad \lambda(I) < \delta. \tag{4.8}$$

The assertion is quite obvious when x is a point of continuity of f. So, let x be a point of discontinuity of f. Select a positive integer p so that

$$|G(y) - G(f(x))| < \tfrac{\varepsilon}{2} \quad \text{whenever} \quad |y - f(x)| < \tfrac{1}{p},$$

and let

$$V = \left\{\, y : |y - f(x)| < \tfrac{1}{2p} \,\right\}.$$

If $x \in I$, then we clearly have

$$\int_{I\cap f^{-1}[V]} \left|G\circ f(t) - G\circ f(x)\right| \psi(t)\, d\mu(t) \leq \frac{\varepsilon}{2}\int_I \psi(t)\, d\mu(t).$$

Thus it remains to be shown that

$$\lim_{I\to x} \frac{\int_{I\setminus f^{-1}[V]} \left|G\circ f(t) - G\circ f(x)\right| \psi(t)\, d\mu(t)}{\int_I \psi(t)\, d\mu(t)} = 0.$$

As

$$\int_{I\setminus f^{-1}[V]} \left|G\circ f(t) - G\circ f(x)\right| \psi(t)\, d\mu(t) \leq 2\, \|G\|_\infty \int_{I\setminus f^{-1}[V]} \psi(t)\, d\mu(t),$$

we shall prove

$$\lim_{I\to x} \frac{\int_{I\setminus f^{-1}[V]} \psi(t)\, d\mu(t)}{\int_I \psi(t)\, d\mu(t)} = 0. \tag{4.9}$$

For convenience we write $W = (0,1) \setminus f^{-1}[V]$.

Using (ii), we select an n such that $x \in F_n$ and $F_n \subset f^{-1}[V]$. If $k \geq n$ and $F_n \supset F_k$, then $E_k \cap W = \emptyset$ from (iii), whence $\int_W \eta_k(t)\, d\mu_{k-1}(t) = 0$ from (v).

So, suppose that $k \geq n$ and $\int_{W \cap I} \eta_k(t)\, d\mu_{k-1}(t) > 0$. From (i) we have $F_n \cap F_k = \emptyset$ and hence $H_k \supset F_n \neq \emptyset$. Thus, by (vii) and formula (4.6), we have

$$\int_{W \cap I} \eta_k(t)\, d\mu_{k-1}(t) \leq \int_{(0,1)} \eta_k(t)\, d\mu_{k-1}(t)$$
$$< \varepsilon_k \leq \tfrac{1}{2^k} \left(\mu(I)\right)^2 \leq \tfrac{1}{2^k}\, \mu(I) \int_I \psi(t)\, d\mu(t).$$

Consequently, by formula (4.7) and the inequality $\eta_n\, \mu_{n-1} \leq \psi\, \mu$, we have

$$\begin{aligned} \int_{W \cap I} \psi(t)\, d\mu(t) &= \sum_{k=n}^{\infty} \int_{W \cap I} \eta_k(t)\, d\mu_{k-1}(t) + \int_{W \cap I} d\mu_{n-1}(t) \\ &\leq \sum_{k=n}^{\infty} \tfrac{1}{2^k}\, \mu(I) \int_I \psi(t)\, d\mu(t) + \int_{W \cap I} d\mu_{n-1}(t) \\ &\leq \left(\mu(I) + \frac{\int_{W \cap I} d\mu_{n-1}(t)}{\int_I \eta_n(t)\, d\mu_{n-1}(t)} \right) \int_I \psi(t)\, d\mu(t). \end{aligned}$$

As $W \cap I \subset I \setminus F_n$, we have from (viii) that formula (4.9) holds. Thereby formula (4.8) is established.

To complete the proof, we use the continuous functions

$$G_1(y) = \frac{y}{1+|y|} \quad \text{for} \quad y \in \mathbb{R} \quad \text{and} \quad G_2(y) = \frac{1}{1+|y|} \quad \text{for} \quad y \in \mathbb{R}.$$

Then, for every x in $(0,1)$,

$$\lim_{I \to x} \frac{\int_I \frac{f(t)}{1+|f(t)|}\, \psi(t)\, d\mu(t)}{\int_I \psi(t)\, d\mu(t)} = \frac{f(x)}{1+|f(x)|}$$

and

$$\lim_{I \to x} \frac{\int_I \frac{1}{1+|f(t)|}\, \psi(t)\, d\mu(t)}{\int_I \psi(t)\, d\mu(t)} = \frac{1}{1+|f(x)|}.$$

Let $\nu = \frac{\psi}{1+|f|}\, \mu$. As

$$\nu\big(I \cap f^{-1}[(c,d)]\big) \geq \frac{1}{1+B}\, \mu\big(I \cap f^{-1}[(c,d)]\big),$$

where $B = |c| + |d|$, we have that f has the ν-Denjoy property. Then

$$\lim_{I \to x} \frac{\int_I f(t)\, d\nu(t)}{\int_I d\nu(t)} = \frac{\frac{f(x)}{1+|f(x)|}}{\frac{1}{1+|f(x)|}} = f(x)$$

yields formula (4.5), and Maximoff's theorem is finally proved.

4.4. Approximate derivatives

It is well-known that a derivative of a function f can be defined at a point x by ignoring a set of dispersion at that point. This derivative is called an approximate derivative and is denoted by $f'_{ap}(x)$. (A precise definition will be given shortly.) We shall apply the material of the earlier sections to show that if an approximate derivative $f'_{ap}(x)$ exists at each point x of $(0,1)$, it must be in $\mathcal{DB}_1$. Our proof is essentially that of [**64**].

Let $f\colon (0,1) \to \mathbb{R}$ and $x \in (0,1)$. A number L is said to be the **approximate derivative of f at x** if for each positive number ε there exists a measurable set E such that

$$E \text{ has density } 1 \text{ at } x \quad \text{and} \quad E \subset \left\{ t : \left| \tfrac{f(t)-f(x)}{t-x} - L \right| < \varepsilon \right\}.$$

4.4.1. f'_{ap} is in the Baire class 1. Suppose that $f\colon (0,1) \to \mathbb{R}$ is such that $f'_{ap}(x)$ exists at every x. It is immediate from the definition of the approximate derivative that f is approximately continuous, whence a Borel measurable function. So, for each w in $\mathbb{R}$,

$$A(w) = \left\{ (u,v) : \tfrac{f(u)-f(v)}{u-v} > w \right\}$$

is a Borel set contained in $(0,1) \times (0,1)$. The Lebesgue measure on $\mathbb{R} \times \mathbb{R}$ will be denoted by λ_2. Recall that $\mathcal{I}$ denotes the collection of all closed intervals contained in $(0,1)$. By defining

$$\varphi(I) = \sup \left\{ w : \lambda_2(A(w) \cap (I \times I)) > \tfrac{1}{2} \lambda_2(I \times I) \right\} \quad \text{for} \quad I \in \mathcal{I},$$

we have an interval function $\varphi\colon \mathcal{I} \to \mathbb{R}$. We shall prove that

$$\lim_{I \to x} \varphi(I) = f'_{ap}(x) \quad \text{for every} \quad x \tag{4.10}$$

from which, in view of Gleyzal's theorem (Theorem 4.7), f'_{ap} will be in the Baire class 1.

Let $x \in (0,1)$ and $\varepsilon > 0$. Then

$$E = \left\{ u : \left| \tfrac{f(u)-f(x)}{u-x} - f'_{ap}(x) \right| < \varepsilon \right\}$$

is a Borel set and x is a point of density 1 of E. For $(u,v) \in E \times E$, we have

$$|f(u) - f(v)| < \varepsilon \left(|u-x| + |v-x| \right).$$

Hence, for any choice of k with $k > 1$, if F is the subset of $\mathbb{R} \times \mathbb{R}$ given by

$$F = \left\{ (u,v) : |u-x| + |v-x| < k\,|u-v| \right\},$$

then

$$(E \times E) \cap F \subset A\big(f'_{ap}(x) - k\varepsilon\big) \setminus A\big(f'_{ap}(x) + k\varepsilon\big). \tag{4.11}$$

A convenient choice is $k = 17$. A straightforward computation will show

$$\begin{aligned} \mathbb{R} \times \mathbb{R} \setminus F \subset \left\{ (u,v) : \tfrac{9}{8}(u-x) \geq v-x \geq u-x \geq 0 \right\} \\ \cup \left\{ (u,v) : \tfrac{9}{8}(v-x) \geq u-x \geq v-x \geq 0 \right\} \\ \cup \left\{ (u,v) : \tfrac{9}{8}(u-x) \leq v-x \leq u-x \leq 0 \right\} \\ \cup \left\{ (u,v) : \tfrac{9}{8}(v-x) \leq u-x \leq u-x \leq 0 \right\}. \end{aligned}$$

Consequently, if $x \in I \in \mathcal{I}$, then

$$\lambda_2\big((\mathbb{R} \times \mathbb{R} \setminus F) \cap (I \times I)\big) \leq \tfrac{2}{9} \lambda_2(I \times I).$$

As

$$\lim_{I \to x} \frac{\lambda_2\big((E \times E) \cap (I \times I)\big)}{\lambda_2(I \times I)} = 1,$$

there is a positive number δ such that

$$\frac{\lambda_2\big((E \times E) \cap (I \times I)\big)}{\lambda_2(I \times I)} > \frac{3}{4} \quad \text{whenever} \quad x \in I \in \mathcal{I} \quad \text{and} \quad \lambda(I) < \delta.$$

Consequently, from (4.11) with $k = 17$, if $x \in I \in \mathcal{I}$ and $\lambda(I) < \delta$, then

$$\begin{aligned}\lambda_2\big(A(f'_{ap}(x) - 17\,\varepsilon) \cap (I \times I)\big) &\geq \lambda_2\big((E \times E) \cap (I \times I) \cap F\big) \\ &> \tfrac{3}{4}\,\lambda_2(I \times I) - \tfrac{2}{9}\,\lambda_2(I \times I) \\ &> \tfrac{1}{2}\,\lambda_2(I \times I)\end{aligned}$$

and

$$\begin{aligned}\lambda_2\big(A(f'_{ap}(x) + 17\,\varepsilon) \cap (I \times I)\big) &\leq \lambda_2\big((I \times I) \setminus ((E \times E) \cap F)\big) \\ &\leq \tfrac{1}{4}\,\lambda_2(I \times I) + \tfrac{2}{9}\,\lambda_2(I \times I) \\ &\leq \tfrac{1}{2}\,\lambda_2(I \times I).\end{aligned}$$

Thus we have shown that

$$f'_{ap}(x) - 17\,\varepsilon \leq \varphi(I) \leq f'_{ap}(x) + 17\,\varepsilon \qquad \text{whenever} \quad x \in I \in \mathcal{I} \quad \text{and} \quad \lambda(I) < \delta.$$

Thereby formula (4.10) follows and the following theorem has been established.

THEOREM 4.15. *$f'_{ap}\colon (0,1) \to \mathbb{R}$ is in the Baire class 1.*

4.4.2. f'_{ap} is in $\mathcal{DB}_1$. As in the last section, $f\colon (0,1) \to \mathbb{R}$ will be assumed to be an approximately differentiable function. We have already shown that f'_{ap} is in the Baire class 1. It remains to be shown that f'_{ap} has the Darboux property.

First, we shall prove a lemma.

LEMMA 4.16. *If $I = [a,b]$ is contained in $(0,1)$ and if $f'_{ap} \geq 0$ on I, then f is monotone nondecreasing on I; consequently, $f'_{ap} = f'$ on (a,b).*

PROOF. Let us first assume that $f'_{ap}(x) > 0$ for every x in I. Define the set

$$E = \{x : x \in I \quad \text{and} \quad f(a) \leq f(x)\}$$

and suppose that α is such that $0 < \alpha < 1$. Let $\mathcal{R}$ be the collection of all nonempty sets C that satisfy

(1) $C \subset E$,
(2) if x and x' are in C with $x < x'$, then $\lambda(E \cap [x,x']) \geq \alpha\,\lambda([x,x'])$.

Since $f'_{ap}(x) > 0$ for every x in I, we have $\mathcal{R} \neq \emptyset$. With set inclusion as the partial order on $\mathcal{R}$, one readily sees that every linearly ordered subset of $\mathcal{R}$ has an upper bound in $\mathcal{R}$. By Zorn's lemma, $\mathcal{R}$ has a maximal element K.

For a maximal element K of $\mathcal{R}$, we claim that $\overline{x} = \sup K$ is a member of K. Indeed, let $x \in K$ with $x < \overline{x}$. Let x_n, $n = 1, 2, \ldots$, be a sequence in K that converges to $\overline{x}$. Then

$$\alpha \leq \lim_{n \to \infty} \frac{\lambda(E \cap [x, x_n])}{\lambda([x, x_n])} = \frac{\lambda(E \cap [x, \overline{x}])}{\lambda([x, \overline{x}])}.$$

As f is approximately continuous, the set E is closed in the density topology. And the above inequality yields that $\overline{x}$ is a limit point in the density topology of the set E. Hence we have shown that $\overline{x}$ is a member of E.

Next, let us show $\overline{x} = b$. To the contrary, suppose $\overline{x} < b$. Then there would exist a x_0 such that $\overline{x} < x_0 < b$ with

$$x_0 \in E \quad \text{and} \quad \frac{\lambda(E \cap [\overline{x}, x_0])}{\lambda([\overline{x}, x_0])} \geq \alpha$$

because $f'_{ap}(\overline{x}) > 0$. Hence $x_0 \in K$, a contradiction. Thus we have shown that $f(a) \leq f(b)$.

For $f'_{ap} \geq 0$ on I, the function defined by

$$h_\varepsilon(x) = f(x) + \varepsilon\,(x - a)$$

has a positive approximate derivative on I whenever ε is a positive number. Hence

$$f(a) \leq f(b) + \varepsilon\,(b - a)$$

from which we infer $f(a) \leq f(b)$.

By considering all closed intervals contained in I, we have that f is monotone nondecreasing on I.

THEOREM 4.17. *$f'_{ap}\colon (0,1) \to \mathbb{R}$ has the Darboux property.*

PROOF. Suppose that f'_{ap} does not have the Darboux property. Then there is a closed interval $[a,b]$ such that f'_{ap} fails to have the intermediate value property. There is no loss in assuming that $f'_{ap}(a) < 0$ and $f'_{ap}(b) > 0$ and that $[a,b] \cap {f'_{ap}}^{-1}[0] = \emptyset$. The sets

$$E = \left\{ x : f'_{ap}(x) < 0 \right\} \quad \text{and} \quad F = \left\{ x : f'_{ap}(x) > 0 \right\}$$

satisfy

$$[a,b] \subset E \cup F \quad \text{and} \quad [a,b] \cap E \cap F = \emptyset, \tag{4.12}$$

As f'_{ap} is in the Baire class 1, the sets E and F are F_σ sets. Let us show that $(a,b) \cap E$ is a set in the Zahorski class M_0. According to Proposition 4.2, we must show that each nondegenerate component of $(a,b) \setminus E$ is a closed interval. Consider such a component C of $(a,b) \setminus E$ and let J be a closed interval contained in C. Then $f'_{ap}(x) > 0$ for each x in J. By the last lemma, f is monotone nondecreasing on J. Consequently, f is also monotone nondecreasing on $\overline{C}$. As f is approximately continuous on $[a,b]$, we have that $f'_{ap}(x) \geq 0$ for every x in $\overline{C}$. Since $\overline{C} \subset [a,b]$, we have from (4.12) that $f'_{ap}(x) > 0$ for every x in $\overline{C}$. Thus we have shown $C = (a,b) \cap \overline{C}$, whence $(a,b) \cap E$ is in the class M_0. Analogously, $(a,b) \cap F$ is in the class M_0. Suppose that $(a,b) \cap E$ is empty. Then $f'_{ap}(x) > 0$ for every x in $[a,b]$ since f is nondecreasing by the above lemma. But we have $f'_{ap}(a) < 0$, a contradiction. Analogously, $(a,b) \cap F$ is not empty. From Proposition 4.3 we infer that $(a,b) \cap E \cap F \neq \emptyset$. But this contradicts (4.12). The theorem is proved.

Summarizing the last two theorems, we have

THEOREM 4.18 (GOFFMAN-NEUGEBAUER). $f'_{ap} \in \mathcal{DB}_1$.

And we have the immediate corollary

COROLLARY 4.19. *If $g\colon (0,1) \to \mathbb{R}$ is an approximate derivative function, then there is a self-homeomorphism h of $(0,1)$ such that $g \circ h$ is a derivative function.*

4.5. Remarks

It has been pointed out that the original proof given by Maximoff of his theorem may not be correct. The first relatively simple proof of the theorem was given by Preiss [**110**]. (Although Preiss's proof seems to use only differentiation from the right, it is not very difficult to correct this oversight.) The paper by Preiss also contains several other interesting properties of the class $\mathcal{DB}_1$. The critical estimates found in our proof are due to Preiss. The present emphasis on the use of Neugebauer's characterization theorem of the derivative function to prove Maximoff's theorem appears here for the first time.

Part 2

Mappings and Measures on $\mathbb{R}^n$

CHAPTER 5

Bi-Lipschitzian Homeomorphisms

The functions that have been considered up to this point have been very nice. Often they were in the Baire class of functions which is nicely preserved on composition with general homeomorphisms. Now the focus will be on the class of Lebesgue measurable functions, a class that is wider than the Baire classes. As will be shown in the first section, composition of Lebesgue measurable functions with general homeomorphisms will destroy Lebesgue measurability. Fortunately, as will be shown for $n = 1$ in the first section and more generally in the chapter, bi-Lipschitzian homeomorphisms do preserve measurability under compositions. (See page 189 for the definition of bi-Lipschitzian maps.) In many problems of analysis, the need to "complete" spaces of continuous functions leads naturally to the consideration of Lebesgue measurable functions. This chapter will focus on those functions which are associated with nonparametric length and area.

The geometric notion of nonparametric, n-dimensional area is used in many parts of mathematics. The value $n = 1$ gives us length and $n = 2$ gives us area. For smooth functions, these notions are well-known to be given by the usual integral formulas. The notion of length has a very natural definition for continuous functions. It is the supremum of all the lengths of inscribed polygonal curves. But for area, the corresponding supremum is infinite even for some of the simplest of surfaces. (See Section A.9 of the Appendix.) Thus the correct definition of the nonparametric, n-dimensional area of a function is not evident. The chapter explores the class of functions (the linearly continuous ones) for which a reasonable notion of n-dimensional area can be defined and determines what sort of self-homeomorphisms will preserve this class of functions. We shall start with the simplest case of length, that is, $n = 1$. The case of $n = 2$ begins to show the difficulties that arise. Here the notion of partial derivatives in the sense of distributions will appear. For higher dimensions, the fact that these functions have an approximation by means of approximately continuous functions will be used in the final step of the question of invariance of the class of functions under bi-Lipschitzian homeomorphisms.

5.1. Lebesgue measurability

Let us begin the chapter with some elementary considerations of real-valued functions of one real variable. The first consists of two counterexamples. And the last presents positive results on measurability preserving homeomorphisms.

5.1.1. Examples. It is well-known that the composition of a function in the Baire classes of functions with a homeomorphism will result in a function in the same Baire class of functions, whence the Lebesgue measurability of such compositions is assured. The first counterexample is a homeomorphism $h\colon \mathbb{R} \to \mathbb{R}$ for which there is a Lebesgue measurable function $f\colon \mathbb{R} \to \mathbb{R}$ such that $f \circ h$ is not Lebesgue measurable. Let h be a self-homeomorphism of $\mathbb{R}$ that sends the Cantor ternary set onto a set P of positive Lebesgue measure. The set P contains a non-Lebesgue measurable set E. The required function f is the composition $\chi_E \circ h$. Clearly, $f = \chi_{h^{-1}[E]}$, whence f is Lebesgue measurable. As $f \circ h$ is χ_E, the non-Lebesgue measurability of $f \circ h$ is clear.

The second counterexample will show that approximately continuous functions are not preserved under compositions with self-homeomorphisms of $\mathbb{R}$. To this end, we shall modify the example found on page 4 of a nonporous set T and a homeomorphism h for which $S = h[T]$ is porous. For the approximate continuity example, the set S will consist of the point 0 and the union of pairwise disjoint, closed intervals I_n, $n = 1, 2, \ldots$, on the right-hand side of 0 and their reflections to the left of 0 such that 0 is a point of density of S; and, T will consist of the point 0 and the union of pairwise disjoint, closed intervals J_n, $n = 1, 2, \ldots$, on the right-hand side of 0 and their reflections to the left of 0 such that 0 is a point of dispersion of T. Obviously, there is a homeomorphism h such that $S = h[T]$. The characteristic function χ_S of S is approximately continuous at $x = 0$ and, from $\chi_S \circ h = \chi_T$, the composition $\chi_S \circ h$ is not approximately continuous at $x = 0$. Unfortunately, χ_S is not approximately continuous at every x. This is easily corrected by 'rounding-off' this function so as to make the new function f continuous at every x that is not 0. The rounding-off can be achieved by selecting a closed set S_0 so that $S_0 \setminus \{0\}$ is contained in the complement of S and 0 is a point of density of $h[S_0]$. Then the function f defined by

$$f(x) = \begin{cases} 1 & \text{for } x = 0, \\ \dfrac{\operatorname{dist}(x, S_0)}{\operatorname{dist}(x, S) + \operatorname{dist}(x, S_0)} & \text{for } x \neq 0, \end{cases}$$

will be suitably rounded-off. To see this, observe that $f(x) = 1$ whenever $x \in S$, and $f(x) = 0$ whenever $x \in S_0$. Consequently, $f \circ h$ is not approximately continuous at $x = 0$. Thus the second counterexample has been constructed.

5.1.2. Preservation of measurability and approximate continuity. An examination of the first of the above examples will show that homeomorphisms that do not preserve sets of Lebesgue measure 0 are bad for the invariance of Lebesgue measurable functions under composition. Indeed, we have the following simple observation.

LEMMA 5.1. *If $g\colon \mathbb{R} \to \mathbb{R}$ is a continuous function with the property that*

$$\lambda_1(g[Z]) = 0 \quad \textit{whenever} \quad \lambda_1(Z) = 0,$$

then $g[M]$ is Lebesgue measurable whenever M is Lebesgue measurable. Consequently, (locally) absolutely continuous functions g possess the property that $g[M]$ is Lebesgue measurable whenever M is Lebesgue measurable.

PROOF. Let M be a Lebesgue measurable set. Then $M = E \cup Z$, where E is an F_σ set and $\lambda_1(Z) = 0$. As E is σ-compact, the continuity of g will yield that $g[E]$ is also an F_σ set. The Lebesgue measurability of $g[M]$ now follows since $g[M] = g[E] \cup g[Z]$. The second part of the lemma is easily shown.

We now have the following result.

THEOREM 5.2. *Let h be a self-homeomorphism of $\mathbb{R}$ such that both h and h^{-1} are (locally) absolutely continuous (in particular, bi-Lipschitzian) and let $f\colon \mathbb{R} \to \mathbb{R}$ be any function. Then $f \circ h$ is Lebesgue measurable if and only if f is Lebesgue measurable. Also, $f \circ h$ is absolutely continuous on $[a, b]$ if and only if f is absolutely continuous on $h\big[[a, b]\big]$.*

PROOF. If f is Lebesgue measurable and h is a self-homeomorphism of $\mathbb{R}$, then we infer from the identity $(f \circ h)^{-1}[U] = h^{-1}\big[f^{-1}[U]\big]$ that the measurability of $f \circ h$ will occur exactly when $h^{-1}[M]$ is Lebesgue measurable for every Lebesgue measurable set M which is of the form $M = f^{-1}[U]$ with U being open. As h^{-1} is locally absolutely continuous, the lemma above completes the proof of the measurability of $f \circ h$. The first assertion obviously follows.

We leave the proof of the second assertion to the reader.

The second of the above examples displays a self-homeomorphism h that does not preserve the class of approximately continuous functions under composition. Notice that this homeomorphism is locally absolutely continuous and that such homeomorphisms do not preserve points of dispersion of subsets of $\mathbb{R}$. In the above theorem we alluded to the fact that Lipschitzian functions preserve sets of Lebesgue measure 0. For homeomorphisms we have a sharper lemma, which will not be proved here since it is proved in Theorem 5.21 for $\mathbb{R}^n$.

LEMMA 5.3. *Let $h\colon \mathbb{R} \to \mathbb{R}$ be a bi-Lipschitzian homeomorphism. If x is a point of dispersion of a measurable set E, then $h(x)$ is a point of dispersion of $h[E]$.*

We now have the invariance of approximately continuous functions under composition with bi-Lipschitzian homeomorphisms.

THEOREM 5.4. *Let h be a bi-Lipschitzian self-homeomorphism of $\mathbb{R}$. Then $h[U]$ is a density topology open set if and only if U is a density topology open set. Hence, for real-valued functions f, the composition $f \circ h$ is approximately continuous if and only if f is approximately continuous.*

PROOF. First we observe that the set $h[U]$ is measurable if and only if U is measurable. Next, for a measurable set U, a point x is a point of density of U if and only if x is a point of dispersion of $\mathbb{R} \setminus U$. Consequently, the first statement of the theorem follows.

The second statement follows from the definition of approximately continuous functions.

5.2. Length of nonparametric curves

For a continuous function $f\colon [0, 1] \to \mathbb{R}$, the **length of f** is defined to be

$$\ell(f, [0, 1]) = \sup \mathrm{E}_1(g, [0, 1]),$$

where the supremum is taken over all piecewise linear, continuous functions g that are inscribed in f and $\mathrm{E}_1(g, [0, 1])$ is the usual elementary length of the nonparametric curve g. For piecewise linear, continuous functions g, the length is given in integral form as

$$\mathrm{E}_1(g, [0, 1]) = \int_0^1 \sqrt{1 + g'^2(x)}\, dx,$$

and

$$\ell(g, [0,1]) = \mathrm{E}_1(g, [0,1]).$$

Moreover, if g_k, $k = 1, 2, \ldots$, is a sequence of piecewise linear, continuous functions that converges uniformly to a piecewise linear, continuous function g on $[0,1]$, then the lower semicontinuity formula

$$\mathrm{E}_1(g, [0,1]) \leq \liminf_{k \to +\infty} \mathrm{E}_1(g_k, [0,1])$$

holds. (See Section A.10 for a discussion on lower semicontinuity.) In fact, the mode of convergence can be relaxed to everywhere convergence and even to almost everywhere convergence in the lower semicontinuity statement above. It is well-known that, for continuous functions f, the length functional is lower semicontinuous. That is,

$$\ell(f, [0,1]) \leq \liminf_{k \to +\infty} \ell(f_k, [0,1])$$

whenever f_k, $k = 1, 2, \ldots$, is a sequence of continuous functions that converges uniformly to f on $[0,1]$. Even more, the above convergence can be weakened to almost everywhere convergence on $[0,1]$ to the continuous function f. We shall discuss the consequences that these modes of convergence have for the length functional.

5.2.1. Finite length. The elementary length functional $\mathrm{E}_1(g, [0,1])$ for a piecewise linear, continuous function g satisfies the inequalities

$$\max\{\, \lambda([0,1]), \mathrm{V}(g, [0,1]) \,\} \leq \mathrm{E}_1(g, [0,1]) \leq \lambda([0,1]) + \mathrm{V}(g, [0,1]).$$

So, if g_k, $k = 1, 2, \ldots$, is a sequence of piecewise linear, continuous functions that converges to f by some mode of convergence, then we will have

$$\liminf_{k \to +\infty} \mathrm{V}(g_k, [0,1]) \leq \liminf_{k \to +\infty} \mathrm{E}_1(g_k, [0,1]) \leq \lambda([0,1]) + \liminf_{k \to +\infty} \mathrm{V}(g_k, [0,1]).$$

It is easy to show that

$$\mathrm{V}(f, [0,1]) \leq \liminf_{k \to +\infty} \mathrm{V}(g_k, [0,1])$$

when the mode is everywhere convergence, and additionally that f is continuous when the mode is uniform convergence.

We just mention the following theorem which combines results of Jordan and Tonelli.

THEOREM 5.5. *For real-valued functions f,*

$$\ell(f, [0,1]) < +\infty \quad \textit{if and only if} \quad \mathrm{V}(f, [0,1]) < +\infty.$$

Moreover, if $\ell(f, [0,1]) < +\infty$, then f is differentiable almost everywhere and

$$\int_0^1 \sqrt{1 + f'^2(x)}\, dx \leq \ell(f, [0,1]),$$

where the equality holds when and only when f is an absolutely continuous function.

Observe that the above theorem has been stated for functions that are not necessarily continuous. The functions f with finite length can have only countably many points of discontinuities and that the discontinuities are of the first kind (that is, the left and right limits both exist). It is obvious that the class of functions of bounded variation is invariant under composition with self-homeomorphisms of $[0,1]$, whence the class of continuous functions of bounded variation is invariant under such compositions.

Let us now introduce Lebesgue measure into the discussion of lower semicontinuity. We shall use almost everywhere convergence as the mode of convergence. For continuous functions f, the functionals $\ell(f,[0,1])$ and $\mathrm{V}(f,[0,1])$ are lower semicontinuous. But, for discontinuous functions f, this fact is no longer true. Indeed, there is a function f of bounded variation and a sequence g_k, $k=1,2,\ldots$, of continuous, piecewise linear functions that converges almost everywhere to f such that $\liminf_{k\to+\infty}\mathrm{E}_1(g_k,[0,1])<\ell(f,[0,1])$. Consequently, $\ell(f,[0,1])$ and $\mathrm{V}(f,[0,1])$ are the wrong lower semicontinuous extensions for the almost everywhere mode of convergence of the elementary length functional and the total variation functional restricted to the class of piecewise linear, continuous functions on $[0,1]$. This discussion leads us to the natural definition of length and total variation in the mode of almost everywhere convergence. We call these functionals the **essential length** and **essential variation** of f. The definitions are

$$\operatorname{ess}\ell(f,[0,1])=\inf\liminf_{k\to+\infty}\mathrm{E}_1(g_k,[0,1])$$

and

$$\operatorname{ess}\mathrm{V}(f,[0,1])=\inf\liminf_{k\to+\infty}\mathrm{V}(g_k,[0,1]),$$

where the infima are taken over all sequences g_k, $k=1,2,\ldots$, of piecewise linear, continuous functions that converge almost everywhere to f. Clearly,

$$\operatorname{ess}\ell(f,[0,1])\le\ell(f,[0,1])$$

and

$$\operatorname{ess}\mathrm{V}(f,[0,1])\le\mathrm{V}(f,[0,1]).$$

It is shown in Section A.3.3 of the Appendix that, for functions f that have bounded essential variations,

$$\operatorname{ess}\ell(f,[0,1])=\ell(\widetilde{f},[0,1])$$

and

$$\operatorname{ess}\mathrm{V}(f,[0,1])=\mathrm{V}(\widetilde{f},[0,1])$$

where $f=\widetilde{f}$ almost everywhere, $\widetilde{f}$ is of bounded variation and, for every x in $[0,1)$,

$$\widetilde{f}(x)=\lim_{t\to x+}\widetilde{f}(t).$$

5.2.2. Distribution derivatives. Let us begin with an example. The characteristic function of a set E, call it f, contained in the usual Cantor ternary set is certainly a measurable function for which $\operatorname{ess}\mathrm{V}(f,[0,1])=0$. It is well-known that there is a self-homeomorphism h of $[0,1]$ that sends a set of positive measure onto the Cantor ternary set. Consequently, by an appropriate choice of the set E, the function $f\circ h$ can be nonmeasurable. Even more, it may happen that $f\circ h$ is measurable with $\operatorname{ess}\mathrm{V}(f\circ h,[0,1])=+\infty$ as witnessed by simply taking E to be the entire Cantor ternary set.

The above example shows that the functional $\operatorname{ess}\mathrm{V}$ is not invariant under Lebesgue equivalence (that is, invariance under compositions with self-homeomorphisms). The task then becomes one of finding a class of self-homeomorphisms h for which this functional is invariant under composition with h. We will find that one such class is the class of all bi-Lipschitzian self-homeomorphisms. But before we show this, let us discuss the role in analysis played by measurable functions with $\operatorname{ess} \mathrm{V}(f, [0,1]) < +\infty$.

For a Lebesgue integrable function f (that is, $\int_0^1 |f(x)|\, dx < +\infty$) one can define a very natural linear functional on the class of all infinitely differentiable functions φ whose supports are contained in the open interval $(0,1)$, namely, the functional defined by

$$f(\varphi) = \int_0^1 \varphi(x) f(x)\, dx.$$

Clearly, this functional is invariant under almost everywhere equality. Using the integration by parts identity, one is naturally lead to the distribution derivative functional of f, denoted by $\mathrm{D}f$, which is defined by

$$\mathrm{D}f(\varphi) = -\int_0^1 \mathrm{D}\varphi(x) f(x)\, dx,$$

where the D inside of the integral is the usual derivative operator. The linear functional $\mathrm{D}f$ may or may not have a measure representation, which means the existence of a signed Borel measure μ on $[0,1]$ such that

$$\mathrm{D}f(\varphi) = \int_0^1 \varphi(x)\, d\mu(x).$$

We have the following theorem.

THEOREM 5.6. *Let f be a Lebesgue integrable function. Then the distribution derivative* $\mathrm{D}f$ *has a measure representation if and only if* $\operatorname{ess} \mathrm{V}(f, [0,1]) < +\infty$.

A finite, signed Borel measure μ also defines a distribution by means of the formula

$$\mu(\varphi) = \int_0^1 \varphi(x)\, d\mu(x).$$

There is no loss in assuming that this measure also satisfies

$$\mu(\{0\}) = 0 \quad \text{and} \quad \mu(\{1\}) = 0.$$

Such a signed measure μ has a decomposition into the difference of two finite measures μ_- and μ_+. Consider the real-valued function given by

$$f(x) = \begin{cases} \mu([0,x)) & \text{for } x \in [0,1), \\ \mu([0,1]) & \text{for } x = 1. \end{cases}$$

This function is of bounded variation with $\operatorname{ess} \mathrm{V}(f, [0,1]) = \mathrm{V}(f, [0,1])$. With the aid of the identity

$$f(x) = \int_0^1 \chi_{[0,x)}(t)\, d\mu(t)$$

on $[0,1]$, a straightforward application of the Fubini theorem will yield the identity

$$\mu = \mathrm{D}f.$$

Thus we find that the notions of measures and functions with bounded essential variations coincide in the distribution sense.

Of course, the measure in the above discussion may have atoms. In the event that the measures are also nonatomic, the functions that correspond to these measures will be continuous functions with finite total variations.

For a measure μ that is absolutely continuous with respect to the Lebesgue measure λ, there is a Lebesgue integrable function g (namely, the Radon-Nykodym derivative) such that $d\mu = g\,dx$. We shall say that a measurable function f is **essentially absolutely continuous** if there exists an absolutely continuous function $\widetilde{f}$ with $f = \widetilde{f}$ λ-almost everywhere. At this point, it will be convenient to introduce the notion **essential derivative** of f, which means that the derivative of f is computed at the point x after deleting an appropriate set of Lebesgue measure 0. For example, the characteristic function of the set of irrational numbers has an essential derivative equal to 0 at each irrational number.

THEOREM 5.7. *Let f be a Lebesgue integrable function. Then there exists a Lebesgue integrable function g such that*

$$\mathrm{D}f(\varphi) = \int_0^1 \varphi(x)g(x)\,dx \quad \textit{for all} \quad \varphi$$

if and only if f is essentially absolutely continuous. Moreover, if f is essentially absolutely continuous, then the essential derivative of f exists almost everywhere and is equal to the above g almost everywhere.

For more on distribution derivatives, see the Appendix.

5.2.3. Invariance under homeomorphism. The notion of essential variation of a function requires that the function be measurable. The essential variation may be computed by using the points of approximate continuity of the function (see page 194). It has been shown in the first section of the chapter that self-homeomorphisms h do not preserve measurability of functions under composition with h. Moreover, it was shown that locally absolutely continuous homeomorphisms do not preserve points of approximate continuity of functions under compositions. But, fortunately, bi-Lipschitzian homeomorphisms do preserve both measurability and points of approximate continuity of measurable functions under compositions. Indeed, we have the following theorem.

THEOREM 5.8. *Let h be a bi-Lipschitzian self-homeomorphism of $[0,1]$ and let f be a real-valued function on $[0,1]$. Then*

$$\operatorname{ess}\mathrm{V}(f,[0,1]) = \operatorname{ess}\mathrm{V}(f\circ h,[0,1]).$$

PROOF. Assume that $\operatorname{ess}\mathrm{V}(f,[0,1])$ is finite. We know that f is equal λ-almost everywhere to a function $\widetilde{f}$ with the same total variation. Let Z be the set on which f and $\widetilde{f}$ differ. For any self-homeomorphism h, we have that $f\circ h$ and $\widetilde{f}\circ h$ differ precisely on the set $h^{-1}[Z]$. As $\mathrm{V}(\widetilde{f},[0,1]) = \mathrm{V}(\widetilde{f}\circ h,[0,1])$, the theorem follows.

We have already pointed out that the class of continuous functions with finite total variations is invariant under compositions with self-homeomorphisms. The absolutely continuous functions form a subclass of this class, but this subclass is not invariant under compositions with self-homeomorphisms. Indeed, let h be a self-homeomorphism of $[0,1]$ with the property that $\lambda(h[K]) > 0$, where K is the Cantor ternary set. This function is not absolutely continuous because $\lambda(K) = 0$. As the identity homeomorphism is absolutely continuous, the composition of the identity function with the above homeomorphism is not absolutely continuous. Fortunately, due to Theorem 5.2, the class of absolutely continuous functions is invariant under composition with bi-Lipschitzian self-homeomorphisms.

5.2.4. Hausdorff measure and essential length. There is a nice connection between the essential nonparametric length of a function and the Hausdorff 1-dimensional measure of the graph of the function. We have learned that if $\operatorname{ess} \mathrm{V}(f, [0,1]) < +\infty$ then there is a function $\widetilde{f}$ that is equal to f almost everywhere with $\mathrm{V}(\widetilde{f}, [0,1]) = \operatorname{ess} \mathrm{V}(f, [0,1])$. It is easily shown that the Hausdorff measure satisfies

$$\mathsf{H}_1(\operatorname{graph} \widetilde{f}\,) \leq \operatorname{ess} \ell(f, [0,1]),$$

where the equality holds if and only if $\widetilde{f}$ is continuous as well as of bounded variation.

We note that the characteristic function of a nowhere dense perfect set with positive Lebesgue measure does not have finite essential variation on $[0,1]$ and yet its graph has Hausdorff measure equal to 1.

5.3. Nonparametric area

This section concerns the more complicated case of functions of several independent variables. For the sake of staying within the theme of the book, the discussion will be more descriptive rather than a presentation of proofs. (References to the relevant papers will be provided in the remarks section at the end of the chapter.) Of course, by area we shall mean the n-dimensional nonparametric area of functions.

5.3.1. Lower semicontinuity of area. The naive approach of computing the nonparametric area of functions by means of inscribed piecewise linear, continuous functions certainly leads to difficulty (see our earlier reference to the Appendix). There are many candidates for the definition of area, but the ones that seem to have won favor are those that have the lower semicontinuity property. There is a natural definition that was originally proposed by Lebesgue. The development presented here extends further the theory of continuous surfaces to that of a natural class of discontinuous surfaces.

For convenience, we shall assume that our functions are defined on $\mathbb{R}^n$ and that I is the n-fold Cartesian product of intervals $[a_i, b_i]$, $i = 1, \ldots, n$. It is well-known that the nonparametric area of a piecewise linear, continuous function g defined on I is given by the integral

$$\mathrm{E}_n(g, I) = \int_I \sqrt{1 + (\mathrm{D}_1 g(x))^2 + \cdots + (\mathrm{D}_n g(x))^2}\, d\lambda_n(x),$$

where D_i denotes the i-th partial derivative operator.

We mention here a very useful fact about the area integral for Lipschitzian functions f.

THEOREM 5.9. *Suppose that f is a Lipschitzian function on $\mathbb{R}^n$ and that h is a bi-Lipschitzian homeomorphism of $\mathbb{R}^n$ into itself. Then*

$$\int_{h^{-1}[I]} \sqrt{1+(\mathrm{D}_1(f\circ h)(x))^2+\cdots+(\mathrm{D}_n(f\circ h)(x))^2}\,|\mathrm{J}(h,x)|\,d\lambda_n(x) = \int_I \sqrt{1+(\mathrm{D}_1 f(x))^2+\cdots+(\mathrm{D}_n f(x))^2}\,d\lambda_n(x),$$

where $\mathrm{J}(h,x)$ is the Jacobian of h at x.

Consequently,

$$\int_I \sqrt{1+(\mathrm{D}_1 f(x))^2+\cdots+(\mathrm{D}_n f(x))^2}\,d\lambda_n(x) \le \big(\mathrm{Lip}(h)\big)^n \int_{h^{-1}[I]} \sqrt{1+(\mathrm{D}_1(f\circ h)(x))^2+\cdots+(\mathrm{D}_n(f\circ h)(x))^2}\,d\lambda_n(x).$$

For each i, where $i=1,\dots,n$, let x_i be the i-th coordinate of x and let $x^{(i)}$ be the remaining coordinates of x. Analogously, we shall use I_i and $I^{(i)}$ to denote the corresponding intervals of the n-fold product I. We shall write $f(x_i,x^{(i)})$ for $f(x)$ when we wish to emphasize the i-th coordinate of x. Now we find that each piecewise linear, continuous function g satisfies

$$\mathrm{V}\big(g(\,\cdot\,,x^{(i)}),I_i\big)=\int_{I_i}|\mathrm{D}_i g(x_i,x^{(i)})|\,dx_i$$

and

$$\int_{I^{(i)}} \mathrm{V}\big(g(\,\cdot\,,x^{(i)}),I_i\big)\,d\lambda_{n-1}(x^{(i)}) = \int_I |\mathrm{D}_i g(x)|\,d\lambda_n(x).$$

As in the previous section, we consider sequences g_k, $k=1,2,\dots$, of piecewise linear, continuous functions that converge almost everywhere to f on I. For such sequences and for each i, we have

$$\liminf_{k\to+\infty}\int_{I^{(i)}} \mathrm{V}\big(g_k(\,\cdot\,,x^{(i)}),I_i\big)\,d\lambda_{n-1}(x^{(i)}) \le \liminf_{k\to+\infty}\mathrm{E}_n(g_k,I).$$

This inequality motivates the definition of the **nonparametric area functional**

$$\mathrm{A}_n(f,I)=\inf\liminf_{k\to+\infty}\mathrm{E}_n(g_k,I)$$

and the **variation functionals**

$$\mathrm{V}_i(f,I)=\inf\int_{I^{(i)}}\liminf_{k\to+\infty}\mathrm{V}\big(g_k(\,\cdot\,,x^{(i)}),I_i\big)\,d\lambda_{n-1}(x^{(i)}),\qquad i=1,\dots,n,$$

where the infima are taken over all sequences g_k, $k=1,2,\dots$, of piecewise linear, continuous functions that converge almost everywhere to f.

It is known that this nonparametric area coincides with the one that results from the mode of convergence being uniform convergence whenever the function f is continuous. Even more, the following theorem connects it to the area integral for Lipschitzian functions f.

THEOREM 5.10. *If f is Lipschitzian on I, then*

$$\mathrm{A}_n(f,I)=\int_I \sqrt{1+(\mathrm{D}_1 f(x))^2+\cdots+(\mathrm{D}_n f(x))^2}\,d\lambda_n(x).$$

Observe that $\mathrm{V}_i(f,I)$ is not defined as an integral. Fortunately, the following equality holds.

$$\mathrm{V}_i(f,I) = \int_{I^{(i)}} \mathrm{V}_E\big(f(\,\cdot\,, x^{(i)}), I_i\big)\, d\lambda_{n-1}(x^{(i)}),$$

where E is the set of all approximate continuity points of f and the variation $\mathrm{V}_E\big(f(\,\cdot\,,x^{(i)}),I_i\big)$ is computed by employing only the points z of E for which $z^{(i)} = x^{(i)}$. It is easily shown that if f and $\widetilde{f}$ differ only on a set of λ_n-measure zero, then they have the same variations on I.

It is now possible to define various classes of functions of bounded variations with the aid of the variation functions $\mathrm{V}_i(f,I)$, $i = 1, \dots, n$. Let f be a measurable function. Then f is said to be of **bounded variation in the sense of Cesari** on I if $\mathrm{V}_i(f,I)$ is finite for every i, abbreviated as BVC (see Cesari [**26**]). In the remainder of the discussion, we shall suppress the reference to the interval I, that is, we will be dealing with the BVC property locally. A BVC function f is said to be **linearly continuous** if, for each i, there is a function $\widetilde{f}$ such that $f = \widetilde{f}$ λ_n-almost everywhere and $\widetilde{f}(\,\cdot\,, x^{(i)})$ is continuous for λ_{n-1}-almost every $x^{(i)}$. Of course, it is not assumed that the same function $\widetilde{f}$ works for all i. Fortunately, the following theorem holds. (See Goffman [**57, 56**].)

THEOREM 5.11. *In the class of* BVC *functions, f is linearly continuous if and only if there exists a function $\widetilde{f}$ such that $\widetilde{f}$ is equal to f λ_n-almost everywhere and such that, for each i, $\widetilde{f}(\,\cdot\,, x^{(i)})$ is continuous for λ_{n-1}-almost every $x^{(i)}$.*

Taking a cue from the notion of linearly continuous functions, we see that the next thing to do is to replace continuity on lines with absolute continuity on lines. Such a function is called **linearly absolutely continuous** or simply **absolutely continuous**. None of the classes of functions that have been singled out in this section are contained in the class of continuous functions. Tonelli has studied these functions under the additional condition of continuity. With this added assumption, the case of linearly continuous functions will disappear.

We close this section with a major theorem. The one-dimensional analogue is fairly simple to prove. The proof for the higher dimensional case is harder (see [**28**, page 24]).

THEOREM 5.12. *For a measurable function f,*

$$\mathrm{A}_n(f,I) < +\infty \quad \textit{if and only if} \quad f \text{ is BVC}.$$

Moreover, for every i,

$$\mathrm{V}_i(f,I) \le \mathrm{A}_n(f,I) \le \lambda_n(I) + \sum_{j=1}^{n} \mathrm{V}_j(f,I).$$

5.3.2. Distribution partial derivatives. The notion of BVC functions is closely associated with that of distribution partial derivatives. Functions f that are integrable on I will define a linear functional on infinitely differentiable functions φ with compact support contained in the interior of I by means of the formula

$$f(\varphi) = \int_I \varphi(x) f(x)\, d\lambda_n(x).$$

For each i, its distribution partial derivative is given by

$$\mathrm{D}_i f(\varphi) = -\int_I \mathrm{D}_i\varphi(x) f(x)\, d\lambda_n(x).$$

Borel measures μ define distributions, namely,

$$\mu(\varphi) = \int_I \varphi(x)\, d\mu(x).$$

The class of functions whose distribution partial derivatives are measures is given by the following theorem.

THEOREM 5.13. *Let f be (locally) integrable. Then there exists a Borel measure μ_i such that $\mathrm{D}_i f = \mu_i$ for each i, $i = 1, \dots, n$, if and only if f is* BVC.

For $n = 1$, it is possible to distinguish which measures came from the continuous BV functions and which came from the absolutely continuous ones. The next theorem shows the analogous distinctions in the higher dimensional case.

THEOREM 5.14 (GOFFMAN-LIU). *Let μ_i, $i = 1, \dots, n$, be the Borel measures that correspond to the distribution partial derivatives of a function f in* BVC. *Then the following three statements hold for Borel sets E.*

(1) *If $\mathsf{H}_{n-1}(E) = 0$, then $\mu_i(E) = 0$ for each i.*
(2) *If f is linearly continuous and $\mathsf{H}_{n-1}(E) < +\infty$, then $\mu_i(E) = 0$ for each i.*
(3) *If f is linearly absolutely continuous and $\mathsf{H}_n(E) = 0$, then $\mu_i(E) = 0$ for each i.*

Moreover,

(4) *if $n - 1 < \alpha < n$, then there is a continuous* BVC *function f and there is a Borel set E such that $\mathsf{H}_\alpha(E) = 0$ and $\mu_1(E) > 0$.*

5.3.3. Hausdorff n-measure of the graph. Just as there is a relationship between the Hausdorff measure of the graph and the nonparametric length of a function, there is an analogous connection in the higher dimensional case [**51**].

THEOREM 5.15 (GOFFMAN). *If f is* BVC *on I, then there is a function $\widetilde{f}$ that is equal to f λ_n-almost everywhere and such that*

$$\mathsf{H}_n(\operatorname{graph} \widetilde{f}\,) \le \mathrm{A}_n(f, I),$$

where the equality holds when and only when f is linearly continuous.

5.4. Invariance under self-homeomorphisms

We turn now to the main point of the chapter. Unlike the BV class for $n = 1$, the class of functions that are BVC on I is not invariant under self-homeomorphisms of I as the example given below will show. We shall show that each of the classes of BVC functions mentioned in the last section are invariant under bi-Lipschitzian self-homeomorphisms of I.

5.4.1. An example. Let f be the function on $\mathbb{R}^2$ given by

$$f(x) = \begin{cases} 0 & \text{if } x_1 < \frac{1}{2}, \\ 1 & \text{if } x_1 \ge \frac{1}{2}. \end{cases}$$

Clearly, f is BVC on I. Suppose that h is a self-homeomorphism of I such that the line segment L in I defined by $x_1 = \frac{1}{2}$ is mapped onto a set of positive planar measure. It is easy to show that $f \circ h$ is not BVC on I.

5.4.2. Measures and bi-Lipschitzian homeomorphisms. Let f be a function that is BVC on I. Associated with each measure μ_i that corresponds to the distribution partial derivative $\mathrm{D}_i f$ there is its total variation measure $\|\mu_i\|$. It will be convenient to define the following positive measure

$$\nu_f = \lambda_n + \sum_{i=1}^{n} \|\mu_i\|.$$

This positive measure has a very nice invariance property.

THEOREM 5.16. *Let K be a Lipschitz constant for both h and h^{-1}, where h is a bi-Lipschitzian self-homeomorphism of I. Then there is a constant M that depends only on K such that, for any* BVC *function f,*

$$\nu_f \leq M\, \nu_{f\circ h}.$$

It now follows that the class of BVC functions is invariant under bi-Lipschitzian self-homeomorphisms of I.

Even more, the class of linearly continuous, BVC functions is invariant under bi-Lipschitzian self-homeomorphisms. This result is shown by first proving the next two theorems.

THEOREM 5.17. *If f is an approximately continuous,* BVC *function, then f is linearly continuous.*

THEOREM 5.18 (GOFFMAN). *Let f be* BVC *on an interval I. Then a necessary and sufficient condition for f to be linearly continuous on I is that there exist for each positive number ε an approximately continuous,* BVC *function g such that*

$$\nu_f(E) < \varepsilon \quad \textit{and} \quad \nu_g(E) < \varepsilon,$$

where $E = \{\, x \in I : f(x) \neq g(x) \,\}$.

We infer from these two theorems the following sharpening of Theorem 5.11.

THEOREM 5.19. *In the class of* BVC *functions, f is linearly continuous if and only if there exists a function $\widetilde{f}$ such that $\widetilde{f}$ is equal to f λ_n-almost everywhere and such that, for almost every hyperplane H of $\mathbb{R}^n$, $\widetilde{f}$ is continuous on almost every line orthogonal to H.*

As the class of approximately continuous functions is invariant under such homeomorphisms, which will be proved in the next section, we finally have the promised invariance theorem for the class of linearly continuous BVC functions.

THEOREM 5.20. *For bi-Lipschitzian self-homeomorphisms h and* BVC *functions f, the composition $f \circ h$ is linearly continuous if and only if f is linearly continuous.*

The last four theorems were first proved by Goffman in [**56**].

5.5. Invariance of approximately continuous functions

Let us prove that the the class of approximately continuous functions is invariant under bi-Lipschitzian homeomorphisms. A function $f\colon \mathbb{R}^n \to \mathbb{R}$ is approximately continuous if each point of the set $f^{-1}[U]$ is a point of density 1 of the set whenever U is an open set (see the Appendix). So, it remains to be shown that bi-Lipschitzian homeomorphisms of $\mathbb{R}^n$ into itself preserve density points and dispersion points of each measurable subset of $\mathbb{R}^n$.

THEOREM 5.21. *If h is a Lipschitzian homeomorphism of $\mathbb{R}^n$ into $\mathbb{R}^n$, then*

$$h\big[B(x,r)\big] \subset B\big(h(x), \operatorname{Lip}(h)\, r\big),$$

and consequently, whenever E is a measurable subset of $\mathbb{R}^n$, the set $h[E]$ is measurable and satisfies

$$\lambda_n(h[E]) \le \big(2\operatorname{Lip}(h)\big)^n \lambda_n(E).$$

Therefore, bi-Lipschitzian maps of $\mathbb{R}^n$ preserve density points and preserve dispersion points of measurable sets.

PROOF. The inclusion statement is obvious. We infer from this inclusion that sets of Lebesgue measure 0 have images that also have Lebesgue measure 0. As each Lebesgue measurable set E is the the union of an F_σ set and a set with Lebesgue measure 0, it is clear that $h[E]$ is Lebesgue measurable if E is. The measure inequality now follows easily from the inclusion statement and properties of the Lebesgue measure.

Finally, let us show that dispersion points of measurable sets are preserved. Let E be a measurable set and x be a point for which the density of E at x is 0. Suppose that ε is a positive number. Then there is a positive number δ such that

$$\lambda_n\big(E \cap B(x,r)\big) < \varepsilon\, \lambda_n\big(B(x,r)\big) \quad \text{whenever} \quad r < \operatorname{Lip}(h^{-1})\, \delta.$$

Since h is a homeomorphism, we have that $h\big[B(x, 2\operatorname{Lip}(h^{-1})\,\delta)\big]$ is an open set that contains the point $h(x)$; whence there is a positive number δ_0 such that

$$B(h(x), s) \subset h\big[B(x, \operatorname{Lip}(h^{-1})\,\delta)\big] \quad \text{whenever} \quad 0 < s < \delta_0 < \delta.$$

We assert that

$$\lambda_n\big(h[E] \cap B(h(x), s)\big) < \varepsilon\,\big(4\operatorname{Lip}(h^{-1})\operatorname{Lip}(h)\big)^n \lambda_n\big(B(h(x), s)\big) \qquad \text{whenever} \quad 0 < s < \delta_0.$$

Indeed, as $\delta_0 < \delta$, the following inequality results.

$$\begin{aligned}\lambda_n\big(E \cap B(x, 2\operatorname{Lip}(h^{-1})\, s)\big) &< \varepsilon\, \lambda_n\big(B(x, 2\operatorname{Lip}(h^{-1})\, s)\big)\\ &= \varepsilon\, \big(2\operatorname{Lip}(h^{-1})\big)^n \lambda_n\big(B(x,s)\big)\\ &= \varepsilon\, \big(2\operatorname{Lip}(h^{-1})\big)^n \lambda_n\big(B(h(x),s)\big).\end{aligned}$$

And from the inclusion

$$E \cap h^{-1}[B(h(x), s)] \subset E \cap B(x, 2\operatorname{Lip}(h^{-1})\, s),$$

we infer the inclusion

$$h[E] \cap B(h(x), s) \subset h[E \cap B(x, 2\operatorname{Lip}(h^{-1})\, s)].$$

To complete the proof of the assertion, we observe that

$$\begin{aligned}\lambda_n\big(h[E] \cap B(h(x), s)\big) &\le \varepsilon\,\big(2\operatorname{Lip}(h^{-1})\big)^n \lambda_n\big(E \cap B(x, 2\operatorname{Lip}(h^{-1})\, s)\big)\\ &= \varepsilon\,\big(4\operatorname{Lip}(h^{-1})\operatorname{Lip}(h)\big)^n \lambda_n\big(B(h(x), s)\big).\end{aligned}$$

Thus we have shown that $h(x)$ is a point of dispersion of $h[E]$.

The proof of the theorem is easily completed by observing that, for measurable sets, points of density of the set are precisely those points that are points of dispersion of the complement of the set.

Note that Theorem 5.4 is the special case of $n = 1$ in the last theorem. It is clear that much more than the invariance of linearly continuous, BVC functions will result from the last theorem. We summarize all of the invariants that have been achieved in the following theorem.

THEOREM 5.22. *Let h be a bi-Lipschitzian homeomorphism of $\mathbb{R}^n$. Then the following statements hold for real-valued functions f on $\mathbb{R}^n$.*

(1) *f is measurable if and only if $f \circ h$ is measurable.*
(2) *f is approximately continuous if and only if $f \circ h$ is approximately continuous.*
(3) *f is* BVC *if and only if $f \circ h$ is* BVC.
(4) *f is* BVC *and linearly continuous if and only if $f \circ h$ is* BVC *and linearly continuous.*
(5) *f is* BVC *and absolutely continuous if and only if $f \circ h$ is* BVC *and absolutely continuous.*

PROOF. The first two statements are consequences of the last theorem. The third statement follows from Theorems 5.9 and 5.10. And the fourth statement (which is Theorem 5.20) is a consequence of the first two statements and Theorems 5.17 and 5.18.

Let us prove the last statement. It is known that BVC functions that are linearly absolutely continuous are those whose distribution partial derivatives are given by measures that are absolutely continuous with respect to Lebesgue measure. Consequently, the statement follows from Theorem 5.16.

5.6. Remarks

Interesting articles on length and area in the continuous setting are the three by Cesari [**30, 29, 31**].

We have presented the material on nonparametric area with very few proofs and references to the literature. The proofs in general are very difficult and long. So we will cite sources where proofs can be found. Very likely, all pertinent references will not have been made. As the references are filled in, we shall at times make additional remarks which the reader may find interesting.

The notion of BVC has been studied by Cesari [**26**], Goffman [**51, 55**], Fleming [**43**], and Krickeberg [**90**], to cite a few authors. The connection between the finiteness of area and BVC (Theorem 5.12) is proved in Cesari [**26**] and Goffman [**51, 55**]. The existence of nice linearly continuous representations (Theorems 5.11 and 5.19) was first proved by Goffman [**57, 56**]; see also Hughs [**77**] and Serrin [**122**]. The connection between Hausdorff measure and area, Theorem 5.15, is found in [**51**].

We have indicated the close connection of BVC to distributions. The reader may wish to consult, for example, the book by Evans and Gariepy [**39**] (the notion of BV functions given in Chapter 5 of their book is defined in a slightly modified form). Theorems 5.9 and 5.10 concerning the area of Lipschitzian functions can be found in [**39**]. Theorem 5.13 can be found in Krickeberg [**90**]. For a proof of the various connections between the Hausdorff measures and the distribution partial derivative measures given in Theorem 5.14, see Goffman and Liu [**60**]. We point out that from this theorem there will result 5 subclasses of the BVC functions when $n > 1$. Indeed, among those that are continuous, there are the linearly absolutely continuous ones. Among the discontinuous ones, there are the linearly continuous ones and the linearly absolutely continuous ones. This is in contrast to the case of $n = 1$ in which there are only 3 subclasses of the BV functions (namely, the continuous, the absolutely continuous and the discontinuous) because the Hausdorff $(n-1)$-measure in this case is H_0, and $\mathsf{H}_0(E) = 0$ if and only if $E = \emptyset$. Finally, Theorem 5.16 is found in [**57**].

The proof of the invariance of the density and dispersion points of a measurable set under bi-Lipschitzian homeomorphisms (Theorem 5.21) is rather simple. A sharper form of this result due to Buczolich [**25**], which requires a substantially more complicated proof, is given in the Appendix (Theorem A.9).

CHAPTER 6

Approximation by Homeomorphisms

The theme of this chapter is the approximation of one-to-one mappings T by means of interpolating homeomorphisms H that agree with T on sets that are relatively large in the sense of outer measures. Such approximations are motivated by Lusin's theorem. The first section is a background discussion of various Lusin-type theorems for real-valued functions. Arbitrary one-to-one maps of the square are shown in the second section to have approximations by homeomorphisms, though not necessarily by an interpolation. And, in the fourth section, one-to-one measurable maps T with measurable inverse maps T^{-1} are approximated by means of interpolations (see that section for the meaning of interpolation in measure). These approximations are established with the aid of extension theorems proved in the third section.

6.1. Background

We begin the background discussion with some consequences of Lusin's theorem on the approximation of measurable functions by continuous functions. For a measurable real-valued function f on the interval $[0,1]$, Lusin's theorem says that for every positive number ε there is a continuous function g on $[0,1]$ such that $f(x) = g(x)$ except on a set of measure less than ε. A standard consequence of this theorem is that there is a function g in the Baire class 2 or less such that the measurable function f and g agree almost everywhere.

6.1.1. Approximation in measure. Not so well-known is a beautiful theorem of Saks and Sierpiński [**116**] on the approximation of arbitrary functions by means of functions in the Baire class 2. Here we shall merely state the theorem and give references and indications of proofs.

THEOREM 6.1 (SAKS-SIERPIŃSKI). *For each real-valued function f defined on $[0,1]$, there exists a Baire class 2 function g such that $|f(x) - g(x)| < \varepsilon$ holds on a set of outer measure 1 for each positive number ε.*

Regarding the proof of this theorem, we shall discuss a proof related to the measurable boundaries of arbitrary real-valued functions. This notion was introduced by H. Blumberg [**13**]. The approach to measurable boundaries given here is due to Goffman and Waterman [**66**] and its application to the proof of the Saks-Sierpiński theorem is due to Goffman and Zink [**72**]. Denote by $\mathcal{M}$ the set of equivalence classes of measurable functions on the interval $[0,1]$, that is, those classes of functions that are equal to each other except on a set of measure 0. The set $\mathcal{M}$ is a lattice. Moreover, it is a complete lattice if the functions are allowed to take the values $\pm\infty$. Indeed, by effecting a homeomorphism of $[-\infty,+\infty]$ onto $[-1,+1]$, the matter is reduced to the bounded case and the easily shown fact that every nonempty bounded subset $\mathcal{S}$ of $\mathcal{M}$ has a least upper bound and a greatest lower bound.

Blumberg has associated with every real-valued function f on $[0,1]$ two measurable functions (extended real-valued) which he calls the **measurable boundaries** of f. If we consider the collection of all equivalence classes whose members are greater than or equal to f almost everywhere, then the greatest lower bound of this collection contains the Blumberg upper measurable boundary u of f. The lower measurable boundary l of f is obtained by considering the measurable functions below f. The functions u and l may be taken to be in the Baire class 2. The proof of the Saks-Sierpiński theorem is almost obvious when f is bounded. A simple trick will handle the unbounded case.

Another relatively simple proof of the Saks-Sierpiński theorem was given in Goffman [**50**]. The proof uses the following lemma.

LEMMA 6.2. *For an interval $[a,b]$ and a positive number ε, if f is a real-valued function and g is a continuous function such that $|f(x)-g(x)|<\varepsilon$ on a set of outer measure greater than $(b-a)-\varepsilon$, then for each positive number η there is a continuous function h such that $|g(x)-h(x)|<\varepsilon$ on a set of measure greater than $(b-a)-\varepsilon$ and such that $|f(x)-h(x)|<\eta$ on a set of outer measure greater than $(b-a)-\eta$.*

The Saks-Sierpiński theorem follows by applying the lemma a countably infinite number of times.

6.1.2. Interpolations in measure.

We mention two more facts in the spirit of the Lusin theorem.

We first observe that a real-valued function f is measurable if and only if it is approximately continuous almost everywhere. Hence the Lusin theorem yields that f is such that for every positive number ε there is a continuous function g such that $f(x)=g(x)$ except on a set of measure less than ε if and only if f is approximately continuous almost everywhere.

An associated theorem first obtained by H. Whitney may be given the following form. *A real-valued function f is approximately differentiable almost everywhere if and only if for every positive number ε there is a continuously differentiable function g such that $f(x)=g(x)$ except on a set of measure less than ε.* The proof is easy in the one-dimensional case but is rather sophisticated in higher dimensions.

6.2. Approximations by homeomorphisms of one-to-one maps

In analogy with the Saks-Sierpiński theorem, we will show that every one-to-one mapping of a square Q onto itself may be approximated by a homeomorphism. Obviously, we do not assume in this section that the maps are continuous because such an assumption would make the theorem trivial.

6.2.1. ε-approximations in measure.

For $\varepsilon>0$, a one-to-one transformation T of a square Q onto itself is said to be **ε-approximated** by another such transformation $\widetilde{T}$ if there are sets A and $\widetilde{A}$ in Q, each of relative outer measure greater than $1-\varepsilon$ with respect to Q, such that, for every p in A, the distance between $T(p)$ and $\widetilde{T}(p)$ is less than ε and, for every p in $\widetilde{A}$, the distance between $T^{-1}(p)$ and $\widetilde{T}^{-1}(p)$ is less than ε.

THEOREM 6.3 (GOFFMAN). *Every one-to-one map of a square Q onto itself may, for each positive number ε, be ε-approximated by a homeomorphism.*

PROOF. We may assume that Q is the unit square. Subdivide Q into nonoverlapping squares Q_i with $\operatorname{diam} Q_i < \varepsilon$, $i = 1, 2, \ldots, n$. For each i, let E_i be a subset of Q_i such that

$$E_i \cap E_j = \emptyset \quad \text{for} \quad i \neq j\,, \quad \text{and} \quad Q = \bigcup_{i=1}^n E_i.$$

Since T is one-to-one, each point p of Q corresponds to precisely one index i with $p \in T[E_i]$ and precisely one index j with $p \in T^{-1}[E_j]$. For each pair (i, j) let $A_{ij} = T[E_i] \cap T^{-1}[E_j]$. Then Q is the union of the collection

$$\{\, A_{ij} : i = 1, 2, \ldots, n,\ j = 1, 2, \ldots, n \,\}$$

of mutually disjoint sets. Employing measurable hulls of the sets A_{ij}, we can find subsets Z_{ij} of A_{ij} with $\lambda(Z_{ij}) = 0$ such that each point p of $A_{ij} \setminus (Z_{ij} \cup \bigcup_{k=1}^n \partial Q_k)$ has arbitrarily small closed squares S which are centered at p and are contained in $Q \setminus \bigcup_{k=1}^n \partial Q_k$ and which have the property that

$$\lambda^*(A_{ij} \cap S) > \left(1 - \tfrac{\varepsilon}{2}\right) \lambda(S).$$

Let

$$E = \left(Q \setminus \bigcup_{i=1}^n \partial Q_i\right) \setminus \bigcup_{i=1}^n \bigcup_{j=1}^n Z_{ij}.$$

The collection of all squares described above forms a Vitali covering of the measurable set E. Let S_k, $k = 1, 2, \ldots, m$, be a finite collection from this cover such that

$$S_k \cap S_{k'} = \emptyset \quad \text{whenever} \quad k \neq k'$$

and

$$\lambda(Q \setminus \bigcup_{k=1}^m S_k) < \tfrac{\varepsilon}{2}\, \lambda(Q).$$

For each k there is a pair (i_k, j_k) such that the center of S_k is in $A_{i_k j_k}$. Define the set A to be

$$A = \bigcup_{k=1}^m A_{i_k j_k} \cap S_k.$$

Now it follows easily that

$$\lambda^*(A) > (1 - \varepsilon)\, \lambda(Q).$$

Also, the interior U_i of $Q_i \setminus \bigcup_{k=1}^m S_k$ is nonempty for $i = 1, 2, \ldots, n$. For each pair (i_k, j_k) select a square R_k inside of U_{i_k}, and a square $\widetilde{R}_{j_k}$ inside of U_{j_k} so that

$$\{\, S_k : k = 1, 2, \ldots, m \,\} \cup \{\, R_k : k = 1, 2, \ldots, m \,\} \cup \{\, \widetilde{R}_k : k = 1, 2, \ldots, m \,\}$$

is a mutually disjoint collection. Let us now define the required homeomorphism $\widetilde{T}$ on the set C defined by the union

$$C = \bigcup_{k=1}^m S_k \cup \bigcup_{k=1}^m R_k \cup \bigcup_{k=1}^m \widetilde{R}_k$$

and on the set ∂Q. Clearly, $\widetilde{T}$ is to be the identity on ∂Q. On the remaining set C, we define $\widetilde{T}$ for each k as follows. The square S_k is mapped homeomorphically onto the square $\widetilde{R}_{j_k}$; the square $\widetilde{R}_{j_k}$ is mapped homeomorphically onto the square R_{i_k}; and, the square R_{i_k} is mapped homeomorphically onto the square S_k (each preserving orientation). Now we have $T[S_k \cap A_{i_k j_k}] \subset Q_{j_k}$ and $\widetilde{T}[S_k \cap A_{i_k j_k}] \subset \widetilde{R}_k \subset Q_{j_k}$. And also, $T^{-1}[S_k \cap A_{i_k j_k}] \subset Q_{i_k}$ and $T^{-1}[S_k \cap A_{i_k j_k}] \subset R_k \subset Q_{i_k}$. It remains now a matter of invoking the well-known Jordan-Schoenflies theorem for finitely connected Jordan regions in the plane to extend this homeomorphism to one on the square Q. Thereby the theorem is proved.

6.2.2. Remarks. Theorem 6.3 first appeared in Goffman [**49**]. Notice that a stronger theorem has been proved here. Indeed, in the proof of the above theorem, we have constructed the homeomorphism $\widetilde{T}$ in such a way that the sets A and $\widetilde{A}$ of the ε-approximation coincide.

The only part of the proof of the theorem given above that relies on the plane is the Jordan-Schoenflies theorem. More to the point is the fact that $\mathbb{R}^2$ is k-homogeneous for each k. By this we mean that there exists a homeomorphism h of $\mathbb{R}^2$ that takes any prescribed ordered k-tuple $\{x_1, x_2, \dots, x_k\}$ of distinct points of $\mathbb{R}^2$ to any other prescribed ordered k-tuple of distinct points $\{y_1, y_2, \dots, y_k\}$. This homogeneity property holds for $\mathbb{R}^n$ with $n > 1$. Of course, in the context of the above proof, the higher dimensional Jordan-Schoenflies theorem is also valid because one deals only with higher dimensional cubes that are contained in a given fixed cube. Consequently, the above theorem holds as well for all dimensions higher than 2. We shall return to the higher dimensional Jordan-Schoenflies theorem in the next section.

However, the theorem does not hold for the 1-dimensional case as the following example will show. The one-to-one transformation T of $[0,1]$ onto itself is given by

$$T(x) = \begin{cases} x + \frac{1}{2} & \text{for} \quad 0 \le x \le \frac{1}{2}, \\ 1 - x & \text{for} \quad \frac{1}{2} < x \le 1. \end{cases}$$

If $\widetilde{T}$ is any homeomorphism of $[0,1]$ onto itself, then it is a monotonic function such that $\widetilde{T}(0)$ is either 0 or 1. In either case, $\widetilde{T}$ cannot be an ε-approximation of T for $\varepsilon = \frac{1}{4}$. The intuitive reason for the failure is that it is impossible to have a homeomorphism of $[0,1]$ onto itself which transforms a disjoint collection of intervals I_1, I_2, I_3 with I_2 between I_1 and I_3 respectively onto three intervals $\widetilde{I}_1$, $\widetilde{I}_2$, $\widetilde{I}_3$ with $\widetilde{I}_3$ between $\widetilde{I}_1$ and $\widetilde{I}_2$.

6.3. Extensions of homeomorphisms

We will be concerned with homeomorphisms defined on I^n that extend given homeomorphisms T of the following kind: a compact subset A of I^n and a compact subset B of I^m are preassigned and T is a homeomorphism defined on A into I^m with $T[A] = B$. Clearly, when $n = m = 1$, such extensions need not exist in general. Of course this problem is related to the Jordan-Schoenflies theorem when $n = m = 2$, where A is a simple closed curve in the plane (a topological 1-sphere). As for dimensions higher than 2, the Jordan-Schoenflies theorem does not extend without additional conditions on the homeomorphism T (see, for example, Brown [**17**]). Indeed, consider the Alexander's horned sphere. For our purposes, the set A will be a compact 0-dimensional set. But, of course, this additional hypothesis is still inadequate due to the existence of the famous Antoine's necklace. (See Blankinship [**11**] for additional material on Antoine's necklace.) Fortunately, the compact 0-dimensional sets that we encounter will have the additional property which we now define.

6.3.1. 0-dimensional sets. Let $n \geq 2$. For each i, denote by Π_i the natural projection of $\mathbb{R}^n$ onto its i-th coordinate space. A compact subset E of $\mathbb{R}^n$ is said to be **sectionally** 0-**dimensional** if $\Pi_i[E]$ is totally disconnected for each coordinate i. Clearly, such a set is topologically 0-dimensional, and $\mathsf{X}_{i=1}^n \Pi_i[E]$ is sectionally 0-dimensional whenever E is. Finally, the union of finitely many sectionally 0-dimensional sets is also sectionally 0-dimensional, and a closed subset of a sectionally 0-dimensional set is a sectionally 0-dimensional set.

LEMMA 6.4. *For $n \geq 2$, let $I = \mathsf{X}_{i=1}^n [a_i, b_i]$ be a product of closed intervals of $\mathbb{R}$. If E is a sectionally 0-dimensional set contained in the interior of I and if ε is a positive number, then there is a finite, disjoint collection $\{ I_k : k = 1, 2, \dots, m \}$ of subsets of the interior of I that are formed by products of closed intervals such that* $\operatorname{diam} I_k < \varepsilon$ *for $k = 1, 2, \dots, m$ and E is contained in the interior of the union of this collection.*

PROOF. There is no loss in assuming that $E = \mathsf{X}_{i=1}^n \Pi_i[E_i]$ where E_i is a compact totally disconnected subset of $\mathbb{R}$. The construction of the collection for this set E is straightforward and will be left to the reader.

In the following lemmas, special types of n-cells in $\mathbb{R}^n$ are used. They are topological n-cells contained in $\mathbb{R}^n$ which are formed by unions of finitely many closed n-dimensional intervals (i.e., products of n closed linear intervals in the coordinate axes). We refer to these n-cells as **projecting** n-**cells**.

LEMMA 6.5. *Let $n \geq 2$. If S is a projecting n-cell and*

$$\{ J_{kp} : p = 1, 2, \dots, p_k,\ k = 1, 2, \dots, m \}$$

is a collection of mutually disjoint closed n-dimensional intervals contained in the interior of S, then there exists a collection

$$\mathcal{S} = \{ S_1, S_2, \dots, S_m \}$$

of mutually disjoint projecting n-cells contained in the interior of S such that J_{kp} is contained in the interior of S_k whenever $p = 1, 2, \dots, p_k$ for each k.

The proof of this lemma can be made by a straightforward induction on p_k and k. We shall omit the proof.

The next lemma is an immediate consequence of Lemmas 6.4 and 6.5.

LEMMA 6.6. *Let $n \geq 2$. If S is a projecting n-cell and*

$$\{ A_k : k = 1, 2, \dots, m \}$$

is a collection of mutually disjoint sectionally 0-dimensional sets contained in the interior of S, then there exists a collection

$$\mathcal{S} = \{ S_1, S_2, \dots, S_m \}$$

of mutually disjoint projecting n-cells contained in the interior of S such that A_k is contained in the interior of S_k for each k.

The following is an application of the higher dimensional Jordan-Schoenflies theorem which applies because we are using projecting n-cells in $\mathbb{R}^n$. We omit the proof.

LEMMA 6.7. *Let $n \geq 2$ and let*

$$\widetilde{\mathcal{I}} = \{ \widetilde{I}_1, \widetilde{I}_2, \ldots, \widetilde{I}_m \}$$

be a collection of mutually disjoint closed n-dimensional intervals contained in the interior of $\widetilde{I} = \mathsf{X}_{i=1}^{n}[a_i, b_i]$. If

$$\mathcal{S} = \{ S_1, S_2, \ldots, S_m \}$$

is a collection of mutually disjoint projecting n-cells contained in the interior of a projecting n-cell S, then there is a homeomorphism H of S onto $\widetilde{I}$ with $H[S_k] = \widetilde{I}_k$ for each k. Moreover, H can be chosen so as to agree with a preassigned orientation preserving homeomorphism defined on ∂S onto ∂I.

The final lemma provides the inductive step of the proof of the extension theorem.

LEMMA 6.8. *Let $n \geq 2$, $\varepsilon > 0$ and $I = \mathsf{X}_{i=1}^{n}[a_i, b_i]$, and suppose that A and B are sectionally 0-dimensional sets contained in the interior of I and that T is a homeomorphism of A onto B.*

If H^* is a homeomorphism of I onto itself, and if

$$\mathcal{S}^* = \{ S_1^*, S_2^*, \ldots, S_m^* \}$$

is a collection of projecting n-cells whose interiors cover A and

$$\widetilde{\mathcal{I}}^* = \{ \widetilde{I}_1^*, \widetilde{I}_2^*, \ldots, \widetilde{I}_m^* \}$$

is a collection of closed n-dimensional intervals whose interiors cover B such that

$$T^{-1}[B \cap \widetilde{I}_k^*] \subset S_k^* \quad \text{and} \quad H^*[S_k^*] = \widetilde{I}_k^* \quad \text{for} \quad k = 1, 2, \ldots, m,$$

then there exists a homeomorphism H of I onto itself and there exist collections

$$\mathcal{S} = \{ S_{kp} : p = 1, 2, \ldots, p_k, \ k = 1, 2, \ldots, m \}$$

of mutually disjoint projecting n-cells and

$$\widetilde{\mathcal{I}} = \{ \widetilde{I}_{kp} : p = 1, 2, \ldots, p_k, \ k = 1, 2, \ldots, m \}$$

of mutually disjoint closed n-dimensional intervals such that

(1) $\operatorname{diam} S_{kp} < \varepsilon$ *and* $\operatorname{diam} \widetilde{I}_{kp} < \varepsilon$ *for every p and k,*
(2) $T^{-1}[B \cap \widetilde{I}_{kp}] \subset S_{kp}$, $H[S_{kp}] = \widetilde{I}_{kp}$ *and* $H[S_{kp}] \subset \widetilde{I}_k^*$ *for every p and k,*
(3) $H(x) = H^*(x)$ *whenever* $x \in I \setminus \bigcup_{k=1}^{m} S_k^*$,
(4) *the interiors of the members of $\mathcal{S}$ form an open cover of A,*
(5) *the interiors of the members of $\widetilde{\mathcal{I}}$ form an open cover of B.*

PROOF. By Lemma 6.4, we select a collection

$$\mathcal{J} = \{ J_{kq} : q = 1, 2, \ldots, q_k, \ k = 1, 2, \ldots, m \}$$

of mutually disjoint closed n-dimensional intervals J_{kq} with $\operatorname{diam} J_{kq} < \varepsilon$ such that their interiors form a cover of A and such that, for each k,

$$J_{kq} \subset S_k^* \quad \text{for} \quad q = 1, 2, \ldots, q_k.$$

We infer from the hypothesis that there exists a positive number δ with $\delta < \varepsilon$ such that

(i) $\operatorname{diam} \widetilde{E} < \delta$ and $\widetilde{E} \cap B \neq \emptyset$ imply $\widetilde{E} \subset \widetilde{I}_k$ for some k,
(ii) $T^{-1}[\widetilde{E} \cap B] \subset J_{kq}$ for some J_{kq} in $\mathcal{J}$ whenever $\operatorname{diam} \widetilde{E} < \delta$ and $\widetilde{E} \cap B \neq \emptyset$.

Again by Lemma 6.4, we select a collection

$$\widetilde{\mathcal{I}} = \{\, \widetilde{I}_{kp} : p = 1, 2, \ldots, p_k,\ k = 1, 2, \ldots, m \,\}$$

of mutually disjoint closed n-dimensional intervals $\widetilde{I}_{kp}$ with $\operatorname{diam} \widetilde{I}_{kp} < \delta$ such that their interiors form a cover of B and such that, for each k,

$$\widetilde{I}_{kp} \subset \widetilde{I}_k^* \quad \text{for} \quad p = 1, 2, \ldots, p_k.$$

For a fixed J_{kq} in $\mathcal{J}$, consider the finitely many sets $\widetilde{I}_{kp}$ such that

$$A_{kp} = T^{-1}[\widetilde{I}_{kp} \cap B] \subset J_{kq}.$$

As T is a homeomorphism, the collection of such sets A_{kp} are disjoint. We apply Lemma 6.6 to get a collection of mutually disjoint projecting n-cells S_{kp} with $A_{kp} \subset S_{kp} \subset J_{kq}$. In this way, a collection

$$\mathcal{S} = \{\, S_{kp} : p = 1, 2, \ldots, p_k,\ k = 1, 2, \ldots, m \,\}$$

is constructed. Moreover, by Lemma 6.7 there are homeomorphisms H_k of S_k onto $\widetilde{I}_k^*$ such that H_k agrees with H^* on ∂S_k and such that

$$H_k[S_{kp}] = \widetilde{I}_{kp} \quad \text{for all} \quad p \quad \text{and} \quad k.$$

The required homeomorphism H is now constructed by cutting and pasting together H^* and H_k, $k = 1, 2, \ldots, m$.

We now state and prove the homeomorphism extension theorem for sectionally 0-dimensional sets.

THEOREM 6.9 (GOFFMAN). *Let $n \geq 2$ and $I = \mathsf{X}_{i=1}^n [a_i, b_i]$. If $T \colon A \to B$ is an onto homeomorphism of sectionally 0-dimensional sets contained in the interior of I, then there exists a homeomorphism H of I onto itself such that $H|A = T$.*

PROOF. The proof is a straightforward induction, where the inductive step is provided by Lemma 6.8. Indeed, there exists a sequence of onto homeomorphisms $H_j \colon I \to I$ such that

$$H_j|A = T \quad \text{and} \quad H_j^{-1}|B = T^{-1}$$

and

$$H_j,\ j = 1, 2, \ldots, \quad \text{and} \quad H_j^{-1},\ j = 1, 2, \ldots,$$

are Cauchy sequences in the uniform norm. Let H and $\widetilde{H}$ be the limits, respectively, of these sequences. Clearly, these limits are continuous maps with $H|A = T$ and $\widetilde{H}|B = T^{-1}$. By proving that $H \circ \widetilde{H}$ and $\widetilde{H} \circ H$ are the identity map, we will complete the demonstration. To this end, consider the inequality

$$\begin{aligned} |H(\widetilde{H}(y)) - y| \leq |H(\widetilde{H}(y)) - H_j(\widetilde{H}(y))| + |H_j(\widetilde{H}(y)) - H_j(H_m^{-1}(y))| \\ + |H_j(H_m^{-1}(y)) - H_m(H_m^{-1}(y))| + |H_m(H_m^{-1}(y)) - y|. \end{aligned}$$

Consequently,

$$|H(\widetilde{H}(y)) - y| \leq \|H - H_j\|_\infty + |H_j(\widetilde{H}(y)) - H_j(H_m^{-1}(y))| + \|H_j - H_m\|_\infty,$$

from which we infer that $H \circ \widetilde{H}$ is the identity map. Analogously, we can show that $\widetilde{H} \circ H$ is the identity map, and the theorem is proved.

Let us now prove some corollaries of the above theorem. It is well-known that each compact, totally disconnected metric space A is homeomorphic to a closed subset of the Cantor set. Let $f_A\colon A \to [0,1]$ denote one such embedding of A into $[0,1]$. With the aid of this embedding, we can construct a "canonical" sectionally 0-dimensional subset of $I = \mathsf{X}_{i=1}^{n}[a_i, b_i]$.

COROLLARY 6.10. *For $n \geq 2$, if A is a sectionally 0-dimensional set contained in the interior of $I = \mathsf{X}_{i=1}^{n}[a_i, b_i]$ and if B is the image of $f_A[A]$ under an affine embedding φ of $[0,1]$ into the interior of I, then there exists a homeomorphism H of I onto itself so that $H|A = \varphi \circ f_A$.*

PROOF. Clearly, B is a sectionally 0-dimensional subset of I.

COROLLARY 6.11. *For $n \geq 2$, if B is a sectionally 0-dimensional set contained in the interior of $I = \mathsf{X}_{i=1}^{n}[a_i, b_i]$ and if $A = f_B[B]$, then there exists a homeomorphism H of $[0,1]$ into I so that $H|A = f_B^{-1}$.*

We come to our second theorem. Its proof is a simple consequence of an affine embedding of I_n into the interior of I_m so that the image of A in I^m is sectionally 0-dimensional. The proof is left to the reader.

THEOREM 6.12 (GOFFMAN). *For $1 \leq n < m$, let I_n and I_m be closed intervals in $\mathbb{R}^n$ and $\mathbb{R}^m$ respectively. If T is a homeomorphism of A onto B, where A is a sectionally 0-dimensional set contained in I_n and B is a sectionally 0-dimensional set contained in the interior of I_m, then there exists a homeomorphism H of I_n into I_m such that $H|A = T$.*

6.3.2. Remarks. It is clear from Remarks 6.2.2 in the previous section that Theorem 6.9 does not hold for $n = m = 1$.

The notion of sectionally 0-dimensional set was introduced in Goffman [**53**]. The proof of Theorem 6.12 does not allow much control on the embedding of the complement of the sectionally 0-dimensional set A, that is, the set $I_n \setminus A$. For $n = 2$ and $m = 3$, there is a construction due to Besicovitch [**10**] in which the Hausdorff 2-dimensional measure of the set $H[I_2 \setminus A]$ can be made as small as one wishes. It is possible to use similar constructions for $2 \leq n < m$ to control the Hausdorff n-dimensional measure of the set $H[I_n \setminus A]$.

In the discussion of a canonical sectionally 0-dimensional set in I_n, we used affine embeddings of $[0,1]$ into I_n. Observe that rotations of sectionally 0-dimensional sets contained in $\mathbb{R}^2$ need not result in sectionally 0-dimensional sets, witness that the square of the Cantor ternary set has this unfortunate property. And finally, in view of the generalized Schoenflies theorem [**17**], every self-homeomorphism of $\mathbb{R}^3$ changes the Antoine necklace into a compact totally disconnected set that is not sectionally 0-dimensional in $\mathbb{R}^3$.

6.4. Measurable one-to-one maps

In the ε-approximation theorem above, Theorem 6.3, it was not possible to interpolate by homeomorphisms because no measurability assumptions were made in the theorem. In order to get a Lusin-type theorem that corresponds to Theorem 6.3 we must assume that the one-to-one mapping and its inverse mapping are measurable. At this point it would be convenient to give such a one-to-one mapping a name. Let $1 \leq n \leq m$. For a one-to-one mapping T of I^n onto I^m, the pair (T, T^{-1}) is said to be **bimeasurable** if both T and T^{-1} are measurable.

Here, we have assumed that I^n and I^m are closed intervals. For fixed n and m, let $\mathcal{M}[I^n, I^m]$ denote the collection of all T such that (T, T^{-1}) are bimeasurable pairs. There is a natural interpolation metric on $\mathcal{M}[I^n, I^m]$ defined by

$$\operatorname{dist}(T, \widetilde{T}) = \lambda_n(\{x : T(x) \neq \widetilde{T}(x)\}) + \lambda_m(\{x : T^{-1}(x) \neq \widetilde{T}^{-1}(x)\}).$$

Clearly, this metric is one that corresponds to interpolation in measure. (The verification that the triangle inequality holds is left to the reader.)

Our task will be to interpolate in measure with a homeomorphism each T in $\mathcal{M}[I^n, I^n]$. That is, for $\varepsilon > 0$, we seek a homeomorphism H of I^n onto I^n such that $T(x) = H(x)$ except on a subset of I^n with measure less than ε and $T^{-1}(y) = H^{-1}(y)$ except on a subset of I^n with measure also less than ε.

Here is a useful lemma.

LEMMA 6.13. *Let $T\colon I^n \to I^m$ be a one-to-one (not necessarily onto) mapping such that both T and T^{-1} are measurable. Then $T[F]$ is measurable for each closed subset F of I^n.*

PROOF. Observe that $T[F] = (T^{-1})^{-1}[F]$. The measurability of T^{-1} completes the proof.

6.4.1. Example. Before going to the solution of the interpolation in measure problem, we shall look at a couple of examples, where $n \geq 2$. With the first example, we show that the measurability of $T\colon I^n \to I^n$ does not imply the measurability of T^{-1}. To this end, let A and B be homeomorphic sectionally 0-dimensional sets contained in the interior of I^n with $\lambda_n(A) = 0$ and $\lambda_n(B) > 0$. By Theorem 6.9, there is a homeomorphism H of I^n onto itself such that $H|A$ is a homeomorphism of A onto B. Let $\varphi\colon A \to B$ be a one-to-one, onto map such that $\varphi[U]$ is not a Lebesgue measurable subset of I^n for some open set U. Then the one-to-one, onto map T defined by $T(x) = H(x)$ for $x \in I^n \setminus A$ and $T(x) = \varphi(x)$ for $x \in A$ does not have a Lebesgue measurable inverse.

For our second example, let A and B be homeomorphic sectionally 0-dimensional sets contained in the interior of I^n. Again we invoke Theorem 6.9 to get a homeomorphism H such that $H|A$ is a homeomorphism of A onto B. Moreover, we may assume that $H|\partial I^n$ is the identity map. Since B is a compact, totally disconnected set, there exists a homeomorphism φ_0 of B onto a subset K contained in ∂I^n. Let φ_1 be any homeomorphism from K onto A. Then the one-to-one, onto map T defined by

$$T(x) = \begin{cases} \varphi_0(x) & \text{for} \quad x \in B, \\ H(x) & \text{for} \quad x \in I^n \setminus (B \cup K), \\ \varphi_1(x) & \text{for} \quad x \in K. \end{cases}$$

gives rise to a pair (T, T^{-1}) that is bimeasurable. Consequently, the existence of bimeasurable pairs that are not homeomorphisms has been established. Observe also that bimeasurable pairs can be defined for one-to-one mappings from the interior of I^n onto the interior of I^n. Clearly, such pairs gives rise to bimeasurable pairs on I^n by a simple extension of the map to be the identity on ∂I^n.

6.4.2. A characterization. Let $\mathcal{H}[I^n]$ denote the collection of all self-homeomorphisms of I^n. We shall characterize the closure of $\mathcal{H}[I^n]$ in the space $\mathcal{M}[I^n, I^n]$. We first establish a necessary condition.

LEMMA 6.14. *For $n \geq 2$, if $T \in \mathcal{M}[I^n, I^n]$ and $H \in \mathcal{H}[I^n]$, then*

$$\lambda_n\big(T[\partial I^n]\big) + \lambda_n\big(T^{-1}[\partial I^n]\big) \leq \operatorname{dist}(T, H).$$

PROOF. By Brouwer's invariance of domain theorem (see [**78**] for example), we have

$$H^{-1}\big[T[\partial I^n] \setminus \partial I^n\big] \cap T^{-1}\big[T[\partial I^n] \setminus \partial I^n\big] \subset (I^n \setminus \partial I^n) \cap \partial I^n = \emptyset.$$

Hence

$$T[\partial I^n] \setminus \partial I^n \subset \{\, x : T^{-1}(x) \neq H^{-1}(x) \,\}$$

and, analogously,

$$T^{-1}[\partial I^n] \setminus \partial I^n \subset \{\, x : T(x) \neq H(x) \,\}.$$

The lemma now follows.

The necessary condition given below is easily proved from this lemma.

THEOREM 6.15. *For $n \geq 2$, if T is in the closure of $\mathcal{H}[I^n]$ in the metric space $\mathcal{M}[I^n, I^n]$, then $\lambda_n\big(T[\partial I^n]\big) = 0$ and $\lambda_n\big(T^{-1}[\partial I^n]\big) = 0$.*

The sufficiency will rely on the next theorem.

THEOREM 6.16 (GOFFMAN). *For $2 \leq n \leq m$, if T is a one-to-one mapping of I^n into I^m such that both T and T^{-1} are measurable, and if ε is a positive number, then there exist closed sets A and B in I^n and $T[I^n]$, respectively, such that*

$$\lambda_n(I^n \setminus A) < \varepsilon \quad \text{and} \quad \lambda_m(T[I^n] \setminus B) < \varepsilon$$

and $T|A$ is a homeomorphism of A onto B. Moreover, the sets A and B may be assumed to be sectionally 0-dimensional.

PROOF. Recall that Lusin's theorem also holds for T and T^{-1}.

Select compact subsets A_0 and B_0 of I^n and $T[I^n]$ with both $T|A_0$ and $T^{-1}|B_0$ continuous and with $\lambda_n(I^n \setminus A_0) < \varepsilon$ and $\lambda_m(T[I^n] \setminus B_0) < \varepsilon$. With $A_1 = T^{-1}[B_0]$ and $B_1 = T[A_0]$, let $A = A_0 \cup A_1$ and $B = B_0 \cup B_1$. The continuity of both $T|A_0$ and $T^{-1}|B_0$ assures that they are homeomorphisms because A_0 and B_0 are compact. Obviously, A_1 and B_1 are compact and $T|A_1$ and $T^{-1}|B_1$ are continuous. Consequently, $T|A$ is a homeomorphisms of A onto B. The inequalities in the first statement of the theorem follows easily.

Let us show that the sets A and B can be made into sectionally 0-dimensional sets. To this end, consider the i-th coordinate axis and denote the natural projection of $\mathbb{R}^n$ onto this axis by $\Pi_{n,i}$. Analogously, $\Pi_{m,i}$ will denote the natural projection of $\mathbb{R}^m$. The collection of real numbers t such that $\lambda_m\big(T[\Pi_{n,i}{}^{-1}(t)]\big) > 0$ for some i or $\lambda_n\big(T^{-1}[\Pi_{m,j}{}^{-1}(t)]\big) > 0$ for some j is a countable set. Hence there is a countable dense set of real numbers t_k, $k = 1, 2, \ldots$, such that

$$\sum_{k=1}^{\infty}\Big(\sum_{i=1}^{n}\lambda_m\big(T[\Pi_{n,i}{}^{-1}(t_k)]\big) + \sum_{j=1}^{m}\lambda_n\big(T^{-1}[\Pi_{m,j}{}^{-1}(t_k)]\big)\Big) = 0.$$

Clearly, the sets

$$\widetilde{A} = A \setminus \textstyle\bigcup_{k=1}^{\infty}\big(\bigcup_{i=1}^{n}\Pi_{n,i}{}^{-1}(t_k) \cup \bigcup_{j=1}^{m} T^{-1}[\Pi_{m,j}{}^{-1}(t_k)]\,\big)$$

and

$$\widetilde{B} = B \setminus \bigcup_{k=1}^{\infty} \left(\bigcup_{i=1}^{n} T[{\Pi_{n,i}}^{-1}(t_k)] \cup \bigcup_{j=1}^{m} {\Pi_{m,j}}^{-1}(t_k) \right)$$

have the same measures as A and B respectively. Also, $T[\widetilde{A}] = \widetilde{B}$ because T is one-to-one. It is a simple matter to find compact subsets of $\widetilde{A}$ and $\widetilde{B}$ with the required properties.

LEMMA 6.17. *Let $n \geq 2$. If $T \in \mathcal{M}[I^n, I^n]$ and $\varepsilon > 0$, then there exist an H in $\mathcal{H}[I^n]$ such that*

$$\operatorname{dist}(T, H) < \lambda_n(T[\partial I^n]) + \lambda_n(T^{-1}[\partial I^n]) + \varepsilon.$$

PROOF. With ε in the statement of the last theorem replaced by $\frac{\varepsilon}{3}$, let A and B be the sectionally 0-dimensional sets provided by the theorem. For each positive number η, let U_η denote the set consisting of the points x in I^n with $\operatorname{dist}(x, \partial I^n) < \eta$. Define the compact sets

$$A_\eta = A \setminus \left(T^{-1}[U_\eta] \cup U_\eta\right) \quad \text{and} \quad B_\eta = B \setminus \left(T[U_\eta] \cup U_\eta\right).$$

That A_η and B_η are compact follows from the fact that $T|A$ and $T^{-1}|B$ are homeomorphisms. Clearly, $T[A_\eta] = B_\eta$ for each η because T is one-to-one. From

$$\bigcup_{\eta>0} A_\eta = A \setminus \left(T^{-1}[\partial I^n] \cup \partial I^n\right) \subset A \setminus \partial I^n$$

and

$$\bigcup_{\eta>0} B_\eta = B \setminus \left(T[\partial I^n] \cup \partial I^n\right) \subset B \setminus \partial I^n,$$

we infer that

$$\lambda_n(I^n \setminus \bigcup_{\eta>0} A_\eta) \leq \lambda_n(I^n \setminus A) + \lambda_n(T^{-1}[\partial I^n])$$

and

$$\lambda_n(I^n \setminus \bigcup_{\eta>0} B_\eta) \leq \lambda_n(I^n \setminus B) + \lambda_n(T[\partial I^n]),$$

whence

$$\lambda_n(I^n \setminus A_\eta) + \lambda_n(I^n \setminus B_\eta) < \lambda_n(T[\partial I^n]) + \lambda_n(T^{-1}[\partial I^n]) + \varepsilon$$

for some η. As A_η and B_η are contained in the interior of I^n, by Theorem 6.9, there is a homeomorphism H in $\mathcal{H}[I^n]$ such that $H|A_\eta = T|A_\eta$. Consequently,

$$\operatorname{dist}(T, H) \leq \lambda_n(I^n \setminus A_\eta) + \lambda_n(I^n \setminus B_\eta) < \lambda_n(T[\partial I^n]) + \lambda_n(T^{-1}[\partial I^n]) + \varepsilon$$

and the theorem is proved.

Here follows the characterization theorem whose proof is immediate from Lemmas 6.14 and 6.17.

THEOREM 6.18. *For $n \geq 2$, a member T of $\mathcal{M}[I^n, I^n]$ is in the closure of $\mathcal{H}[I^n]$ if and only if $\lambda_n(T[\partial I^n]) + \lambda_n(T^{-1}[\partial I^n]) = 0$.*

An immediate consequence of this characterization is the theorem for open intervals I^n.

THEOREM 6.19 (GOFFMAN). *For $n \geq 2$, if T is a one-to-one mapping of an open interval I^n onto itself such that both T and T^{-1} are measurable and if ε is a positive number, then there is a homeomorphism H of I^n onto itself so that* $\operatorname{dist}(T, H) < \varepsilon$.

PROOF. Extend T to the closure of I^n in the obvious way by letting it be the identity on ∂I^n.

6.4.3. Mappings into higher dimensions. Here, we shall assume that the dimensions satisfy $1 \leq n < m$.

THEOREM 6.20 (GOFFMAN). *If ε is a positive number and T is a one-to-one mapping of a closed interval I^n onto an open interval I^m such that both T and T^{-1} are measurable, then there is a homeomorphism H of I^n into I^m such that*

$$\lambda_n(\{x : T(x) \neq H(x)\}) < \varepsilon \quad \textit{and} \quad \lambda_m(\{y : T^{-1}(y) \neq H^{-1}(y)\}) < \varepsilon.$$

PROOF. We shall give the proof for the case $2 \leq n < m$; the case $n = 1 < m$ is an easy modification of this case. Select suitable compact sets A in I^n and B in I^m so that $T|A$ is a homeomorphism of A onto B and $\lambda_n(I^n \setminus A) < \varepsilon$ and $\lambda_m(I^m \setminus B) < \varepsilon$. The existence of such sets are assured by Theorem 6.16. Now, by Theorem 6.12, there exists a homeomorphism H of I^n into I^m such that $T|A = H|A$. The inequalities follow easily.

6.4.4. Remarks. It has already been remarked that the above theorems are false when $n = m = 1$. We state the next result for $n = m = 1$ without proof.

THEOREM 6.21. *If f is a one-to-one function of the open interval $(0, 1)$ onto itself such that f and f^{-1} are measurable, then there is a one-to-one function g of $(0, 1)$ onto itself such that both g and g^{-1} are in the Baire class 2 and both $f = g$ and $f^{-1} = g^{-1}$ hold λ-almost everywhere.*

Theorems 6.16, 6.19 and 6.20 are from Goffman [**53**]. And Theorem 6.18 is new.

CHAPTER 7

Measures on $\mathbb{R}^n$

The discussion in this chapter concerns the behavior of Lebesgue measure under homeomorphisms. In particular, an interesting connection between nonatomic Borel measures and Lebesgue measure is the following conjecture which was made by S. L. Ulam, and was established by J. von Neumann (see the remarks at the end of the chapter).

THEOREM (VON NEUMANN). *Let μ be a Borel measure on an interval I^n of $\mathbb{R}^n$ such that $\mu(I^n) = \lambda_n(I^n)$, where λ_n is the Lebesgue measure on $\mathbb{R}^n$. In order that there exists a self-homeomorphism h of I^n such that*

$$\lambda_n(E) = \mu(h[E]) \quad \textit{for every Borel set} \quad E,$$

it is necessary and sufficient that μ satisfies

(i) $\mu(\partial I^n) = 0$,

(ii) *μ be nonatomic,*

(iii) *$\mu(U) > 0$ whenever U is a nonempty, relatively open subset of I^n.*

Moreover, h may be assumed to be the identity map on ∂I^n.

By a Borel measure we mean a regular Borel measure.

A sharpening of this theorem to a measure preserving homotopy version is the goal of the chapter. The case $n = 1$ is easily proved and its proof does not shed any light on the difficulties encountered in the higher dimensional cases. This proof is given in the short second section. The third section is devoted to the not so obvious, yet simple, constructions of deformations used in the proof of the main theorem found in the fourth section. There are some obvious unanswered questions that have been placed at the end of the chapter.

7.1. Preliminaries

It will be useful to extend our discussion to the level of compact topological n-cells X and finite Borel measures μ defined on them. To a pair (μ, X) and a homeomorphism $h\colon X \to Y$ there corresponds a natural pair $(h_{\#}\mu, Y)$ defined by

$$h_{\#}\mu(E) = \mu(h^{-1}[E]) \quad \text{for each Borel set} \quad E \quad \text{of} \quad Y.$$

We shall find it convenient in this chapter to denote composition of functions in multiplicative notation. Note that, for the composition gh,

$$(gh)_{\#}\mu = g_{\#}h_{\#}\mu.$$

Two pairs (μ, X) and (ν, Y) are said to be **Lebesgue equivalent** if there is a homeomorphism $h\colon X \to Y$ such that $h_{\#}\mu = \nu$.

In view of von Neumann's theorem, the following definition is appropriate. A pair (μ, X) is said to be **Lebesgue-like** if μ satisfies

(i) $\mu(\partial X) = 0$,
(ii) μ is nonatomic,
(iii) $\mu(U) > 0$ whenever U is a nonempty open subset of X.

As usual, we shall use E^o to denote the topological interior of a set E in the space X. In the context of a topological n-cell X, we remind the reader that the set ∂X is not the topological boundary of X which, of course, is the empty set.

The notion of Lebesgue-like measures has been introduced already in Chapter 4 for $n = 1$ (see page 43). Another name, positive measure, was also mentioned there.

Let us state the main theorem.

THEOREM 7.1. *Let μ be a Borel measure on an interval I^n of $\mathbb{R}^n$ such that $\mu(I^n) = \lambda_n(I^n)$. Then there is a continuous deformation D_t, $0 \le t \le 1$, of I^n by self-homeomorphisms that satisfy*

$$D_0(x) = x \text{ for } x \text{ in } I^n \quad \text{and} \quad D_t(x) = x \text{ for } (x,t) \text{ in } (\partial I^n) \times [0,1],$$

and such that

$$D_{0\#}\mu(E) = \mu(E) \quad \text{and} \quad D_{1\#}\mu(E) = \lambda_n(E)$$

holds for every Borel set E if and only if μ is a Lebesgue-like Borel measure on I^n.

7.1.1. A direction lemma. A key element in the proof of the main theorem is the deceptively simple lemma about nonatomic Borel measures proved below.

A hyperplane E in $\mathbb{R}^n$ is the solution set of the equation $(x - b)\cdot e = 0$ where e is in the unit sphere $\mathbb{S}^{n-1}$ and b is in $\mathbb{R}^n$. Of course, e is a normal vector to E. As usual, by a k-flat we mean a nonempty intersection of $n - k$ hyperplanes whose normals are linearly independent. Each k-flat determines a family consisting of all hyperplanes that contains the k-flat; and, this family yields a set of unit normals that forms a closed, nowhere dense subset of $\mathbb{S}^{n-1}$.

For a finite, Borel measure μ on $\mathbb{R}^n$ (atoms being permitted) and for each positive integer k with $k < n$, let $\mathcal{F}_k$ be the collection of all k-flats F with the properties that $\mu(F) > 0$ and that F contains no j-flat with positive μ measure when $j < k$. One can easily show that $\mathcal{F}_1$ is a countable set. Indeed, let us assume the contrary. Then there is an uncountably infinite number of 1-flats F with the property that $\mu(F) > 0$ and $\mu(\{p\}) = 0$ for each point p in F. Clearly, we may assume that there is a positive number ε with $\mu(F) \ge \varepsilon$ for these uncountably many 1-flats F. Now select a sequence F_k, $k = 1, 2, \ldots$, of distinct members from this collection. Clearly, $\mu\big(F_k \cap (\bigcup_{j<k} F_j)\big) = 0$ for each k. So we will have $k\,\varepsilon \le \mu\big(\bigcup_{j\le k} F_j\big) \le \mu(\mathbb{R}^n)$ for each k, a contradiction. Analogously, one can show that each $\mathcal{F}_k$ is a countable set. Let

$$\mathcal{F}_\mu = \bigcup\nolimits_{j=1}^{n-1} \mathcal{F}_j.$$

Then, by the Baire category theorem, the set

$$\mathcal{E}_\mu = \bigcup\nolimits_{F \in \mathcal{F}_\mu} \{\, e \in \mathbb{S}^{n-1} : e \text{ is normal to } F \,\}$$

is of first category in $\mathbb{S}^{n-1}$. We have the following lemma.

LEMMA 7.2. *If μ is a σ-finite, nonatomic Borel measure on $\mathbb{R}^n$, then, except for points e in a subset of the first Baire category in $\mathbb{S}^{n-1}$,*

$$\mu(E) = 0 \text{ for every hyperplane } E \text{ for which } e \text{ is a normal.}$$

PROOF. First assume that μ is a finite measure on $\mathbb{R}^n$. Let $e \in \mathbb{S}^{n-1} \setminus \mathcal{E}_\mu$. Suppose that $\mu(E) > 0$ for some hyperplane E with e as its normal. From $e \in \mathbb{S}^{n-1} \setminus \mathcal{E}_\mu$ we infer $E \notin \mathcal{F}_{n-1}$. Let E_{n-2} be an $(n-2)$-flat contained in E such that some j-flat contained in E_{n-2} has positive measure, where $j < n-1$. Clearly, $E_{n-2} \notin \mathcal{F}_{n-2}$. After finitely many steps, we will get a 1-flat E_1 contained in E such that some 0-flat of E_1 has positive measure. This shows that μ has an atom. Hence, if μ is a finite, nonatomic measure, then $\mathbb{S}^{n-1} \setminus \mathcal{E}_\mu$ is a set of second category of $\mathbb{S}^{n-1}$ for which the conclusion of the theorem holds.

Turning to the σ-finite measure μ, we write μ as the sum $\sum_{k=1}^{\infty} \mu_k$, where the summands are finite measures. Then

$$\mathcal{E} = \bigcup_{k=1}^{\infty} \mathcal{E}_{\mu_k}$$

is a subset of the first Baire category in S^{n-1}. If e is in $S^{n-1} \setminus \mathcal{E}$ and E is a hyperplane with normal e, then

$$\mu(E) = \sum_{k=1}^{\infty} \mu_k(E) = 0.$$

This completes the proof.

7.2. The one variable case

The proof of the main theorem must be separated into two cases, namely, $n = 1$ and $n \geq 2$. The need for this separation will become evident later since hyperplanes in $\mathbb{R}^1$ are singleton sets.

We first consider the case $n = 1$. The proof is straightforward, but it does not generalize to the higher dimensional case.

7.2.1. Lebesgue-Stieltjes measures. It is well-known that a finite, nonatomic measure μ on $[0,1]$ is uniquely determined by the continuous function $h(x) = \mu([0,x])$, $x \in [0,1]$. When the measure μ is positive on each nondegenerate interval, this function is a homeomorphism. It is a simple calculation to show that the Lebesgue measure λ_1 on $[0,1]$ is equal to $h_{\#}\mu$ when μ is Lebesgue-like and $\mu([0,1]) = 1$ (see also the proof of Proposition 4.10 on page 46). Furthermore, it is easily shown that

$$D_t(x) = t\,x + (1-t)\,h(x), \quad (x,t) \in [0,1] \times [0,1],$$

is the required continuous deformation of h by homeomorphisms.

7.3. Constructions of deformations

The proof of the theorem in higher dimensions will require some special constructions. The first one concerns deformations that preserve Lebesgue-like measures and 'capture' preassigned compact subsets in the interior of topological n-cells. The second one, called the precise partitioning lemma, allows the partitioning of a topological n-cell into two topological n-cells with precisely preassigned values of the Lebesgue-like measure defined on the n-cell. It is the direction lemma that permits us to accomplish the second deformation.

7.3.1. A deformation. Before embarking on the constructions of the two special deformations, let us give a lemma that will prove useful.

LEMMA 7.3. *Let $f\colon I^m \to (-1,+1)$ be a continuous function with*

$$f^{-1}[0] = \partial I^m.$$

Then

$$A = \{\, (x,y) : x \in I^m,\ f(x) \le y \le +1 \,\}$$

and

$$B = \{\, (x,y) : x \in I^m,\ -1 \le y \le f(x) \,\}$$

are $(m+1)$-cells for which the common m-face $A \cap B$ satisfies

$$\operatorname{graph}(f) = A \cap B.$$

Moreover, there is a continuous deformation H_t, $0 \le t \le 1$, of $I^m \times [-1,+1]$ by self-homeomorphisms such that, for each x in I^m and for each t in $[0,1]$, the map $H_t(x,\cdot)$ is a self-homeomorphism of $\{x\} \times [-1,+1]$ which satisfies

$$H_0(x,\cdot) \quad \textit{is the identity map}$$

and

$$H_1(x,-1) = (x,-1), \quad H_1\big(x, f(x)\big) = (x,0), \quad H_1(x,+1) = (x,+1).$$

Consequently,

1. *A is deformed onto $I^m \times [0,+1]$,*
2. *B is deformed onto $I^m \times [-1,0]$,*
3. *$\operatorname{graph}(f)$ is deformed onto $I^m \times \{0\}$,*

and

4. *H_0 is the identity map on $I^m \times [-1,+1]$ and H_t is the identity map on $\partial(I^m \times [-1,+1])$ for each t.*

We leave the proof to the reader.

Let us construct some functions that have special features. For the $(m+1)$-cell $I^m \times [-1,+1]$, let μ be a Lebesgue-like measure on it and let M be a compact set with

$$M \subset \big(I^m \times [-1,+1]\big) \setminus \partial(I^m \times [-1,+1]).$$

Consider a continuous function $g\colon I^m \to [0,+1)$ that satisfies

$$g^{-1}[0] = \partial I^m.$$

For positive real numbers r, denote by g^r the usual function given by

$$g^r(x) = \big(g(x)\big)^r \quad \text{for} \quad x \in I^m.$$

(This is not to be confused with the multiplicative notation for k-fold compositions h^k of homeomorphisms h, where k is an integer.)

For all r that are sufficiently near 0 we have

$$M \subset \{\, (x,y) : x \in I^m,\ -g^r(x) \le y \le g^r(x) \,\}. \tag{7.1}$$

Note that

$$\operatorname{graph}(g^{r_1}) \cap \operatorname{graph}(g^{r_2}) = \big(\partial I^m\big) \times \{\, 0 \,\} \quad \text{for} \quad 0 < r_1 < r_2 < 1.$$

As μ is Lebesgue-like, we have

$$\mu(\operatorname{graph}(g^r)) = \mu(\operatorname{graph}(-g^r)) = 0 \tag{7.2}$$

for a dense set of r in $(0,1)$. In the constructions to follow, we will let f be either g^r or $-g^r$ for which both (7.1) and (7.2) hold.

7.3.2. Capture deformation. Let us construct a deformation P_t, $0 \le t \le 1$, that 'captures' M in the upper $(m+1)$-cell $I^m \times [0,+1]$. Use $-g^r$ of the previous section for f in Lemma 7.3. In addition to the properties (1), (2) and (3) of that lemma, the deformation P_t will have the further properties

(5) $P_1[M] \subset (I^m \times [0,+1]) \setminus \partial(I^m \times [0,+1])$,
(6) $P_{t\#}\,\mu(P_t[M]) = \mu(M)$ and $P_{t\#}\,\mu(P_t[\operatorname{graph}(f)]) = \mu(\operatorname{graph}(f)) = 0$ for $0 \le t \le 1$,

and

(7) $P_{1\#}\,\mu$ is Lebesgue-like on both $I^m \times [-1,0]$ and $I^m \times [0,+1]$.

7.3.3. Precise partitioning. This section provides the inductive step of the final construction. We begin with the 'precise partitioning' lemma.

LEMMA 7.4. *Let μ be a Lebesgue-like Borel measure on $I^m \times [-1,+1]$ and let p be such that $0 < p < \mu(I^m \times [-1,+1])$. Then there exists a continuous deformation H_t, $0 \le t \le 1$, of $I^m \times [-1,+1]$ by self-homeomorphisms such that*

(α) *H_0 is the identity map on $I^m \times [-1,+1]$, whence $H_{0\#}\mu = \mu$, and H_t is the identity map on $\partial(I^m \times [-1,+1])$ for each t,*
(β) *$H_{1\#}\mu$ is Lebesgue-like on each of $I^m \times [-1,0]$ and $I^m \times [0,+1]$, and*
$$H_{1\#}\mu(I^m \times [-1,0]) = p.$$

PROOF. Let M be a compact subset of $I^m \times [-1,+1] \setminus \partial(I^m \times [-1,+1])$ such that

$$\mu((I^m \times [-1,0]) \setminus M) < p < \mu(M).$$

Employing the capture deformation of the previous section, we have that $P_1[M]$ is contained in the interior of $I^m \times [0,+1]$ and

$$P_{1\#}\mu(I^m \times [-1,0]) < p < P_{1\#}\mu(P_1[M]).$$

From our earlier constructions we have a continuous function $\overline{f}\colon I^m \to [0,1]$ with

$$\overline{f}^{-1}[0] = \partial I^m \quad \text{and} \quad P_{1\#}\mu(\operatorname{graph}(\overline{f})) = 0$$

such that

$$P_1[M] \subset \{\, (x,y) : x \in I^m,\ 0 < y \le \overline{f}(x) \,\}. \tag{7.3}$$

Let y_0 and y_1 be a positive numbers for which

$$P_{1\#}\mu(I^m \times [-1,y_0]) < p < P_{1\#}\mu(I^m \times [-1,y_1]) < P_{1\#}\mu(I^m \times [-1,+1]). \tag{7.4}$$

We infer from the direction lemma (Lemma 7.2) that there is an e in $\mathbb{S}^m$ for which every hyperplane E of $\mathbb{R}^{m+1}$ whose normal is e has $P_{1\#}\mu(E) = 0$ and, moreover, one of the hyperplanes separates $I^m \times \{+1\}$ and $I^m \times [-1,y_1]$ and another one separates $I^m \times [-1,0]$ and $I^m \times [y_0,+1]$. Now each hyperplane E whose normal is e corresponds to a continuous real-valued function defined on I^m whose graph is contained in E. Moreover, this family of continuous functions is a continuous one-parameter family $\overline{f}_s$, $s \in \mathbb{R}$. Consequently, we have a continuous, one parameter family of continuous functions

$$f_s = \overline{f} \wedge \overline{f}_s, \quad s \in \mathbb{R},$$

with

$$P_{1\#}\mu\big(\operatorname{graph}(f_s)\big) = 0 \quad \text{for every} \quad s.$$

In view of (7.3) and (7.4), there is an s such that the function $f = f_s$ satisfies the hypothesis of Lemma 7.3 as well as the condition

$$P_{1\#}\mu\big(\{\,(x,y) : x \in I^m,\ -1 \le y \le f(x)\,\}\big) = p.$$

Denote by N_t, $0 \le t \le 1$, the deformation provided by that lemma. The desired deformation H_t will result on adding in the homotopy sense the two deformations P_t and N_t.

7.3.4. Measure deformations. This section provides the inductive step of the final construction.

Consider an n-cell I^n, where $n > 1$, and form 2^n smaller n-cells by dividing each edge of I^n in two, and let μ be a Lebesgue-like Borel measure on I^n. If each of the smaller n-cells is assigned a positive number in any manner one wishes and if the sum of these numbers is $\mu(I^n)$, then one can easily devise a scheme that uses the above lemma and the addition in the homotopy sense of a finite number of deformations to create a deformation H_t such that

(α) H_0 is the identity map on I^n, whence $H_{0\#}\mu = \mu$, and H_t is the identity map on ∂I^n,

(β) $H_{1\#}\mu$ is Lebesgue-like on each of the smaller n-cells, and has precisely the value on it that was assigned to it.

In fact, one can repeat this process any number of times. This fact will prove useful in the proof of the next lemma.

We are ready for the lemma that facilitates the inductive step in the proof of the main theorem. The following terminology will lead to a clean presentation of this lemma.

Let X be a compact topological n-cell. An **n-cell partition** of X is a finite collection $\mathcal{P} = \{\,\sigma_i : i = 1, 2, \dots, k\,\}$ of compact n-cells σ_i such that

$$\bigcup_{i=1}^{k} \sigma_i = X \quad \text{and} \quad \sigma_i{}^o \cap \sigma_j{}^o = \emptyset \quad \text{whenever} \quad i \ne j.$$

For such a partition $\mathcal{P}$, the collection $\partial\mathcal{P}$, called the **boundary collection**, is defined to be $\{\,\partial\sigma_i : i = 1, 2, \dots, k\,\}$. As usual, the **mesh** of $\mathcal{P}$ is defined to be

$$\operatorname{mesh}\mathcal{P} = \max\{\,\operatorname{diam}\sigma_i : i = 1, 2, \dots, k\,\}.$$

If $h \colon X \to Y$ is a homeomorphism and $\mathcal{P} = \{\,\sigma_i : i = 1, 2, \dots, k\,\}$ is an n-cell partition of Y, then

$$h^{\#}\mathcal{P} = \{\,h^{-1}[\sigma_i] : i = 1, 2, \dots, k\,\}$$

is an n-cell partition of X. We shall also agree to write

$$h^{\#}\partial\mathcal{P} = \partial h^{\#}\mathcal{P}.$$

Observe that, for a composition gh of homeomorphisms, we have

$$(gh)^{\#}\mathcal{P} = h^{\#}g^{\#}\mathcal{P}.$$

For Lebesgue-like pairs (μ, X) and (ν, Y), a homeomorphism $h\colon X \to Y$ and an n-cell partition $\mathcal{P}$ of Y are said to be (μ,ν)-**admissible** if, for each σ in $\mathcal{P}$,

$$(h_{\#}\mu, \sigma) \quad \text{and} \quad (\nu, \sigma) \quad \text{are Lebesgue-like pairs with} \quad h_{\#}\mu(\sigma) = \nu(\sigma).$$

LEMMA 7.5. *For Lebesgue-like pairs (μ, X) and (ν, Y), let $h\colon X \to Y$ be a homeomorphism and $\mathcal{P}_0$ be an n-cell partition of Y that are (μ,ν)-admissible. Then, for each positive number ε, there is a continuous deformation D_t, $0 \le t \le 1$, by homeomorphisms of X onto Y and an n-cell partition $\mathcal{P}$ of Y such that*

$$D_0 = h, \tag{7.5}$$

$$\mathcal{P} \ \textit{refines}\ \mathcal{P}_0 \quad \textit{and} \quad \operatorname{mesh}\mathcal{P} < \varepsilon, \tag{7.6}$$

$$D_1 : X \to Y \quad \textit{and} \quad \mathcal{P} \quad \textit{are} \quad (\mu,\nu)\textit{-admissible}, \tag{7.7}$$

$$D_1{}^{\#}\mathcal{P} \quad \textit{partitions} \quad X \quad \textit{with} \quad \operatorname{mesh} D_1{}^{\#}\mathcal{P} < \varepsilon, \tag{7.8}$$

and, for every t,

$$D_t(x) = h(x) \quad \textit{whenever} \quad x \in h^{-1}[\partial\sigma] \in h^{\#}\partial\mathcal{P}_0, \tag{7.9}$$

$$D_t : X \to Y \quad \textit{and} \quad \mathcal{P}_0 \quad \textit{are} \quad (\mu,\nu)\textit{-admissible}. \tag{7.10}$$

PROOF. Recall that we have assumed $n > 1$. For $\sigma \in \mathcal{P}_0$, let $\eta\colon \sigma \to I^n$ be a homeomorphism. Then $(\eta\, h)_{\#}\mu$ and $\eta_{\#}\nu$ are Lebesgue-like Borel measures on I^n. There is n-cell partition $\overline{\mathcal{P}}_{\sigma 1}$ of I^n formed by finitely many hyperplanes of $\mathbb{R}^n$ that are parallel to the coordinate hyperplanes of $\mathbb{R}^n$ such that $\operatorname{diam} \eta^{-1}[\sigma'] < \varepsilon$ for each σ' in $\overline{\mathcal{P}}_{\sigma 1}$ and such that the $\eta_{\#}\nu$ measures of the hyperplanes are 0. We may assume that the partition is formed by 2^k points on each edge of I^n. To each σ' in $\overline{\mathcal{P}}_{\sigma 1}$, assign the positive number $\eta_{\#}\nu(\sigma')$. Let $H_{\sigma t}$ be the continuous deformation of I^n described earlier for which

$$(H_{\sigma 0}\eta\, h)_{\#}\mu = (\eta\, h)_{\#}\mu, \quad \text{and} \quad H_{\sigma t}(x) = x \quad \text{whenever} \quad (x,t) \in (\partial I^n) \times [0,1],$$

and, for each σ' in $\overline{\mathcal{P}}_{\sigma 1}$,

$$(H_{\sigma 1}\eta\, h)_{\#}\mu \quad \text{is Lebesgue-like on} \quad \sigma' \quad \text{with} \quad (H_{\sigma 1}\eta\, h)_{\#}\mu(\sigma') = \eta_{\#}\nu(\sigma').$$

Then $\widehat{D}_{\sigma t} = \eta^{-1}H_{\sigma t}\eta\, h$ is a deformation by homeomorphisms of $h^{\#}\sigma$ onto σ such that $\widehat{D}_{\sigma t} = h|(h^{\#}\sigma)$ on $h^{-1}[\partial\sigma]$ and such that $\mathcal{P}_{\sigma 1} = \eta^{\#}\overline{\mathcal{P}}_{\sigma 1}$ is an n-cell partition of σ. Let $\widehat{D}_t$ be the sum in the homotopy sense of all of these deformations and let $\mathcal{P}_1 = \bigcup_{\sigma\in\mathcal{P}_0} \mathcal{P}_{\sigma 1}$.

Now $\widehat{D}_t$ and $\mathcal{P}_1$ satisfy all of the conditions listed in the lemma with the possible exception of (7.8). To adjust for this, we interchange the roles of (μ, X) and (ν, Y) and define $h^* = \widehat{D}_1^{-1}$ and $\mathcal{P}_0^* = \widehat{D}_1^{\#}\mathcal{P}_1$. Then a deformation $\widehat{D}_t^*\colon Y \to X$ and an n-cell partition $\mathcal{P}_1^*$ of X exist, analogous to $\widehat{D}_t\colon Y \to X$ and $\mathcal{P}_1$, so that the corresponding conditions (7.5), (7.6), (7.7), (7.8), (7.9) and (7.10) are satisfied by $\widehat{D}_t^*$ and $\mathcal{P}_1^*$. The deformation $\widehat{D}_1\, \widehat{D}_t^*\, \widehat{D}_1\colon X \to Y$ and the partition $\mathcal{P} = h^{*\#}\mathcal{P}_1^*$ of Y clearly satisfy

$$\mathcal{P} \quad \text{refines} \quad \mathcal{P}_1, \quad \big(\widehat{D}_1\, \widehat{D}_1^*\, \widehat{D}_1\big)^{\#}\mathcal{P} = \mathcal{P}_1^*, \quad \widehat{D}_1\, \widehat{D}_0^*\, \widehat{D}_1 = \widehat{D}_1,$$

$$\text{and} \quad \big(\widehat{D}_1\, \widehat{D}_t^*\, \widehat{D}_1\big)(x) = h(x) \quad \text{for} \quad x \in h^{-1}[\partial\sigma] \in h^{\#}\partial\mathcal{P}_0.$$

Hence, $\widehat{D}_t$ and $\widehat{D}_1\, \widehat{D}_t^*\, \widehat{D}_1$ can be added to form a deformation D_t by homeomorphisms of X onto Y such that (7.5), (7.6), (7.7), (7.8), (7.9) and (7.10) are satisfied. The straightforward verification of this assertion is left to the reader.

7.4. Deformation theorem

We are now in the position to prove Theorem 7.1. It will be a corollary of the lemma proved below.

7.4.1. The proof. Suppose that (μ, X) and (ν, Y) are 2 Lebesgue-like pairs with $\mu(X) = \nu(Y)$ and that $h\colon X \to Y$ is a homeomorphism. For positive integers k, let $I_{(k)} = [a_{k-1}, a_k]$ where a_k, $k = 1, 2, \ldots$, is a sequence that is strictly increasing to 1 with $a_0 = 0$. Using Lemma 3.2 with $\mathcal{P}_0 = \{Y\}$ and $\varepsilon = \frac{1}{2}$, we select a deformation $D_{(1)\,t}$, $t \in I_{(1)}$, by homeomorphisms of X onto Y and an n-cell partition $\mathcal{P}_{(1)}$ of Y such that (7.5), (7.6), (7.7), (7.8), (7.9) and (7.10) hold, where the obvious change from $t \in [0,1]$ to $t \in I_{(1)}$ has been made. Assuming that $D_{(k-1)\,t}$, $t \in I_{(k-1)}$, and $\mathcal{P}_{(k-1)}$ have been selected, we next use Lemma 7.5 to select a deformation $D_{(k)\,t}$, $t \in I_{(k)}$, and an n-cell partition $\mathcal{P}_{(k)}$ of Y that satisfies

$$D_{(k-1)\,a_{k-1}}(x) = D_{(k)\,a_{k-1}}(x), \quad x \in X, \tag{7.11}$$

$$\mathcal{P}_{(k)} \text{ refines } \mathcal{P}_{(k-1)} \text{ and } \operatorname{mesh} \mathcal{P}_{(k)} < \tfrac{1}{2^k}, \tag{7.12}$$

$$D_{(k)\,a_k}\colon X \to Y \text{ and } \mathcal{P}_{(k)} \text{ are } (\mu,\nu)\text{-admissible}, \tag{7.13}$$

$$D_{(k)\,a_k}{}^{\#}\mathcal{P}_{(k)} \text{ partitions } X \text{ with } \operatorname{mesh} D_{(k)\,a_k}{}^{\#}\mathcal{P}_{(k)} < \tfrac{1}{2^k}, \tag{7.14}$$

and, for every t in $I_{(k)}$,

$$\begin{aligned} &D_{(k)\,t}(x) = D_{(k-1)\,a_{k-1}}(x) \quad \text{whenever} \\ &\qquad x \in D_{(k-1)\,a_{k-1}}{}^{-1}[\partial\sigma] \in D_{(k-1)\,a_{k-1}}{}^{\#}\partial\mathcal{P}_{(k-1)}, \end{aligned} \tag{7.15}$$

$$D_{(k)\,t}\colon X \to Y \text{ and } \mathcal{P}_{(k-1)} \text{ are } (\mu,\nu)\text{-admissible}. \tag{7.16}$$

In view of condition (7.11), we can paste these deformations together to yield D_t, $t \in [0,1)$. From (7.12) we infer that $D_t(x)$ is uniformly continuous on $[0,1) \times X$. Hence we may assume that $D_t(x)$ is defined and continuous on $[0,1] \times X$. Next, we infer from (7.14) and (7.15) that $D_1\colon X \to Y$ is one-to-one and onto Y. Finally, from (7.13), (7.15) and (7.16) we infer that $\lim_{t\to 1} D_{t\#}\mu(\sigma) = \nu(\sigma)$ whenever $\sigma \in \mathcal{P}(k)$, whence $D_{1\#}\mu(\sigma) = \nu(\sigma)$.

It is now a simple matter to show that $D_{1\#}\mu(U) = \nu(U)$ for every open set U of Y. Indeed, let U be a nonempty open set in the space Y. For each k, let $\mathcal{M}_k$ be the collection of all σ in $\mathcal{P}_{(k)}$ that are contained in U. Then

$$\sum_{\sigma\in\mathcal{M}_k} D_{1\#}\mu(\sigma) = \sum_{\sigma\in\mathcal{M}_k} \nu(\sigma)$$

for each k. Since $\mathcal{P}_{(k+1)}$ refines $\mathcal{P}_{(k)}$ for each k, we have

$$U = \textstyle\bigcup_{k=1}^{\infty} \bigcup_{\sigma\in\mathcal{M}_k} \sigma,$$

whence $D_{1\#}\mu(U) = \nu(U)$.

As ν and $D_{1\#}\mu$ are Borel measures, we have established the following lemma.

LEMMA 7.6. *Let (μ, X) and (ν, Y) be Lebesgue-like pairs with $\mu(X) = \nu(Y)$. If $h\colon X \to Y$ is a homeomorphism, then there exists a continuous deformation D_t, $t \in [0,1]$, by homeomorphisms of X onto Y such that*

$$D_0 = h \quad \text{and} \quad D_t(x) = h(x) \quad \text{whenever} \quad (x,t) \in (\partial X) \times [0,1],$$

and such that

$$D_{0\#}\mu = h_{\#}\mu \quad \text{and} \quad D_{1\#}\mu = \nu.$$

Our main deformation theorem is a corollary of the lemma.

PROOF OF THE MAIN THEOREM. Let μ be a Borel measure on the interval I^n with $\mu(I^n) = \lambda_n(I^n)$.

Suppose that a deformation D_t, $0 \le t \le 1$, of I^n that satisfies the conditions given in the theorem exists. Then

$$\mu = D_{1\#}^{-1}\lambda.$$

Hence,

$$\mu(\partial I^n) = \lambda_n(D_1[\partial I^n]) = \lambda_n(\partial I^n) = 0,$$

since $D_1[\partial I^n] = \partial I^n$ by the invariance of domain theorem from topology (see [**78**]). And, for each x,

$$\mu(\{x\}) = \lambda_n(D_1[\{x\}]) = 0.$$

Finally, for each nonempty, relatively open set U of I^n,

$$\mu(U) = \lambda_n(D_1[U]) > 0$$

because $D_1[U]$ is also a nonempty, relatively open set of I^n by the invariance of domain theorem. Thereby we have shown that μ is a Lebesgue-like measure on I^n.

For the converse, we have already separately proved the theorem for $n = 1$. So we assume that $n > 1$. Here we apply the last lemma with $\nu = \lambda_n$ and with h being the identity map on I^n. Then the existence of the desired deformation is assured.

7.4.2. Example. Let E_0 be the union of a countable collection of line segments I_k contained in $I^n \setminus \partial I^n$. For each k, let p_k be a positive number such that $\sum_{k=1}^{\infty} p_k = \lambda_n(I^n)$. With H_1 denoting the Hausdorff 1-measure on $\mathbb{R}^n$, let μ be defined by

$$\mu = \sum_{k=1}^{\infty} \frac{p_k}{\mathsf{H}_1(I_k)} \mathsf{H}_1 \llcorner I_k,$$

where $\mathsf{H}_1 \llcorner I_k(E) = \mathsf{H}_1(I_k \cap E)$ for Borel sets E. Then μ is Lebesgue-like if and only if E_0 is dense in I^n. Of course, this example is most interesting when $n \ge 2$. Note that piecewise linear self-homeomorphisms of I^n preserve the essential nature of the construction of this measure.

7.5. Remarks

In 1941, J. C. Oxtoby and S. L. Ulam [**107**] gave the second proof of what we have called the von Neumann theorem [**128**]. It has already been mentioned that Ulam had conjectured the theorem. The conjecture was made in 1936 and von Neumann's proof is actually unpublished. The citation [**128**] only states that von Neumann had proved the conjecture and that the proof had not been published.

C. Goffman and G. Pedrick [**65**] gave a new and simple third proof of the theorem in 1975 that used the direction lemma. In his review of the new proof (see [**65**]), Oxtoby obliquely suggested that the proof of a crucial lemma appeared to depend on the fact that n was 2. The proof of the deformation theorem, Theorem 7.1, overcomes this difficulty, thereby yielding a truly simple proof. The key to the proof is the direction lemma, Lemma 7.2, which first appeared in [**65**]. The reader may find that many proofs given here are foreshadowed in [**65**]. Apropos to the final statement of the Oxtoby's review, we remark that the deformation in Lemma 7.4

can be made piecewise linear. But, of course, the above example shows that the deformation in Lemma 7.6 cannot be piecewise linear even though the map h may have been piecewise linear.

7.5.1. Problems. We leave the reader with the following unsolved problem.

Problem 1: Characterize the Lebesgue equivalence classes of nonatomic Borel measures on I^n. Characterize the homotopy classes of nonatomic Borel measures on I^n.

More generally,

Problem 2: Characterize, in a simple way, the Lebesgue equivalence classes of Borel measures on I^n. Characterize, in a simple way, the homotopy classes of Borel measures on I^n.

CHAPTER 8

Blumberg's Theorem

In 1922 [**12**], H. Blumberg proved the following remarkable continuity fact about every real-valued function on $\mathbb{R}$.

THEOREM (BLUMBERG). *For each function $f\colon \mathbb{R} \to \mathbb{R}$ there exists a dense subset D of $\mathbb{R}$ such that the restriction $f|D$ is continuous.*

It will be shown that a similar homeomorphism result does not hold for arbitrary one-to-one transformations. Indeed, an example is given of a one-to-one mapping f of the interval $(0,1)$ onto itself for which there is no dense subset D such that $f[D]$ is also dense and $f|D : D \to f[D]$ is a homeomorphism. But first, a detailed discussion of Blumberg's theorem as well as a short presentation of the notion of Blumberg spaces will be given. Finally, some positive results due to Neugebauer will conclude the chapter.

8.1. Blumberg's theorem for metric spaces

We characterize the metric spaces for which Blumberg's theorem holds. As the statement of the characterization theorem requires the use of properties that depend on the Baire categories, we shall begin with the appropriate definitions.

Let E be a subset of a metric space X. A point x of X is said to be of the **second category relative to** E if every sphere that is centered at x contains a subset of E that is of second category in X. Otherwise, x is said to be of the **first category relative to** E. (The reader should note that the point x in the above definitions has not been restricted to be in E.) Finally, X is said to be of **homogeneous second category** if every point of X is of the second category relative to X.

Here is a simple proposition.

PROPOSITION 8.1. *For subsets E of a metric space X, the set $\mathrm{S}(E)$ consisting of all of the points that are of second category relative to E is a closed set contained in the closure of E. And, if $E_1 \subset E_2 \subset X$, then $\mathrm{S}(E_1) \subset \mathrm{S}(E_2)$.*

The characterization theorem was first obtained by Bradford and Goffman [**15**].

THEOREM 8.2 (BRADFORD-GOFFMAN). *Let X be a metric space. If X is of homogeneous second category, then for each real-valued function f on X there is a dense subset D of X such that the restricted function $f|D$ is continuous on D. Conversely, if X is not of homogeneous second category, then there is a real-valued function f on X such that $f|D$ is not continuous for every dense subset D of X.*

8.1.1. Proof of the converse. Let X be a metric space that is not of homogeneous second category. That is, there is a sphere K that is of the first category in X. We write

$$K = \bigcup_{n=1}^{\infty} K_n,$$

where the union is over a countable collection of disjoint, nowhere dense sets of X.

The required function $f\colon X \to \mathbb{R}$ is defined to be

$$f(x) = \begin{cases} n & \text{for } x \in K_n,\ n = 1, 2, \ldots, \\ 0 & \text{for } x \notin K. \end{cases}$$

Let D be any dense subset of X and let $x \in D \cap K$. Then $x \in D \cap K_n$ for some n. Let K' be any sphere centered at x. As K_n is nowhere dense, there is a point x' with $x' \in (D \setminus K_n) \cap K'$. Since $|f(x) - f(x')| \geq 1$, we infer that $f|D$ is not continuous at x.

Clearly, we may assume the existence of a bounded function with the requisite property.

8.1.2. Proof of the statement. This part of the proof the characterization theorem is more difficult than that of the converse. The proof uses the following fact that was first observed by Banach [**7**]. For a subset E of a metric space X, the set which consists of all points of E that are of the first category relative to E is a first category set in X. This fact is obtained by a transfinite construction of disjoint spheres that are centered at points of E and meet E in a set of first category so that the complement of the union of these spheres contains all points of E that are interior to the set points of the second category relative to E.

The following definitions will prove useful. Let E be a subset of a metric space X. Then a point x of X is said to be **heavy relative to** E if there is a sphere that is centered at x and that consists entirely of points of the second category relative to E. A real-valued function on X is said to be **almost continuous at** x if, for every open set G that contains $f(x)$, the point x is a heavy point relative to the set $f^{-1}[G]$.

Here is another simple proposition.

PROPOSITION 8.3. *For subsets E of a metric space X, the set $\mathrm{H}(E)$ consisting of all points that are heavy relative to E is the interior of the closed set $\mathrm{S}(E)$ that consists of all points that are of second category relative to E. Consequently, $\mathrm{S}(E) \setminus \mathrm{H}(E)$ is nowhere dense. And, if $E_1 \subset E_2 \subset X$, then $\mathrm{H}(E_1) \subset \mathrm{H}(E_2)$.*

We have several auxiliary lemmas.

LEMMA 8.4. *Each subset E of a metric space X can be written as the union of disjoint sets A and B with the properties that each point of A is heavy relative to E and the set B is of the first category in X.*

PROOF. Let B_1 consist of all of the points of E that are of the first category relative to E. According to Banach, B_1 is a set of the first category in X. Let B_2 consist of the points of E which are of the second category relative to E but are not heavy relative to E. That is, $B_2 = E \cap \big(\mathrm{S}(E) \setminus \mathrm{H}(E)\big)$ in the notation of the last proposition. Clearly, B_2 is nowhere dense in X. Define B to be the first category set $B_1 \cup B_2$ and define A to be $E \setminus B$. We infer from $E \setminus B = E \cap \mathrm{H}(E)$ that A is a dense subset of $\mathrm{H}(E)$, whence every x in A is contained in a sphere that consists entirely of points of second category relative to E.

LEMMA 8.5. *If f is a real-valued function defined on a metric space X, then f is almost continuous at every point of X except at a set of the first category.*

PROOF. Let G_n, $n = 1, 2, \ldots$, be an enumeration of the collection of all open intervals of $\mathbb{R}$ whose end points are rational numbers. For each n let E_n be the set of points of $f^{-1}[G_n]$ that are not heavy points relative to $f^{-1}[G_n]$. Since the sets E_n are of the first category by Lemma 8.4, we have that $E = \bigcup_{n=1}^{\infty} E_n$ is also of the first category. Certainly, $x \in E$ whenever f is not almost continuous at x.

The proof of the characterization theorem is by means of an inductive construction. The next lemma provides the inductive step. We shall use $B(x, r)$ to denote the open sphere with center at x and radius r.

LEMMA 8.6. *Let $\varepsilon > 0$ and let R be the set of points at which a real-valued function f on a metric space X of homogeneous second category is almost continuous. Suppose that, for a set N of the first category and for an indexed collection $\{ (x_\alpha, r_\alpha) : \alpha \in A \}$ of points in $(R \setminus N) \times \{ r : 0 < r \}$, the following hold:*

(1) *$\mathcal{U} = \{ B(x_\alpha, r_\alpha) : \alpha \in A \}$ is mutually disjoint,*
(2) *$\bigcup_{\alpha \in A} B(x_\alpha, r_\alpha)$ is dense in X,*

and, for each α in A,

(3) *$x_\alpha \in V_\alpha = B(x_\alpha, r_\alpha) \setminus N \subset R$,*
(4) *every point of $B(x_\alpha, r_\alpha)$ is a heavy point relative to V_α.*

Then there is a first category set N' and an indexed collection $\{ (x_{\alpha'}, r'_{\alpha'}) : \alpha' \in A' \}$ of points in $(R \setminus N') \times \{ r : 0 < r < \varepsilon \}$ with $N \subset N'$ and $A \subset A'$ such that the following hold:

(1′) *$\mathcal{U}' = \{ B(x_{\alpha'}, r'_{\alpha'}) : \alpha' \in A' \}$ is mutually disjoint and refines $\mathcal{U}$,*
(2′) *$\bigcup_{\alpha' \in A'} B(x_{\alpha'}, r'_{\alpha'})$ is dense in X,*

and, for each α' in A',

(3′) *$x_{\alpha'} \in V'_{\alpha'} = B(x_{\alpha'}, r'_{\alpha'}) \setminus N' \subset R$,*
(4′) *every point of $B(x_{\alpha'}, r'_{\alpha'})$ is a heavy point relative to $V'_{\alpha'}$,*
(5′) *$|f(x) - f(x_{\alpha'})| < \varepsilon$ whenever $x \in V'_{\alpha'}$.*

PROOF. Clearly, condition (5′) is the one of interest. From (1) we see that $\{ x_\alpha : \alpha \in A \}$ is a subset of R that is discrete in the open subspace $\bigcup_{\alpha \in A} B(x_\alpha, r_\alpha)$. As f is almost continuous at each point x_α, there is a number r'_α with

$$0 < r'_\alpha < \min\{r_\alpha/2, \varepsilon\}$$

such that

$$|f(x) - f(x_{\alpha'})| < \varepsilon \quad \text{whenever} \quad x \in R \cap B(x_\alpha, r'_\alpha),$$

and every point of $B(x_\alpha, r'_{\alpha'})$ is a heavy point relative to $R \cap B(x_\alpha, r'_\alpha)$. Observe that $\{ B(x_\alpha, r'_\alpha) : \alpha \in A \}$ is a discrete collection of spheres in the open subspace $\bigcup_{\alpha \in A} B(x_\alpha, r_\alpha)$. We must take care of the points of the open set

$$U = \bigcup_{\alpha \in A} B(x_\alpha, r_\alpha) \setminus \overline{\bigcup_{\alpha \in A} B(x_\alpha, r'_\alpha)}.$$

Since X is of homogeneous second category, we infer from Lemma 8.5 that the set R is dense in X. The required collections will be shown to exist by a transfinite construction in the open set U. Let $\{ x'_\gamma : \gamma < \xi \}$ be a well ordering of $(R \setminus N) \cap U$. (Here, we are assuming that the indexing set A does not have ordinal numbers as members.) Let $x_1 = x'_1$. From the definition of almost continuous we have a positive number r'_1 such that

(a_1) $|f(x) - f(x_1)| < \varepsilon$ whenever $x \in R \cap B(x_1, r'_1)$,

(b_1) every point of $B(x_1, r'_1)$ is a heavy point relative to $R \cap B(x_1, r_1)$,
(c_1) $B(x_1, r'_1)$ is contained in some member of $\mathcal{U}$.

Let $x_2 = x'_{\gamma_2}$, where

$$\gamma_2 = \min\{\, \gamma : x_\gamma \notin \overline{B(x_1, r'_1)}\,\}.$$

There exists a positive number r'_2 such that

(a_2) $|f(x) - f(x_2)| < \varepsilon$ whenever $x \in R \cap B(x_2, r'_2)$,
(b_2) every point of $B(x_2, r'_2)$ is a heavy point relative to $R \cap B(x_2, r_2)$,
(c_2) $B(x_2, r'_2)$ is contained in some member of $\mathcal{U}$,
(d_2) $B(x_2, r'_2) \cap \overline{B(x_1, r'_1)} = \emptyset$.

Now suppose that η is an ordinal number with $\eta < \xi$ such that for every ordinal number γ with $\gamma < \eta$ the pair (x_γ, r'_γ) has been selected so that

(a_γ) $|f(x) - f(x_\gamma)| < \varepsilon$ whenever $x \in R \cap B(x_\gamma, r'_\gamma)$,
(b_γ) every point of $B(x_\gamma, r'_\gamma)$ is a heavy point relative to $R \cap B(x_\gamma, r_\gamma)$,
(c_γ) $B(x_\gamma, r'_\gamma)$ is contained in some member of $\mathcal{U}$,
(d_γ) $B(x_\gamma, r'_\gamma) \cap \overline{\bigcup_{\delta<\gamma} B(x_\delta, r'_\delta)} = \emptyset$.

Let $x_\eta = x'_{\gamma_\eta}$, where

$$\gamma_\eta = \min\{\, \gamma : x_\gamma \notin \overline{\textstyle\bigcup_{\delta<\eta} B(x_\delta, r'_\delta)}\,\}.$$

We can now select r'_η in the same way as r'_2 was. It is clear that $\bigcup_{\gamma<\xi} B(x_\gamma, r'_\gamma)$ is dense in U. (Here, we have assumed that the most efficient ordinal number ξ has been used to well order the set $(R \setminus N) \cap U$.) With

$$A' = A \cup \{\, \gamma : \gamma < \xi\,\} \quad \text{and} \quad N' = N \cup \big(X \setminus \textstyle\bigcup_{\alpha' \in A'} B(x_{\alpha'}, r'_{\alpha'})\big)$$

one can easily verify that the conditions (1′)–(5′) are satisfied.

In passing, we remark that the collection $\{\, x_{\alpha'} : \alpha' \in A'\,\}$ in Lemma 8.6 has the property that, for each x in X, there is an α' in A' for which $\mathrm{d}(x, x_{\alpha'}) < 2\varepsilon$ holds.

Proof of the theorem. By Lemma 8.5, the set R is a dense set of the second category because X is of homogeneous second category. So select a member x_0 from R. As there is no loss in assuming that the metric is bounded by $\frac{1}{2}$ we may assume $X = B(x_0, 1)$. Beginning with

$$N_0 = X \setminus R, \quad \{\, (x_{\alpha_0}, 1) : \alpha_0 \in A_0 = \{0\}\,\} \quad \text{and} \quad \mathcal{U}_0 = \{B(x_0, 1)\},$$

we can inductively construct for each positive integer n the sets N_n of first category and indexed collections

$$\{\, (x_{\alpha_n}, r_{n,\alpha_n}) : \alpha_n \in A_n\,\}$$

contained in $(R \setminus N_n) \times \{\, r : 0 < r < \frac{1}{n}\,\}$ such that

(0″) $N_{n-1} \subset N_n$ and $A_{n-1} \subset A_n$,
(1″) $\mathcal{U}_n = \{\, B(x_{\alpha_n}, r_{n,\alpha_n}) : \alpha_n \in A_n\,\}$ is a mutually disjoint collection that refines $\mathcal{U}_{n-1}$,
(2″) $\bigcup_{\alpha_n \in A_n} B(x_{\alpha_n}, r_{n,\alpha_n})$ is dense in X,

and, for each α_n in A_n,

(3″) $x_{\alpha_n} \in V_{n,\alpha_n} = B(x_{\alpha_n}, r_{n,\alpha_n}) \setminus N_n \subset R$,
(4″) every point of $B(x_{\alpha_n}, r_{n,\alpha_n})$ is a heavy point relative to V_{n,α_n},
(5″) $|f(x) - f(x_{\alpha_n})| < \frac{1}{n}$ whenever $x \in V_{n,\alpha_n}$.

Let

$$D = \{\, x : \text{ for some } n,\ x = x_{\alpha_n} \text{ for some } \alpha_n \text{ in } A_n \,\}.$$

In view of the passing remark above, we see from (2″) that D is dense in X. Let $x \in D$ and $\varepsilon > 0$. There is an n such that $\varepsilon > \frac{1}{n}$ and $x = x_{\alpha_n}$ for some α_n in A_n. From (0″), (1″), (3″) and (5″) we infer that

$$|f(x') - f(x)| < \varepsilon \quad \text{whenever} \quad x' \in D \cap B(x, r_{n,\alpha_n}).$$

Thereby we have shown that $f|D$ is continuous at x.

8.2. Non-Blumberg Baire spaces

Let us discuss the property found in Blumberg's theorem. A topological space X is called a **Blumberg** space if every real-valued function f on X possesses a dense subset D of X for which $f|D$ is continuous. In the above characterization theorem, the Baire categories played an important role. A topological space X is called a **Baire** space if every sequence of open dense sets has a dense intersection. (See the Appendix, page 187.) Baire first proved that $\mathbb{R}$ had this property.

It is quite clear that the proof of the converse of the above characterization theorem can be generalized to yield the fact that every Blumberg space is a Baire space. We give two examples of Baire spaces whose cardinalities are $\mathfrak{c} = 2^{\aleph_0}$ and fail to be Blumberg spaces.

8.2.1. Levy's example.

The example of Levy [**94**] is from the theory of the rings of continuous functions. A good general reference on rings of continuous functions is the book by Gillman and Jerison [**47**].

Levy's example is an η_1-set X whose cardinality is $\mathfrak{c}$. That is, X is a totally ordered set with the property that, for every pair of countable subsets A and B with $a < b$ whenever $a \in A$ and $b \in B$, there is an x such that $a < x < b$ for every a in A and every b in B. That such an η_1-set exists is shown in [**47**, page 187]. The topology for X will be the interval topology.

To see that X is a Baire space, let G_n, $n = 1, 2, \ldots$, be a sequence of dense open sets and let (a, b) be an open interval of X. We can inductively find a nested sequence of open intervals (a_n, b_n) such that $(a_n, b_n) \subset (a, b) \cap G_n$ for each n. With $A = \{\, a_n : n = 1, 2, \ldots \}$ and $B = \{\, b_n : n = 1, 2, \ldots \}$, there is a point x such that $a_m < x < b_n$ for every m and n. Clearly, $x \in (a, b) \cap \bigcap_{n=1}^{\infty} G_n$, whence X is a Baire space.

Before proving that X is not a Blumberg space, we shall give a brief discussion of P-spaces, a notion from the theory of rings of continuous functions. A completely regular space X is a **P-space** if each prime ideal in the ring $\mathrm{C}(X)$ of continuous real-valued functions is a maximal ideal. P-spaces are characterized as those completely regular spaces for which the zero-sets of continuous real-valued functions are open sets [**47**, page 62]. Hence, every singleton zero-set of a continuous real-valued function is necessarily an isolated point since it would be simultaneously open and closed. According to [**47**, page 63], subspaces of P-spaces are also P-spaces.

It is proved in [**47**, page 193] that our η_1-set X is a P-space with no isolated points. Hence every dense subspace of X is a P-space with no isolated points. Let $f\colon X \to \mathbb{R}$ be a one-to-one onto function. Clearly such a function exists because X and $\mathbb{R}$ have the same cardinality. Let D be a subspace of X such that $f|D$ is continuous. For $y \in (f|D)[D]$, we have that $(f|D)^{-1}[y]$ is a singleton zero-set in the subspace D. That is, D consists of only isolated points. Hence D is not dense. Thus we have shown that X is not a Blumberg space.

An interesting consequence of this proof is the fact that this topology on X does not contain a coarser topology $\mathcal{T}'$ such that $(X, \mathcal{T}')$ is homeomorphic to $\mathbb{R}$ with the Euclidean topology because the identity map from the finer to the coarser topology is continuous. In a sense, this example is a little too esoteric for the purposes of this book; the next example is more to the point.

8.2.2. White's example. Assuming the continuum hypothesis, White proved in [**140**] that $\mathbb{R}$ with the density topology $\mathcal{T}_d$ (see Section A.4 of the Appendix) fails to be a Blumberg space. It is clear that the Euclidean topology on $\mathbb{R}$ is coarser than $\mathcal{T}_d$. We shall reproduce his proof here.

Let us first prove that the density topology on $\mathbb{R}$ yields a Baire space. Observe that a density-open set is dense in $\mathbb{R}$ with respect to the density topology if and only if $\mathbb{R} \setminus E$ has Lebesgue measure 0. Hence it follows very easily that the density topology on $\mathbb{R}$ yields a Baire space.

To prove that $\mathbb{R}$ with the density topology is not a Blumberg space, we will need to study the Hausdorff 1-measure on arbitrary subsets E of $\mathbb{R}$. (See Section A.6.1 of the Appendix for the definition of the Hausdorff 1-measure.) We shall use H_1^* and H_1^{E*} for the outer Hausdorff 1-measures on $\mathbb{R}$ and E respectively. It is clear that $\mathsf{H}_1^*[A] = \mathsf{H}_1^{E*}[A]$ whenever $A \subset E$. Since the Hausdorff measures are Carathéodory metric outer measures, closed sets in the respective metric spaces $\mathbb{R}$ and E are measurable. Moreover, they are regular Borel measures. Now observe that, if E is a set with $\mathsf{H}_1^*[E] > 0$ and $g \colon E \to \mathbb{R}$ is an H_1^E-measurable function, there is by Lusin's theorem a set F that is closed in the subspace E with $\mathsf{H}_1^E[F] > 0$ such that $g|F$ is continuous (of course with the Euclidean topologies on its domain and range). Clearly, such a set F is uncountable.

Let us derive properties of the relative density topology for subsets E of $\mathbb{R}$. If U is a density-open set of $\mathbb{R}$, then it is an H_1-measurable set. Hence it is the union of a Borel set of $\mathbb{R}$ and a set of H_1-measure 0. Consequently, $U \cap E$ is an H_1^E-measurable set. Now let $f \colon \mathbb{R} \to \mathbb{R}$ be a function such that $f|E$ is continuous when E is endowed with the relative density topology. Then $f|E$ is also H_1^E-measurable. As the Blumberg property requires that E be dense in $\mathbb{R}$ with respect to the density topology, we have that $\mathsf{H}_1^*[E] > 0$, whence there is an uncountable set F contained in E such that $f|F$ is continuous when F is endowed with the Euclidean topology.

White now makes the simple observation that it is known that under the Continuum Hypothesis there is a function $f \colon \mathbb{R} \to \mathbb{R}$ with the property that $f|D$ is continuous only when D is a countable subset of $\mathbb{R}$ (see [**54**, page 148]). Therefore, $\mathbb{R}$ with the density topology can not be a Blumberg space.

8.3. Homeomorphism analogues

Turning to the main point of the chapter, namely homeomorphisms, we begin with an example of a one-to-one mapping between $(0, 1)$ and itself which is not a homeomorphism between dense subsets [**52**]. This example will show that the natural analogue of Blumberg's theorem fails. The next natural question then becomes that of finding a large class of one-to-one functions for which the natural analogue of Blumberg's theorem will hold. Here we find that the notion of quasi-continuity (defined originally by Kempisty [**89**]) has played a role. For this, we shall follow a development of Neugebauer [**100**].

8.3.1. The counterexample. Our task is to construct a one-to-one function f which simultaneously possesses the property that $f|D$ is not a homeomorphism and the property that $f[D]$ is dense in $(0,1)$ whenever D is a dense subset of $(0,1)$.

With $I=(0,1)$, for each natural number n, consider the open subintervals

$$I_{nm}=(\tfrac{m}{2^n},\tfrac{m+1}{2^n}) \quad \text{for} \quad m=0,1,\dots,2^n-1.$$

Next, let

$$0<a_1<\cdots<a_k<\cdots<1$$

be an increasing sequence that converges to 1. These points will form disjoint semi-open intervals, progressing to the right, which we label as

$$\widetilde{I}_{11}=(0,a_1],\ \widetilde{I}_{12}=(a_1,a_2],\ \widetilde{I}_{21}=(a_2,a_3],\ \widetilde{I}_{22}=(a_3,a_4],\dots,$$

in an 'increasing' manner so that there is an $\widetilde{I}_{nm}$ for every natural number n and for every integer m with $m=0,1,\dots,2^n-1$. By increasing we mean that $\widetilde{I}_{n_1m_1}$ is to the right of $\widetilde{I}_{n_2m_2}$ whenever $n_1>n_2$ or whenever $n_1=n_2$ and $m_1>m_2$. The union of the intervals $\widetilde{I}_{nm}$ is clearly I.

In order to achieve simplicity of notation in the construction of our one-to-one function $f\colon(0,1)\to(0,1)$, we shall denote the domain by I and the range by J. We shall also denote the corresponding subintervals of J in the above construction by J_{nm} and $\widetilde{J}_{nm}$.

For each interval I_{nm} and J_{nm}, select nonempty, perfect, nowhere dense sets S_{nm} and T_{nm} such that

$$S_{nm}\subset I_{nm} \quad \text{and} \quad T_{nm}\subset J_{nm},$$

and such that each of

$$\{\,S_{nm}:m=0,1,\dots,2^n-1,\ n=1,2,\dots\}$$

and

$$\{\,T_{nm}:m=0,1,\dots,2^n-1,\ n=1,2,\dots\}$$

is a mutually disjoint collection. Denote their unions as

$$S=\bigcup\nolimits_{n,m}S_{nm}$$

and

$$T=\bigcup\nolimits_{n,m}T_{nm}.$$

Note that S and its complement $I\setminus S$ meet every subinterval of I in a set of cardinal number $\mathfrak{c}$ and similarly for T, $J\setminus T$ and J.

For every n and m let

$$\widetilde{S}_{nm}=\widetilde{I}_{nm}\setminus S \quad \text{and} \quad \widetilde{T}_{nm}=\widetilde{J}_{nm}\setminus T,$$

and let

$$\widetilde{S}=\bigcup\nolimits_{n,m}\widetilde{S}_{nm} \quad \text{and} \quad \widetilde{T}=\bigcup\nolimits_{n,m}\widetilde{T}_{nm}.$$

It is clear that each of the sets S_{nm}, $\widetilde{S}_{nm}$, T_{nm} and $\widetilde{T}_{nm}$ have cardinal number $\mathfrak{c}$ and that each of

$$\{\,\widetilde{S}_{nm}:m=0,1,\dots,2^n-1,\ n=1,2,\dots\}$$

and

$$\{\widetilde{T}_{nm} : m = 0, 1, \ldots, 2^n - 1,\ n = 1, 2, \ldots\}$$

is a mutually disjoint collection. A simple calculation shows that

$$I = S \cup \widetilde{S} \quad \text{and} \quad J = T \cup \widetilde{T}.$$

By piecing together arbitrary one-to-one correspondences between S_{nm} and $\widetilde{T}_{nm}$ and between $\widetilde{S}_{nm}$ and T_{nm}, we construct a one-to-one function $f\colon I \to J$. To see that f is the required one, we make two observations.

First observe that if D is a subset of S that is dense in some subinterval I' of I then $f|D$ is discontinuous at every point of $D \cap I'$. For suppose that x is member of $D \cap I'$. As every S_{nm} is nowhere dense, there is a sequence x_i, $i = 1, 2, \ldots$, in $D \cap I'$ converging to x such that $x_i \in S_{n_i m_i} \subset I_{n_i m_i}$ and such that the intervals $J_{n_i m_i}$ increase to the right towards 1. Since $f(x_i) \in J_{n_i m_i}$, we have that $\lim_{i \to \infty} f(x_i) = 1 > f(x)$ as desired. By an analogous argument, if D is a subset of I such that $E = f[D]$ is a subset of T that is dense in some subinterval J' of J then $f^{-1}|E$ is discontinuous at every point of $E \cap J'$.

Second, suppose that D is a subset of $\widetilde{S}$ that is dense in I. Then $f[D]$ is dense in J. To see this, we have from $D \cap \widetilde{I}_{nm} \neq \emptyset$ that $D \cap \widetilde{S}_{nm} \neq \emptyset$ for every n and m. From this we have that $f[D] \cap T_{nm} \neq \emptyset$, whence $f[D] \cap J_{nm} \neq \emptyset$, for every n and m.

To complete our task, we let D be a dense subset of I and define $D_1 = D \cap S$ and $D_2 = D \cap \widetilde{S}$. If D_1 is not nowhere dense, then we infer from the first observation that $f|D_1$ has points of discontinuities. Consequently, $f|D$ will have a point of discontinuity. If D_1 is nowhere dense, then D_2 is dense in I. And then, by the second observation, we have that $E_2 = f[D_2]$ is a subset of T that is dense in J. The first observation yields that $f^{-1}|E_2$ has points of discontinuities. Thus, $f^{-1}|E$ will have a point of discontinuity, where $E = f[D]$. Our task is completed.

It is clear from the proof that f has the following properties. If D is a dense subset of I such that $f|D$ is continuous, then $E = f[D]$ is dense in J and $f^{-1}|E$ is not continuous. If E is a dense subset of J such that $f^{-1}|E$ is continuous, then $D = f^{-1}[E]$ is dense in I and $f|D$ is not continuous.

8.3.2. Positive results. We now present Neugebauer's positive results concerning the class of one-to-one quasi-continuous mappings [**100**].

For ease of exposition we shall make two definitions. For a real-valued function f on $(0, 1)$, a dense subset D for which $f|D$ is continuous will be called a **Blumberg set** for f. And, for a one-to-one, onto function $f\colon (0, 1) \to (0, 1)$, a subset D of $(0, 1)$ is called a **simultaneous** Blumberg set if D is a Blumberg set for f and $E = f[D]$ is a Blumberg set for f^{-1}. In passing, we remark that the above example fails to have simultaneous Blumberg sets.

It is shown by Neugebauer that the class of quasi-continuous, one-to-one, onto functions is a natural one in which a positive result on the existence of simultaneous Blumberg sets can be found (see Theorem 8.12).

Let us begin with a general discussion of quasi-continuous functions defined on $(0, 1)$. A function $f\colon (0, 1) \to \mathbb{R}$ is said to be **quasi-continuous** if, for each x in $(0, 1)$ and each pair of open sets U and V with $x \in U$ and $f(x) \in V$, the interior of $U \cap f^{-1}[V]$ is not empty. For an arbitrary function f defined on $(0, 1)$, the **continuity set**, denote by $C(f)$, is the set consisting of all the points of continuity of f.

PROPOSITION 8.7. *$C(f)$ is a G_δ Blumberg set for a every quasi-continuous function $f\colon (0,1) \to \mathbb{R}$.*

PROOF. The set of points at which the oscillation of f is no smaller than $\frac{1}{n}$ is a closed set. Consequently, for a quasi-continuous function f on $(0,1)$, the set $C(f)$ is a dense G_δ set.

We have the following lemma.

LEMMA 8.8. *If $f\colon (0,1) \to \mathbb{R}$ is quasi-continuous and U is a nonempty open subset of $(0,1)$, then the set $f[D \cap U]$ is dense in $f[U]$ for each dense subset D of $(0,1)$.*

PROOF. Suppose that $f\colon (0,1) \to \mathbb{R}$ is quasi-continuous and D is a dense subset of $(0,1)$. Let U be a nonempty open subset of $(0,1)$. For $y \in f[U]$, let V be an open set with $y \in V$. As f is quasi-continuous, the interior of $U \cap f^{-1}[V]$ is not empty. The denseness of D yields $D \cap U \cap f^{-1}[V] \neq \emptyset$, whence $f[D \cap U] \cap V \neq \emptyset$. Hence $f[D \cap U]$ is dense in $f[U]$.

Obviously, a connection between quasi-continuity and Blumberg sets must be found. To this end, we define the notion of a strong Blumberg set. A subset D of $(0,1)$ is said to be a **strong Blumberg set** for a function f if D is a Blumberg set with the property that, for every nonempty open subset U of $(0,1)$, the set $f[D \cap U]$ is dense in $f[U]$. The following proposition is an immediate consequence of Lemma 8.8.

PROPOSITION 8.9. *For quasi-continuous functions on $(0,1)$, Blumberg sets are strong Blumberg sets and conversely.*

There is the following characterization of quasi-continuity.

THEOREM 8.10. *For a function $f\colon (0,1) \to \mathbb{R}$, the following statements are equivalent.*

(1) *f is quasi-continuous on $(0,1)$.*
(2) *$C(f)$ is a strong Blumberg set for f.*
(3) *f possesses a strong Blumberg set.*

PROOF. Suppose that f is quasi-continuous. Then the two propositions show that (1) implies (2).

Clearly, (2) implies (3).

Finally, suppose that f has a strong Blumberg set D. For $x \in (0,1)$ and for open sets U and V such that $x \in U$ and $f(x) \in V$, let J be a open subinterval of $(0,1)$ such that $f(x) \in J$ and $\overline{J} \subset V$. As U is open, there is an open subinterval I of $(0,1)$ such that $x \in I \subset U$. Since D is a strong Blumberg set, the set $f[D \cap I]$ is dense in $f[I]$. Observe that $f(x)$ is a member of $f[I]$. Hence there is a point x' in $D \cap I$ such that $f(x') \in J$. We now use the continuity of $f|D$ at x' to find an open subinterval I' of I such that $f[D \cap I'] \subset J$. Now we have

$$f[I'] \subset \overline{f[D \cap I']} \subset \overline{J} \subset V.$$

That is, $I' \subset I \cap f^{-1}[V] \subset U \cap f^{-1}[V]$ and quasi-continuity is proved.

We have the following connection between quasi-continuous functions on $(0,1)$ and Baire spaces.

LEMMA 8.11. *Let $f\colon (0,1) \to (0,1)$ be a one-to-one and onto function with both f and f^{-1} being quasi-continuous. Then the set $(0,1) \setminus f^{-1}[V]$ is nowhere dense for each dense open set V of $(0,1)$. Consequently, $f^{-1}[G]$ is a Baire space for each dense G_δ set G of $(0,1)$.*

PROOF. Let V be a dense open subset of $(0,1)$ and let U be a nonempty open subset of $(0,1)$. By Lemma 8.8, $f^{-1}[V]$ is dense in $(0,1)$. So, $U \cap f^{-1}[V]$ is not empty. As f is quasi-continuous, the set $U \cap f^{-1}[V]$ contains a nonempty open subset of $(0,1)$, whence $(0,1) \setminus f^{-1}[V]$ is nowhere dense in $(0,1)$.

The second statement of the lemma is a simple consequence of the first.

THEOREM 8.12 (NEUGEBAUER). *Every one-to-one function f of $(0,1)$ onto itself, where f and f^{-1} are quasi-continuous, possesses a simultaneous Blumberg set.*

PROOF. By Proposition 8.7, $C(f^{-1})$ is a G_δ Blumberg set for f^{-1}. From the last lemma we infer that $G = f^{-1}\big[C(f^{-1})\big]$ is a Baire space that is dense in $(0,1)$. The main theorem (Theorem 8.2) of Section 1 provides the existence of a Blumberg set D contained in G for the function $f|G$. As f is quasi-continuous, we have by Lemma 8.8 that $f[D]$ is a dense subset of $(0,1)$ with $f[D] \subset C(f^{-1})$. Obviously, the set D is a simultaneous Blumberg set for f.

The next theorem characterizes those quasi-continuous, one-to-one, onto functions $f\colon (0,1) \to (0,1)$ that possess a quasi-continuous inverse f^{-1}. The proof will rely on the next lemma.

LEMMA 8.13. *If $f\colon (0,1) \to \mathbb{R}$ is a quasi-continuous function and D is a dense subset of $(0,1)$, then $D \cap f^{-1}[V]$ is dense in $f^{-1}[V]$ for every open subset V of $\mathbb{R}$ with $V \cap f\big[(0,1)\big] \neq \emptyset$.*

PROOF. Let U be any open set such that $U \cap f^{-1}[V] \neq \emptyset$. The quasi-continuity of f yields a nonempty open set W contained in $U \cap f^{-1}[V]$. Since D is dense in $(0,1)$, we have that $D \cap W \neq \emptyset$. Hence

$$\emptyset \neq D \cap W \subset D \cap U \cap f^{-1}[V] = \big(D \cap f^{-1}[V]\big) \cap \big(U \cap f^{-1}[V]\big),$$

and the denseness of $D \cap f^{-1}[V]$ in $f^{-1}[V]$ is established.

THEOREM 8.14 (NEUGEBAUER). *If $f\colon (0,1) \to (0,1)$ is a one-to-one, onto, quasi-continuous function, then f possesses a simultaneous Blumberg set when and only when f^{-1} is quasi-continuous.*

PROOF. Theorem 8.12 provides the proof in one direction. For the converse, let D be any simultaneous Blumberg set for f. According to Theorem 8.10, we must show that $E = f[D]$ is a strong Blumberg set for f^{-1}. This will follow immediately from the last lemma because f is one-to-one and onto.

Finally, we remark that it has been shown in [**100**], by means of a construction, that there are quasi-continuous, one-to-one functions f of $(0,1)$ onto itself that possess no simultaneous Blumberg sets.

Part 3

Fourier Series

CHAPTER 9

Improving the Behavior of Fourier Series

In this chapter we discuss the extent to which the behavior of the Fourier series of a continuous function can be improved by a change of variable. We begin in the first section with an introduction to the convergence theory of Fourier series, some related function spaces, some elementary results on the modulus of continuity, and some estimates of the magnitude of the Fourier coefficients. For further details we refer the reader to the classical text of Zygmund [**147**].

In Section 9.2 we discuss the Pál-Bohr theorem which asserts that for every f in $\mathrm{C}(T)$ there is a change of variable, i.e., a self-homeomorphism g of T such that the Fourier series of $f \circ g$ converges uniformly. We eschew the usual complex analysis proof of this result in favor of a real analysis argument of Kahane and Katznelson which shows that one change of variable will suffice for all f in any given compact subset of $\mathrm{C}(T)$. In the chapters dealing with Fourier series, we follow the usual notation in this area and denote a self-homeomorphism by g rather than the h of earlier the chapters. In Section 9.3 we will demonstrate a generalization of the Pál-Bohr theorem due to Jurkat and Waterman. Section 9.4 is devoted to producing absolute convergence by a change of variable and presenting the negative result on absolute convergence due to Olevskii.

9.1. Preliminaries

The group $\mathbb{R}/2\pi\mathbb{Z}$, the **circle group**, will be denoted by T. $\mathrm{C}(T)$ denotes the continuous real-valued functions on T and $\mathrm{L}^1(T)$ the integrable functions on T. The (trigonometric) Fourier coefficients of functions f in $\mathrm{L}^1(T)$ are

$$a_k = a_k(f) = \frac{1}{\pi}\int_T f(t)\,\cos kt\,dt \quad \text{and} \quad b_k = b_k(f) = \frac{1}{\pi}\int_T f(t)\,\sin kt\,dt$$

for $k = 0, 1, 2, \ldots$, and the complex Fourier coefficients are

$$\hat{f}(k) = \frac{1}{2\pi}\int_{-\pi}^{\pi} f(t)\,e^{-ikt}\,dt$$

for $k = 0, \pm 1, \pm 2, \ldots$. The Fourier series of f is

$$S(f) = S(x; f) = \frac{a_0}{2} + \sum_{k=1}^{\infty}(a_k\,\cos kx + b_k\,\sin kx) = \sum_{k=-\infty}^{\infty}\hat{f}(k)\,e^{ikx}$$

and its n-th partial sum is

$$S_n(f) = S_n(x; f) = \frac{a_0}{2} + \sum_{k=1}^{n}(a_k\,\cos kx + b_k\,\sin kx) = \sum_{k=-n}^{n}\hat{f}(k)\,e^{ikx}.$$

For the n-th partial sum we have the integral formula

$$S_n(x;f) = \frac{1}{\pi}\int_T f(t)\Big(\frac{1}{2} + \sum_{\nu=1}^{n} \cos\nu(t-x)\Big)\, dt = \frac{1}{\pi}\int_{-\pi}^{\pi} f(t)\, D_n(t-x)\, dt,$$

where

$$D_n(t) = \frac{1}{2} + \sum_{\nu=1}^{n} \cos\nu t$$

is the **Dirichlet kernel**. The last integral above can be written in convolution notation as

$$f * D_n(x),$$

where

$$f * g(x) = \frac{1}{\pi}\int_T f(t)\, g(t-x)\, dt$$

is the convolution of the two functions f and g.

For the Dirichlet kernel $D_n(t)$, we have the formula

$$D_n(t) = \frac{\sin(n+\frac{1}{2})t}{2\,\sin\frac{1}{2}t}.$$

It is often more convenient to use the fact that

$$S_n(x;f) = \frac{1}{\pi}\int_{-\varepsilon}^{\varepsilon} f(x+t)\,\frac{\sin nt}{t}\, dt + o(1)$$

and

$$S_n(x;f) - f(x) = \frac{1}{\pi}\int_{0}^{\varepsilon} \varphi_x(t)\,\frac{\sin nt}{t}\, dt + o(1)$$

whenever $\varepsilon \in (0,\pi]$, where

$$\varphi_x(t) = f(x+t) + f(x-t) - 2f(x).$$

In the first case, the $o(1)$ term is uniformly in x; in the second case, it is uniformly in x in any set where f is bounded.

For $f \in \mathrm{L}^1(T)$, the Fejér-Lebesgue theorem asserts that $S(f)$ is **$(C,1)$-summable** to f a.e., that is, we have

$$\sigma_n(x;f) = \frac{1}{n+1}\big(S_0(x;f) + \cdots + S_n(x;f)\big) \to f(x) \quad \text{a.e.}$$

Further

$$\sigma_n(x;f) \to f(x)$$

uniformly on closed intervals of points of continuity of f and

$$\sigma_n(x;f) \to \frac{1}{2}\big(f(x+) + f(x-)\big)$$

wherever the right and left limits exist. The $(C,1)$ summability method is **regular**, i.e., for convergent series, the $(C,1)$ sum (mean) is the usual sum.

If $\sum_{n=0}^{\infty} u_n$ is a numerical series, then

$$\sum_{n=0}^{\infty} u_n r^n, \quad 0 \le r < 1$$

is known as the **Abel mean** of the series. The series is **Abel summable** to S if

$$\lim_{r\to 1-}\sum_{n=0}^{\infty} u_n r^n = S.$$

Abel summability is also a regular method and $(C,1)$ summability implies Abel summability to the same sum [**147**, Chapter III].

For functions f of bounded variation, the Dirichlet-Jordan theorem asserts that $S_n(x;f)$ is everywhere convergent to $\frac{1}{2}\big(f(x+)+f(x-)\big)$ and convergence is uniform on closed intervals of points of continuity of f. In view of these results, the class $\mathrm{R}(T)$ of **regulated** functions will be of importance. Regulated functions are those which have right and left limits at each point and satisfy

$$f(x) = \frac{1}{2}\big(f(x+)+f(x-)\big).$$

Let

$$\mathrm{E}_0(T) = \{\, f : S_n(f),\ n=1,2,\ldots, \text{ is uniformly bounded}\,\},$$

and, for each f in $\mathrm{E}_0(T)$, define

$$\|f\|_{\mathrm{U}} = \sup_{n,\,x} |S_n(x;f)|.$$

Clearly, $\mathrm{E}_0(T)$ is a linear space with norm $\|\cdot\|_{\mathrm{U}}$. We define three linear subspaces of $\mathrm{E}_0(T)$,

$$\mathrm{U}(T) = \{\, f : S_n(f),\ n=1,2,\ldots, \text{ converges uniformly}\,\},$$
$$\mathrm{E}(T) = \mathrm{E}_0(T)\cap\{\, f : S_n(f),\ n=1,2,\ldots, \text{ converges everywhere}\,\}$$

and

$$\mathrm{E}_{\mathrm{C}}(T) = \mathrm{E}(T)\cap \mathrm{C}(T).$$

It is well-known that the space $\mathrm{U}(T)$ is a Banach space with norm $\|\cdot\|_{\mathrm{U}}$, and it is clear that $\mathrm{E}_0(T) \supseteqq \mathrm{E}(T) \supseteqq \mathrm{E}_{\mathrm{C}}(T) \supseteqq \mathrm{U}(T)$.

The following result does not appear to be known.

PROPOSITION 9.1. *$\mathrm{E}_0(T)$, $\mathrm{E}(T)$, $\mathrm{E}_{\mathrm{C}}(T)$ and $\mathrm{U}(T)$ are Banach spaces with the same norm, namely $\|\cdot\|_{\mathrm{U}}$, and*

$$\mathrm{E}_0(T) \supsetneqq \mathrm{E}(T) \supsetneqq \mathrm{E}_{\mathrm{C}}(T) \supsetneqq \mathrm{U}(T).$$

PROOF. By the Fejér-Lebesgue theorem, for each f in $\mathrm{E}_0(T)$ and for almost every x, we have

$$|f(x)| = \lim_{n\to\infty} |\sigma_n(x;f)| \le \|S_n(f)\|_\infty \le \|f\|_{\mathrm{U}}.$$

Thus if f_k, $k=1,2,\ldots$, is a $\|\cdot\|_{\mathrm{U}}$-Cauchy sequence of functions in $\mathrm{E}_0(T)$, then it is a $\|\cdot\|_\infty$-Cauchy sequence in $\mathrm{L}^\infty(T)$. So there is an f in $\mathrm{L}^\infty(T)$ such that $\|f-f_k\|_\infty \to 0$ as $k\to\infty$. Further, for any n and any x,

$$|S_n(x;f)| \le |S_n(x;f-f_k)| + |S_n(x;f_k)| \le \frac{1}{\pi}\|f-f_k\|_\infty \int_T |D_n(t)|\,dt + \|f_k\|_{\mathrm{U}}.$$

For each n, we may make the first term on the right as small as we wish by choosing k large and the second term does not exceed the finite number $\sup_{k\in\mathbb{Z}} \|f_k\|_{\mathrm{U}}$. That is, $f\in \mathrm{E}_0(T)$. The completeness of the space $\mathrm{E}_0(T)$ will follow easily.

For f in $\mathrm{E}(T)$, we may redefine f on a set of Lebesgue measure 0 so that

$$f(x) = \lim_{n\to\infty} \sigma_n(x; f) \quad \text{for every} \quad x.$$

Then for f in any of the spaces $\mathrm{E}(T)$, $\mathrm{E_C}(T)$ or $\mathrm{U}(T)$, we have, by the above argument

$$|f(x)| \le \|f\|_{\mathrm{U}} \quad \text{for every} \quad x.$$

By the same argument as above, $\|\cdot\|_{\mathrm{U}}$-convergence implies convergence in the supremum norm $\|\cdot\|_\infty$ now meaning (with some abuse of notation)

$$\|f\|_\infty = \sup\{\, |f(x)| : x \in T\,\}.$$

Thus a Cauchy convergent sequence in the spaces $\mathrm{E_C}(T)$ or $\mathrm{U}(T)$ converges to a continuous function.

If $f_k \in \mathrm{E}(T)$ or $f_k \in \mathrm{E_C}(T)$, $k = 1, 2, \ldots$, are such that $\|f - f_k\|_{\mathrm{U}} \to 0$ as $k \to \infty$, then for every m, n, k and every x

$$\begin{aligned} |S_m(x;f) - S_n(x;f)| &\le \|S_m(f) - S_m(f_k)\|_\infty + |S_m(x;f_k) - S_n(x;f_k)| \\ &\quad + \|S_n(f_k) - S_n(f)\|_\infty \\ &\le 2\,\|f - f_k\|_{\mathrm{U}} + |S_m(x;f_k) - S_n(x;f_k)|. \end{aligned}$$

Note that we can replace the pointwise evaluations in this inequality by $\|\cdot\|_\infty$ evaluations when $f_k \in \mathrm{U}(T)$. By choosing k large enough, the right hand side will be small for all large m and n, implying that the Fourier series of f converges everywhere whenever $f_k \in \mathrm{E}(T)$ or $f_k \in \mathrm{E_C}(T)$ and converges uniformly whenever $f_k \in \mathrm{U}(T)$. The first assertion of the proposition has now been established.

In [**8**, Chapter 1, Section 45] an example, due to Fejér, is given of a function in $\mathrm{E}_0(T)$ whose Fourier series diverges at some point. Also in [**8**, Chapter 1, Section 44] there is an example of a function which is in $\mathrm{E_C}(T)$ but not in $\mathrm{U}(T)$. Every function of bounded variation has an everywhere convergent Fourier series by the Dirichlet-Jordan theorem, and

$$|S_n(x;f)| \le \|f\|_\infty + C\ \mathrm{V}(f,T),$$

which is an easy consequence of Fejér's theorem on $(C, 1)$-summability and the estimate of the order of magnitude of the Fourier coefficients of a function of bounded variation [**147**, volume I, page 90, Theorem 3.7]. Thus a discontinuous function of bounded variation is in $\mathrm{E}(T)$ but not in $\mathrm{E_C}(T)$. The proof of Proposition 9.1 is now complete.

Associated with each real-valued function f on T is the function

$$\omega(\delta; f) = \begin{cases} 0 & \text{for } \delta = 0 \\ \sup\{\, |f(t_1) - f(t_2)| : |t_1 - t_2| < \delta\,\} & \text{for } \delta > 0, \end{cases}$$

called the **modulus of continuity** of f. Clearly $\omega(\delta; f)$ is increasing and tends to zero with δ whenever $f \in \mathrm{C}(T)$. A function f such that $\omega(\delta; f) \le C\,\delta^\alpha$ for all δ is said to be in the **Lipschitz class** Λ_α.

PROPOSITION 9.2. (i) *$\omega(\,\cdot\,; f)$ is continuous whenever $f \in \mathrm{C}(T)$,*
(ii) *If K is a compact subset of $\mathrm{C}(T)$ and*

$$\omega_K(\delta) = \sup\{\, \omega(\delta; f) : f \in K\,\} \quad \textit{for} \quad 0 \le \delta,$$

then $\omega_K(\delta)$ is continuous, increasing, and tends to zero with δ.

PROOF. Since (i) is a special case of (ii), we prove (ii). The following are clear

(a) If $\beta > \alpha$, then $\omega(\beta; f) \le \omega(\alpha; f) + \omega(\beta - \alpha; f)$ for any f in K,
(b) $\omega(\eta; f) \to 0$ as $\eta \to 0$ for any f in K,
(c) $\omega_K(\delta)$ is increasing.

Given $\varepsilon > 0$ there is a finite collection of functions f_i, $i = 1, 2, \ldots, n$, in K so that the union of ε-neighborhoods of the f_i covers K. For $\delta > 0$, we have

$$\omega_K(\delta) \le \max_i \omega(\delta; f_i) + 2\,\varepsilon,$$

which, with (b) above, implies

$$\omega_K(\delta) \to 0 \quad \text{as} \quad \delta \to 0.$$

If $\delta' > \delta > 0$, then

$$\max_i \omega(\delta; f_i) - 2\,\varepsilon \le \omega_K(\delta) \le \omega_K(\delta') \le \max_i \omega(\delta'; f_i) + 2\,\varepsilon,$$

which, with (a) and (c) above, implies

$$0 \le \omega_K(\delta') - \omega_K(\delta) \le \max_i \big(\omega(\delta'; f_i) - \omega(\delta; f_i)\big) + 4\,\varepsilon \le \max_i \omega(\delta' - \delta; f_i) + 4\,\varepsilon,$$

and so

$$\omega_K(\delta') - \omega_K(\delta) \to 0 \quad \text{as} \quad \delta' - \delta \to 0,$$

establishing the continuity of ω_K and completing the proof of the proposition.

We now make some elementary estimates of the magnitude of the Fourier coefficients. The **integral modulus of continuity** (in $\mathrm{L}^p(T)$) of f is

$$\omega_p(\delta; f) = \sup_{0 \le h \le \delta} \frac{1}{2\,\pi} \left\{ \int_T |f(x+h) - f(x)|^p \, dx \right\}^{\frac{1}{p}}.$$

It is easily seen that the Fourier coefficients $\hat{f}(\nu)$ of f satisfy the inequalities

$$|\hat{f}(\nu)| \le \frac{1}{2}\,\omega(\pi/\nu; f) \quad \text{and} \quad |\hat{f}(\nu)| \le \frac{1}{2}\,\omega_1(\pi/\nu; f) \quad \text{when} \quad \nu \ne 0,$$

for

$$\begin{aligned} \hat{f}(\nu) &= \frac{1}{2\,\pi} \int_T f(x)\, e^{-i\nu x} dx \\ &= -\frac{1}{2\,\pi} \int_T f(x + \pi/\nu)\, e^{-i\nu x}\, dx = \frac{1}{4\,\pi} \int_T \big(f(x) - f(x+\pi/\nu)\big)\, e^{-i\nu x}\, dx \end{aligned}$$

or

$$|\hat{f}(\nu)| \le \frac{1}{2} \cdot \frac{1}{2\,\pi} \int_T |f(x) - f(x + \pi/\nu)|\, dx.$$

For functions f of bounded variation we have

$$|\hat{f}(\nu)| \le \frac{\mathrm{V}(f,T)}{2\,\pi\,|\nu|} \quad \text{when} \quad \nu \ne 0,$$

since

$$\left| \frac{1}{2\,\pi} \int_T f(t)\, e^{-i\nu t}\, dt \right| = \left| \frac{1}{2\,\pi\,\nu} \int_T e^{-i\nu t}\, df(t) \right| \le \frac{\mathrm{V}(f,T)}{2\,\pi\,|\nu|}.$$

LEMMA 9.3. *Let T be partitioned into k intervals of length no less than a positive number η and let $0 < M < \infty$. Suppose that E is the set of functions f in $\mathrm{C}(T)$ which are linear on each interval of the partition and for which $\|f\|_\infty \le M$. Then for any positive ε there is an N depending on η, k, M, and ε such that*

$$\|S_n(f) - f\|_\infty < \varepsilon \quad \textit{whenever} \quad n \ge N \quad \textit{and} \quad f \in E.$$

PROOF. Let $x \in T$ and let $S_n(x) = S_n(x; f)$. The **modified** n-th partial sum of f is

$$S_n^*(x) = S_n^*(x; f) = \frac{1}{2}\big(S_n(x) + S_{n-1}(x)\big) = \frac{1}{\pi}\int_T f(x+t) D_n^*(t)\, dt,$$

where

$$D_n^*(t) = \frac{\sin nt}{2\tan\frac{1}{2}t}.$$

Note that

$$\mathrm{V}(f,T) \le 2\,k\,\|f\|_\infty$$

and

$$|S_n(x) - S_n^*(x)| \le \frac{1}{2}\,|\hat{f}(n) + \hat{f}(-n)| \le \frac{\mathrm{V}(f,T)}{2\,\pi n} \le \frac{k\,\|f\|_\infty}{\pi n}.$$

Now

$$\begin{aligned} &S_n^*(x) - f(x) \\ &\quad = \frac{1}{\pi}\int_0^\pi \big(f(x+t) - f(x)\big) D_n^*(t)\, dt + \frac{1}{\pi}\int_0^\pi \big(f(x-t) - f(x)\big) D_n^*(t)\, dt. \end{aligned}$$

We shall consider only the first integral since the second can be estimated in an analogous manner. We can write the function $f(x+t) - f(x)$ in the variable t as the difference of two nondecreasing functions on $[0,\pi]$, its positive and negative variations. Extend these functions to T by setting them equal to zero on $(-\pi, 0)$. We see then that it suffices to estimate this integral under the assumption that $f(x+t) - f(x)$ is nondecreasing on $[0,\pi]$, continuous on $(-\pi,\pi)$ and with a jump at π of magnitude no more than the positive variation of f on T. Then

$$\frac{1}{\pi}\int_0^\pi \big(f(x+t) - f(x)\big) D_n^*(t)\, dt = \frac{1}{\pi}\int_0^\delta (\cdots)\, dt + \frac{1}{\pi}\int_\delta^\pi (\cdots)\, dt = \mathrm{I} + \mathrm{II}$$

for any δ in $(0,\pi)$. By the second mean-value theorem, for some δ_1 in $(0,\delta)$, we have

$$|\,\mathrm{I}\,| \le \frac{1}{\pi}\,|f(x+\delta) - f(x)|\left|\int_{\delta_1}^\delta D_n^*(t)\, dt\right| \le |f(x+\delta) - f(x)| \le \frac{2\,\|f\|_\infty\,\delta}{\eta}.$$

Now II is the n-th Fourier sine coefficient of the function

$$h(t) = \frac{1}{2}\big(f(x+t) - f(x)\big)\,\chi_{(\delta,\pi)}(t)\cot\frac{t}{2}$$

and so

$$|\,\mathrm{II}\,| \le \frac{\mathrm{V}(h,T)}{\pi n} \le \frac{1}{\pi n}\Big(\mathrm{V}(f,T)\cot\frac{\delta}{2} + \|f\|_\infty \cot\frac{\delta}{2}\Big).$$

Choose δ so that

$$|\,\mathrm{I}\,| \le \frac{\varepsilon}{2},$$

a choice which depends on η and $\|f\|_\infty$. Then we may choose N so large that

$$|S_n(x) - S_n^*(x)| + |\,\mathrm{II}\,| \leq \frac{\varepsilon}{2} \quad \text{whenever} \quad n > N,$$

a choice depending on k, $\|f\|_\infty$, and δ, establishing the inequality

$$|S_n(x; f) - f(x)| < \varepsilon \quad \text{whenever} \quad x \in T.$$

9.2. Uniform convergence

The theorem of Fejér-Lebesgue tells us that if f is in $\mathrm{C}(T)$ then $S(f)$ is uniformly $(C,1)$-summable. However $S(f)$ need converge only a.e. It is natural to ask if for any f in $\mathrm{C}(T)$ there is a change of variable g such that $f \circ g \in \mathrm{U}(T)$. This question was answered affirmatively by Pál [**108**] and Bohr [**14**]. Pál showed that uniform convergence of $S(f \circ g)$ could be produced in an arbitrarily large proper subinterval of T and Bohr extended the result to T. The argument was later refined by Salem (see [**119**] and [**147**, volume I, page 294]).

The classical proofs of this result rely on complex variable methods. The salient features of the proof include an application of the Riemann mapping theorem, an extension theorem of Carathéodory and a theorem of Fejér which asserts that if $f(z) = \sum_{n=0}^{\infty} a_n z^n$ is a conformal mapping of $|z| < 1$ onto a simply connected bounded open set, then the series converges uniformly in $|z| \leq 1$. It is remarkable that a theorem which appears to fall in the provenance of real analysis should appear to depend on complex analysis for its proof. The need for a real analysis proof was pointed out in a survey paper of Goffman and Waterman [**68**] and the argument was supplied by Saakyan [**113, 114**] and by Kahane and Katznelson [**84, 85**] who also affirmatively answered another open question at the same time, namely, given two real functions, f_1 and f_2 in $\mathrm{C}(T)$, is there a change of variable g such that $f_1 \circ g$ and $f_2 \circ g$ are in $\mathrm{U}(T)$?

We proceed then to the statement and proof of this theorem, following the method of Kahane and Katznelson.

THEOREM 9.4 (KAHANE-KATZNELSON). *If K is a compact subset of* $\mathrm{C}(T)$, *then there exists a change of variable g such that $f \circ g \in \mathrm{U}(T)$ for every f in K.*

PROOF. For a compact subset K of $\mathrm{C}(T)$ let ω_K be its corresponding modulus of continuity. Then $M = \max\{\, \|f\|_\infty : f \in K\}$ is finite and the modulus of continuity of each f in K satisfies $\omega(\delta; f) \leq \omega_K(\delta)$ (see page 114).

The desired homeomorphism g is constructed in stages. The first stage is constructed by dividing the $I^0 = T = [-\pi, \pi]$ into two collections of intervals

$$I_1^1,\ I_2^1,\ I_3^1,\ I_4^1 \quad \text{and} \quad J_1^1,\ J_2^1,\ J_3^1,\ J_4^1$$

by choosing four points

$$t_1^1 = -\pi,\ t_2^1 = -\pi + \varepsilon_1,\ t_3^1 = 0 \quad \text{and} \quad t_4^1 = \varepsilon_1,$$

to form the intervals I_j^1, $j = 1, 2, 3, 4$, and by choosing four points

$$x_1^1 = -\pi,\ x_2^1 = -\eta_1,\ x_3^1 = 0 \quad \text{and} \quad x_4^1 = \pi - \eta_1.$$

to form J_j^1, $j = 1, 2, 3, 4$. Then define

$$g(x_j^1) = t_j^1 \quad \text{for each} \quad j,$$

an order-preserving map.

At the n-th stage we will have 4^n pairs of intervals

$$I_j^n \quad \text{and} \quad J_j^n, \quad j = 1, 2, \ldots, 4^n.$$

The order-preserving map g will satisfy $g(x) = t$ whenever x and t are the corresponding end points of J_j^n and I_j^n for each j.

To define the $(n+1)$-th stage, we divide each I_j^n and J_j^n into four subintervals in a manner analogous to the first stage by using the midpoints of I_j^n and J_j^n and numbers ε_n and η_n, thereby yielding the new interval I_j^{n+1} and J_j^{n+1}. The order-preserving map g is defined on the end points of these new intervals in the obvious manner. In this way, g is an order-preserving map defined on a dense subset of T onto another dense subset of T. Clearly h can be extended to a unique homeomorphism of T.

Using these intervals, we shall define a sequence F_n, $n = 0, 1, 2, \ldots$, of continuous, piecewise linear functions on T. The function F_n is to be linear on the intervals J_j^n with values at the end points being that of f at the corresponding end points of I_j^1. The values of ε_n and η_n will be determined by the compact set K. The construction will also define an integer m_n that depends only on the compact set K. For convenience, we shall let

$$\varepsilon_0 = \eta_0 = 2\,\pi.$$

For positive integers n, the numbers ε_n, η_n and m_n are to satisfy

$$\varepsilon_n \le 4^{-2n}\,\varepsilon_{n-1} \quad \text{and} \quad \omega_K(\varepsilon_n) \le 4^{-2n},$$

$$\eta_n \le 4^{-2n}\,\eta_{n-1} \quad \text{and} \quad 4^n\,m_n\,\omega_K(\pi\,2^{-(n+1)})\,\eta_n < 2^{-n}.$$

The choice of η_n depends on the prior choice of m_n.

We can easily choose ε_1 to satisfy these conditions. Defined the function F_0 to be the constant function $f(0)$. Then

$$\|S_m(F_0) - F_0\|_\infty = 0 \quad \text{for every} \quad m.$$

We select m_1 to be 0. Clearly we can select η_1 to meet the above conditions. Corresponding to each function f in K is a continuous, piecewise linear function F_1. With E as the collection of functions in Lemma 9.3 provided by the partition x_j^1, $j = 1, 2, 3, 4$, and the positive number M determined at the outset of the proof, there is a positive integer m_2 with $m_2 > m_1$ such that

$$\|S_m(G) - G\|_\infty < \frac{1}{2^1} \quad \text{whenever} \quad m \ge m_2 \quad \text{and} \quad G \in E.$$

The manner in which ε_2 and η_2 are chosen is obvious. The general case n is straightforward. The integer m_n satisfies the condition

$$\|S_m(G) - G\|_\infty < \frac{1}{2^{n-1}} \quad \text{whenever} \quad m \ge m_n \quad \text{and} \quad G \in E, \tag{9.1}$$

where E is the collection of functions in Lemma 9.3 provided by the partition x_j^n, $j = 1, 2, \ldots, 4^n$, and the positive number M.

Define

$$G_n = F_{n+1} - F_n \quad \text{for} \quad n = 0, 1, 2, \ldots.$$

Then

$$F - F_0 = \sum_{n=0}^{\infty} G_n = \sum_{n=0}^{\infty} (G_n^1 + G_n^2),$$

where

$$G_n = G_n^1 + G_n^2$$

is the decomposition of G_n defined as follows: G_n^1 equals G_n on the intervals J_j^n of length larger than η_{n-1} and equals 0 on the remaining intervals J_j^n. If $|J_j^n| > \eta_{n-1}$, then $|h[J_j^n]| = |I_j^n| = \varepsilon_{n-1}$ and thus we have

$$\max_{x \in J_j^n} |G_n(x)| < \omega(\varepsilon_{n-1}) \leq 4^{2-2n} \quad \text{and} \quad \|G_n^1\|_\infty < 4^{2-2n}.$$

From this we can conclude that the variation of G_n^1 is less than

$$6 \cdot 4^{2-2n} \cdot 2 \cdot 4^{n-1} = 3 \cdot 4^{2-n};$$

implying that $\sum_{n=0}^\infty G_n^1$ is of bounded variation and, since it is continuous, by the Dirichlet-Jordan theorem, it is in $\mathrm{U}(T)$. From

$$F = F_0 + \sum_{\nu=0}^\infty G_\nu^1 + \sum_{\nu=0}^n G_\nu^2 + \sum_{\nu=n+1}^\infty G_\nu^2,$$

it is enough to show that the $\|\cdot\|_{\mathrm{U}}$-norm of the remainder term on the right-hand side of the above identity tends to 0 as n tends to ∞.

For a fixed n and all m, let us turn to estimates of $\|S_m(G_k^2)\|_\infty$ with $k \geq n$. The function G_k^2 may be thought of as the sum of $2 \cdot 4^{k-1}$ functions γ_j^k with support in the interval J_j^k of length η_{k-1} and with distance between the supports of two such functions at least $\frac{\eta_{k-2}}{2} - \eta_{k-1}$. Clearly we have

$$\int_T |\gamma_j^k| \, dt \leq \eta_{k-1} \omega_K(\pi \, 2^{-k}) \tag{9.2}$$

and the variation of a γ_j^k does not exceed $6 \, \omega_K(\pi \, 2^{-k})$. Then, at any t in T, we have that the m-th partial sum of the Fourier series of a γ_j^k whose support is at a distance from t exceeding $\frac{1}{4} \, \eta_{k-2}$ will satisfy

$$|S_m(t; \gamma_j^k)| = \left| \frac{1}{\pi} \int_{-\pi}^{\pi} \gamma_j^k(u) \, \frac{\sin(m+\frac{1}{2})(u-t)}{\sin \frac{1}{2}(u-t)} \, du \right| \leq \frac{4\eta_{k-1}}{\eta_{k-2}} \, \omega_K(\pi \, 2^{-k}).$$

There is only one other γ_j^k and the partial sums of its Fourier series are bounded by twice the variation of γ_j^k. Thus

$$\|G_k^2\|_{\mathrm{U}} \leq 4^{k+1} \, \frac{\eta_{k-1}}{\eta_{k-2}} \, \omega_K(\pi \, 2^{-k}) + 12 \, \omega_K(\pi \, 2^{-k}) = o(1) \quad \text{as} \quad k \to \infty. \tag{9.3}$$

From (9.1) we have

$$\|S_m(G_{k-2}^2) - G_{k-2}^2\|_\infty \leq \frac{1}{2^{k-2}} \quad \text{whenever} \quad m \geq m_{k-1}. \tag{9.4}$$

We infer from inequality (9.2) that the Fourier coefficients of G_k^2 do not exceed

$$\pi^{-1} \, 4^{k-1} \, \eta_{k-1} \, \omega_K(\pi \, 2^{-k});$$

so, for $m < m_{k-1}$,

$$\|S_m(G_k^2)\|_\infty \leq \pi^{-1} \, (2 \, m_{k-1} + 1) \, 4^{k-1} \, \eta_{k-1} \, \omega_K(\pi \, 2^{-k}) < 2^{-(k-1)}. \tag{9.5}$$

Now we shall estimate $\| \sum_{\nu=n+1}^{\infty} G_\nu^2 \|_{\mathrm{U}}$ for $n > 3$. Suppose that $m \geq m_n$. Let k be such that $m_{k-1} \leq m < m_k$. Clearly, $n \leq k-1$. We have

$$\begin{aligned}\Big\| \sum_{\nu=n+1}^{\infty} S_m(G_\nu^2) \Big\|_\infty &\leq \Big\| \sum_{\nu=n+1}^{k-2} S_m(G_\nu^2) \Big\|_\infty \\ &\qquad + \big\| S_m(G_{k-1}^2) \big\|_\infty + \sum_{\nu=k}^{\infty} \big\| S_m(G_\nu^2) \big\|_\infty \\ &= \mathrm{I}_n + \mathrm{II}_n + \mathrm{III}_n.\end{aligned}$$

We have

$$\begin{aligned}\mathrm{I}_n &\leq \Big\| \sum_{\nu=n+1}^{k-2} G_\nu^2 \Big\|_\infty + \sum_{\nu=n+1}^{k-2} 2^{-(\nu-1)} && \text{by (9.1)},\\ \mathrm{II}_n &\leq \| G_{k-1}^2 \|_{\mathrm{U}} \leq \max_{\nu \geq n} \| G_\nu^2 \|_{\mathrm{U}} && \text{by (9.3)},\end{aligned}$$

and

$$\mathrm{III}_n \leq 2^{-k} \leq 2^{-(n+1)} \qquad \text{by (9.5)}.$$

So, for $m \geq m_n$, we have

$$\begin{aligned}&\Big\| S_m\Big(\sum_{\nu=n+1}^{\infty} G_\nu^2 \Big) \Big\|_\infty \\ &\qquad \leq \max_{\nu \geq n} \| G_\nu^2 \|_{\mathrm{U}} + \frac{1}{2^{n-1}} + \max_{\nu > n} \big\| F_\nu - F_n \big\|_\infty + \sum_{\nu=n+1}^{\infty} \| G_\nu^1 \|_\infty,\end{aligned}$$

which tends to 0 as n tends to ∞. It remains to consider $m < m_n$. For such m we have by (9.5) that

$$\Big\| S_m\Big(\sum_{\nu=n+1}^{\infty} G_\nu^2 \Big) \Big\|_\infty \leq \sum_{\nu=n+1}^{\infty} \| S_m(G_\nu^2) \|_\infty \leq \sum_{\nu=n+1}^{\infty} 2^{-(\nu-1)}.$$

Thereby we have shown that the $\| \cdot \|_{\mathrm{U}}$-norm of the remainder tends to 0 as n tends to ∞. The completeness of the Banach space $\mathrm{U}(T)$ finishes the proof of the theorem.

9.3. Conjugate functions and the Pál-Bohr theorem

A result of Jurkat and Waterman [**81, 82**], which we present in this section, shows that if f is in $\mathrm{C}(T)$, then there is a change of variable g such that substantially more than the uniform convergence of $S(f \circ g)$ is accomplished. To understand this result, some information concerning the notions of **conjugate series** and **conjugate functions** will be required.

For an integrable function f, the **conjugate series** to its Fourier series $S(f)$ is

$$\widetilde{S}(f) = \widetilde{S}(x; f) = \sum_{k=1}^{\infty} (a_k \sin kx - \cos kx) = -i \sum_{-\infty}^{\infty} \operatorname{sign}(n) \hat{f}(n)\, e^{int}.$$

If $\widetilde{S}(f)$ is a Fourier series of a function g (i.e., $\widetilde{S}(f) = S(g)$), then $f + ig$ *cannot have a simple discontinuity at any point.* Thus if f and g can have only simple discontinuities (e.g., if they are of bounded variation), *then they are continuous* [**147**, volume I, page 89].

The conjugate function $\tilde{f}$ of f is defined by

$$\tilde{f}(x) = -\frac{1}{\pi}\,\text{P.V.}\int_{-\pi}^{\pi}\frac{f(x+t)}{2\tan\frac{1}{2}t}\,dt = -\frac{1}{\pi}\lim_{\varepsilon\to 0+}\int_{\epsilon}^{\pi}\frac{f(x+t)-f(x-t)}{2\tan\frac{1}{2}t}\,dt.$$

The proof of the existence of this integral is by no means simple, but it exists a.e. if $f \in \mathrm{L}^1(T)$ [**147**, volume I, page 131], and $\tilde{f} \in \mathrm{L}^1(T)$ if $f \in \mathrm{L}^p(T)$, $p > 1$, or even if $|f|\log^+|f| \in \mathrm{L}^1(T)$ [**147**, volume I, page 254]. If $\tilde{f} \in \mathrm{L}^1(T)$, then $S(\tilde{f}) = \widetilde{S}(f)$.

If $f(r,t)$ and $\tilde{f}(r,t)$ denote the Abel means of $S(f)$ and $\widetilde{S}(f)$, then

$$f(r,t) + i\,\tilde{f}(r,t)$$

is an analytic function.

We now return to complex variables methods to obtain the following improvement of the Pál-Bohr theorem.

THEOREM 9.5 (JURKAT-WATERMAN). *If $f \in \mathrm{C}(T)$, then there is a change of variable g so that the conjugate of $f \circ g$ is continuous and of bounded variation.*

An immediate consequence of this is the following corollary, which will imply the Pál-Bohr theorem.

COROLLARY 9.6. *If $f \in \mathrm{C}(T)$, then there is a change of variable g so that the Fourier series*

$$S(f\circ g) = \frac{a_0}{2} + \sum_{k=1}^{\infty}(a_k\cos kx + b_k\sin kx),$$

satisfies, as $n \to \infty$,

(i) $$n\,(|a_n| + |b_n|) = O(1);$$

(ii) $$\frac{1}{n}\sum_{k=1}^{n} k\,(a_k^2 + b_k^2)^{\frac{1}{2}} = o(1).$$

Indeed, given the theorem, applying the usual estimate of the Fourier coefficients of a function of bounded variation to the conjugate of $f \circ g$ yields (i). The estimate (ii) is Wiener's necessary and sufficient condition for a function of bounded variation to be continuous applied to the conjugate of $f \circ g$ [**8**, volume I, page 212], [**147**, volume I, page108].

To see that the corollary yields the Pál-Bohr theorem, observe that the continuity of $f \circ g$ implies that $\sigma_n(x; f \circ g)$, the $(C,1)$ means of $S(f \circ g)$, converge uniformly to $f \circ g$. For numerical series $\sum_{n=0}^{\infty} u_n$, the $(C,1)$ means σ_n and the partial sums s_n satisfy

$$s_n - \sigma_n = \sum_{k=0}^{n} u_k - \frac{1}{n+1}\sum_{k=0}^{n}(k+1)\,u_k = \frac{1}{n+1}\sum_{k=1}^{n} k\,u_k.$$

With condition (ii), we see that the uniform $(C,1)$ summability of $S(f \circ g)$ implies uniform convergence of $S(f \circ g)$, which is the Pál-Bohr theorem.

Another proof of this result has been given by Oniani [**104**]. Saakyan [**115**] has shown that the Pál-Bohr theorem has an analogue for multiple trigonometric series.

We come now to the promised stronger consequence of the theorem.

COROLLARY 9.7. *If f is in $\mathrm{C}(T)$ and is of bounded variation, then there is a change of variable g such that $S(f \circ g)$ converges absolutely.*

According to the theorem, there is a change of variable g such that $\widetilde{f \circ g}$ is of bounded variation. Since $f \circ g$ is also of bounded variation, the F. and M. Riesz theorem yields that $f \circ g$ and $\widetilde{f \circ g}$ are absolutely continuous. It follows from a theorem of Hardy that $S(f \circ g)$ converges absolutely (see [**147**, volume I, pages 285–288] and [**86**, pages 88–91]).

REMARK 9.8. From the fact that the conjugate of $h = f \circ g$ is continuous and of bounded variation, one can deduce that h and $\tilde{h}$ belong to the Lipschitz classes $\lambda^p_{1/p}$ for $1 < p < \infty$ and also to Λ^1_*. See [**147**, volume I, page 45] for the definition of these classes.

The proof of our theorem is based on the following result which is an analogous statement for f in $\mathrm{C}(\mathbb{R})$ with compact support.

LEMMA 9.9. *If f is a continuous, real-valued function on $\mathbb{R}$ with bounded support, then there is a strictly increasing continuous function g, mapping $\mathbb{R}$ onto $\mathbb{R}$, such that the Hilbert transform of $f \circ g$*

$$H[f \circ g](x) = \frac{1}{\pi}\,\mathrm{P.V.}\int_{\mathbb{R}} \frac{f \circ g(t)}{x - t}\,dt, \qquad x \in \mathbb{R}, \tag{H}$$

is continuous and of bounded variation on $\mathbb{R}$.

REMARK 9.10. The Hilbert transform is defined by (H) almost everywhere, and the conclusion of the lemma applies to $H[f \circ g](x)$ suitably extended to all of $\mathbb{R}$. If we show that $H[f \circ g](x)$, denoted by $H(x)$, is of bounded variation on an interval whose interior contains the support of $f \circ g$ then that H is of bounded variation on $\mathbb{R}$ follows from a trivial estimate of $H'(x)$ for x outside that interval.

PROOF OF THE LEMMA. It clearly suffices to consider the case where the support of f is contained in the interior of a bounded interval $[0, a]$. Let $M = 1 + \sup f$. Consider the simple closed curve γ in the complex plane composed of the following pieces:

$$\begin{array}{lll}
(i) & x + i\,f(x), & -1 \le x \le a, \\
(ii) & a + i\,y, & 0 \le y \le M, \\
(iii) & x + i\,M, & -1 \le x \le a, \\
(iv) & -1 + i\,y, & 0 \le y \le M.
\end{array}$$

The pieces (ii), (iii) and (iv) can be parametrized by $x \in [a, a+1]$ in a piecewise linear fashion so that γ is given by a continuous function

$$\varphi(x) + i\,\psi(x), \qquad x \in [-1, a+1], \tag{9.6}$$

which extends (i) and, in particular, φ is piecewise linear on the whole interval. We extend (9.6) periodically.

By the Riemann mapping theorem there is a conformal mapping Φ of $|z| < 1$ onto the inside of γ so that the boundary mapping is an orientation-preserving homeomorphism and $\Phi(1) = 0$. Since $f(x) = 0$ in a neighborhood of 0 (and of a as well), γ is linear at 0 and, by the Schwarz reflection principle, Φ is analytic at $z = 1$. Now let

$$z = (w - i)/(w + i), \qquad w = u + i\,v,$$

so that $|z| < 1$ corresponds to the upper half-plane $v = \mathrm{Im}\,(w) > 0$, and $z = 1$ corresponds to $w = \infty$. Let $F(w) = F_1(w) + i\,F_2(w)$ be the composed map.

Then F is analytic at ∞ and $F(\infty) = 0$, implying $F(w) = O(1/w)$ as $w \to \infty$. From the boundary mapping equation

$$\varphi(x) + i\,\psi(x) = F(u), \qquad u \in \mathbb{R},$$

we obtain a strictly increasing, continuous function $x = g(u)$ with $g(-\infty) = 0$ and $g(+\infty) = a + 2$. Clearly $F(w)$ is of class H^p in the upper half-plane for every $p > 1$ and $F(u)$ is almost everywhere the nontangential limit of $F(u + i\,v)$. Thus

$$-H[\psi \circ g](u) = -H[F_2](u) = F_1(u) = \varphi \circ g(u)$$

which is a continuous function of bounded variation. Choose an interval $[\alpha, \beta]$ so that $g[(\alpha, \beta)]$ contains the support of f and $\beta < b = g^{-1}(a)$. Then, for $x < b$,

$$H[\psi \circ g](x) = H\big[(f \circ g)\,\chi_{(\alpha,\beta)}\big](x) + \frac{1}{\pi}\int_b^\infty \frac{F_2(u)}{x-u}\,du. \tag{9.7}$$

Let J be a closed interval such that $[\alpha, \beta] \subset J^o$ but $b \notin J$. Clearly the second term on the right in (9.7) is continuously differentiable on J and, therefore, the first term on the right in (9.7) must be continuous and of bounded variation on J. If we now redefine g outside of $[\alpha, \beta]$ so that g is a strictly increasing function mapping $\mathbb{R}$ onto $\mathbb{R}$, then

$$H[f \circ g](x) = H\big[(F \circ g)\,\chi_{(\alpha,\beta)}\big](x)$$

and is a continuous function of bounded variation on J. In light of our previous remark, this completes the proof of the lemma.

PROOF OF THE THEOREM. Since adding a constant to f will not alter the problem, we may assume $f(-\pi) = 0 = f(\pi)$. Let f_1 be the real-valued function on $\mathbb{R}$ equal to f on $[-\pi, \pi]$ and 0 elsewhere. We use the change of variable g_1 as in the lemma and let $[a, b] = g_1\big[[-\pi, \pi]\big]$. Let u be the linear function with $u(a) = -\pi$ and $u(b) = \pi$, and denote $g_1 \circ u$ by g. Then both $H[f_1 \circ g_1]$ and

$$\frac{1}{\pi}\,\mathrm{P.V.}\int_{-\pi}^{\pi} \frac{f \circ g(t)}{x-t}\,dt$$

are continuous and of bounded variation on $\mathbb{R}$. Thus

$$\frac{1}{\pi}\,\mathrm{P.V.}\int_{-\pi}^{\pi} f \circ g(t)\left[\frac{1}{x-t} + \frac{1}{x+2\pi-t} + \frac{1}{x-2\pi-t}\right] dt \tag{9.8}$$

is continuous and of bounded variation on $[-\pi, \pi]$. However, since (9.8) differs from the conjugate of $f \circ g$, in the form

$$\widetilde{f \circ g}(x) = \frac{1}{2\pi}\,\mathrm{P.V.}\int_{-\pi}^{\pi} \frac{f \circ g(t)}{\tan\frac{1}{2}(x-t)}\,dt,$$

by a continuously differentiable function on $[-\pi, \pi]$, the conjugate $\widetilde{f \circ g}(x)$ also is continuous and of bounded variation on $[-\pi, \pi]$.

9.4. Absolute convergence

A function f whose Fourier series[1] $\sum \hat{f}(n)\, e^{int}$ converges absolutely is equal almost everywhere to the continuous function which is the uniform limit of the partial sums of that series. Therefore we may consider the class $\mathrm{A}(T)$ of such functions to be a subset of $\mathrm{C}(T)$. $\mathrm{A}(T)$ is a Banach space under the norm

$$\|f\|_{\mathrm{A}(T)} = \sum |\hat{f}(n)|,$$

for if f_k, $n = 1, 2, \ldots$, is a Cauchy sequence in this norm then $\hat{f}_k(n) \to c_n$ as $k \to \infty$ uniformly in n and $\sum |c_n| < \infty$. If $f(t) = \lim_{N\to\infty} \sum_{n=-N}^{N} c_n\, e^{int}$, then it is clear that $f \in \mathrm{A}(T)$ and $\|f_n - f\|_{\mathrm{A}(T)} \to 0$.

Corollary 9.7 to the Jurkat-Waterman theorem states that if $f \in \mathrm{BV} \cap \mathrm{C}(T)$, then there is a change of variable g such that $f \circ g \in \mathrm{A}(T)$. In contrast to this corollary, it has been shown by Olevskii [**102, 103**] that without the assumption of properties additional to continuity, there may be no such change of variable. At about the same time, Kahane and Katznelson [**83**] obtained a weaker result, namely that, in general, there is no single change of variable which will bring two functions in $\mathrm{C}(T)$ into $\mathrm{A}(T)$.

We state and prove the Olevskii theorem.

THEOREM 9.11 (OLEVSKII). *There exists an f in $\mathrm{C}(T)$ such that $f \circ g \notin \mathrm{A}(T)$ for every change of variable g.*

The proof of this result depends on three lemmas which lead to the observation that if a function vanishes for $t < 0$ and, for $t > 0$, consists of high frequency pulsations with unit amplitude, then the $\mathrm{A}(T)$ norm of this function must be large. The proof hinges on this observation. The argument will use the dual space $\mathrm{A}^*(T)$ of the space $\mathrm{A}(T)$.

The following lemma is due to H. Davenport [**37**] who used it to improve the well-known result of P. S. Cohen [**36**]. It is elementary in nature but highly nontransparent.

LEMMA 9.12. *If $(z_1, z_2, \ldots, z_n)$ is an n-tuple, $n \geq 3$, of complex numbers with $|z_j| \leq 1$ for all j, then*

$$\left| 1 - \frac{2}{n^2} - \frac{\sum_{k<j} z_k \bar{z}_j}{n^3} \right| + \frac{\left|\sum_{k=1}^{n} z_k\right|}{n^{5/2}} < 1.$$

PROOF. Let $\sum_{k<j} z_k \bar{z}_j = P + i\,Q$, where P and Q are the real and imaginary parts. Then

$$0 \leq \left| \sum_{k=1}^{n} z_k \right|^2 \leq n + 2\,P$$

and

$$P^2 + Q^2 \leq \Big(\sum_{k<j} |z_k z_j| \Big)^2 \leq \left| \sum_{j=1}^{n} (j-1)\, |z_j| \right|^2 \leq \Big(\frac{n\,(n-1)}{2} \Big)^2 < \frac{n^4}{4}.$$

[1]It is customary to denote $\sum_{n=-\infty}^{\infty}$ simply by $\sum$ when discussing Fourier series written in exponential form.

Then the left side of the desired inequality is the left side of

$$\begin{aligned}\left\{\left(1-\frac{2}{n^2}-\frac{P}{n^3}\right)^2+\frac{Q^2}{n^6}\right\}^{\frac{1}{2}} &+\frac{1}{n^{5/2}}(n+2P)^{\frac{1}{2}} \\ &<\left\{\left(1-\frac{2}{n^2}\right)^2-2\left(1-\frac{2}{n^2}\right)\frac{P}{n^3}+\frac{1}{4n^2}\right\}^{\frac{1}{2}}+\frac{1}{n^{5/2}}(n+2P)^{\frac{1}{2}} \\ &=A^{\frac{1}{2}}+B^{\frac{1}{2}}.\end{aligned}$$

Estimating A, using the fact that $P \geq -\frac{n}{2}$, we have

$$\begin{aligned}A=1-\frac{15}{4n^2}+\frac{4}{n^4}-\frac{2P}{n^3}+\frac{4P}{n^5} &\leq 1-\frac{15}{4n^2}-\frac{P}{n^3}+\frac{4}{n^4}+\frac{1}{2n^2}-\frac{2}{n^4} \\ &=\left(1-\frac{3}{n^2}-\frac{P}{n^3}\right)+\left(\frac{-1}{4n^2}+\frac{2}{n^4}\right)<1-\frac{3}{n^2}-\frac{P}{n^3}.\end{aligned}$$

Then

$$\begin{aligned}A^{\frac{1}{2}}+B^{\frac{1}{2}} &< \left(1-\frac{3}{2n^2}-\frac{P}{2n^3}\right)+\frac{(n+2P)^{\frac{1}{2}}}{n^{5/2}} \\ &=1-\frac{1}{n^2}\left\{\frac{1}{4}+\left(\frac{(n+2P)^{\frac{1}{2}}}{2n^{1/2}}-1\right)^2\right\}<1\end{aligned}$$

as was to be shown.

The next result is of a combinatorial character.

LEMMA 9.13. *Suppose f is of bounded variation on an interval I with variation $\mathrm{V}(f,I)$, denoted by V. Given a positive number ε, a positive integer n, and points x_i in I, $i=1,2,\ldots,m$, and points*

$$t_1<\tau_1<t_2<\tau_2<\cdots<t_N<\tau_N$$

such that

$$x_i+t_1-\tau_N \in I \quad \text{for all} \quad i,$$

where

$$N>n\,(4m\,\varepsilon^{-1}V+1), \tag{9.9}$$

then there are integers

$$0<\nu_1<\nu_2<\cdots<\nu_n$$

such that for all i

$$\begin{aligned}|f(x_i+t_{\nu_k}-t_{\nu_j})+f(x_i+\tau_{\nu_k}-\tau_{\nu_j}) \\ -f(x_i+t_{\nu_k}-\tau_{\nu_j})-f(x_i+\tau_{\nu_k}-t_{\nu_j})|<\varepsilon\end{aligned} \tag{$*$}$$

whenever $1 \leq k \leq j \leq n$.

PROOF. Let

$$E(\nu,i)=E'(\nu,i)\cup E''(\nu,i)$$

where

$$E'(\nu,i)=\left\{p:\nu<p\leq N,\quad |f(x_i+t_\nu-t_p)-f(x_i+t_\nu-\tau_p)|\geq\frac{\varepsilon}{2}\right\}$$

and

$$E''(\nu,i)=\left\{p:\nu<p\leq N,\quad |f(x_i+\tau_\nu-t_p)-f(x_i+\tau_\nu-\tau_p)|\geq\frac{\varepsilon}{2}\right\}.$$

Since

$$\frac{\varepsilon}{2}\,\operatorname{card} E'(\nu,i) \le \sum_{p>\nu} |f(x_i+t_\nu-t_p)-f(x_i+t_\nu-\tau_p)| \le V$$

and

$$\frac{\varepsilon}{2}\,\operatorname{card} E''(\nu,i) \le \sum_{p>\nu} |f(x_i+\tau_\nu-t_p)-f(x_i+\tau_\nu-\tau_p)| \le V,$$

we see that

$$\operatorname{card} E(\nu,i) \le 4\,\varepsilon^{-1}\,V.$$

Let us set

$$\nu_1=1,\quad E_0=\{\,1,2,\ldots,N\,\},\quad E_1=E_0\setminus\big(\textstyle\bigcup_i E(\nu_1,i)\cup\{\nu_1\}\big).$$

Then

$$\operatorname{card} E_1 > N-(4\,m\,\varepsilon^{-1}\,V+1).$$

Suppose now that $\nu_1,\ldots,\nu_{k-1}$ and $E_0\supset E_1\supset\cdots\supset E_{k-1}$ are defined and

$$\operatorname{card} E_{k-1} > N-(k-1)(4\,m\,\varepsilon^{-1}\,V+1).$$

Let $\nu_k=\min E_{k-1}$ and set

$$E_k=E_{k-1}\setminus\big(\textstyle\bigcup_i E(\nu_k,i)\cup\{\nu_k\}\big).$$

Clearly

$$\operatorname{card} E_k > N-k\,(4\,m\,\varepsilon^{-1}\,V+1).$$

We may continue the inductive construction until $k=n$ because of the inequality (9.9). We will then have an increasing sequence of integers ν_k, $k=1,2,\ldots,n$, such that

$$\nu_j\notin\textstyle\bigcup_i E(\nu_k,i)\quad\text{whenever}\quad j>k,$$

which implies $(*)$.

LEMMA 9.14. *For* $0<\gamma\le\pi$, *let* φ *be a continuous, increasing function on* $[0,\gamma]$ *with* $\varphi(0)=0$ *and* $\varphi(\gamma)=\pi$. *If*

$$f(t)=\begin{cases}0, & \gamma\le t\le\pi,\\ \sin 2N\varphi(t), & 0\le t<\gamma,\\ 0, & -\pi\le t<0,\end{cases}$$

where N *is an integer with* $N>40^4$, *then there is an absolute, positive constant* K *such that, for some* μ *in the space of signed, Radon measures* $\mathrm{M}(T)$,

$$\int_T f\,d\mu > K\,\log^{1/12} N\quad\textit{and}\quad \|\mu\|_{\mathrm{A}^*(T)}\le 1,$$

where

$$\|\mu\|_{\mathrm{A}^*(T)}=\sup_{n\in\mathbb{Z}}|\hat\mu(n)|.$$

If I *is an interval containing the support of* μ *and if* $h\in\mathrm{C}(T)$ *with* $\|\chi_I\,h\|_\infty<e^{-N}$, *then*

$$\|\lambda f+h\|_{\mathrm{A}(T)}\ge K\,|\lambda|\log^{1/12} N-1$$

for each constant λ.

PROOF. To each μ in $\mathrm{M}(T)$ there corresponds a linear functional Φ_μ on $\mathrm{A}(T)$ given by

$$\Phi_\mu(G) = \int_T G\,d\mu = \int_T \sum \hat{G}(n)\,e^{in\theta}\,d\mu(\theta) = 2\pi \sum \hat{G}(n)\,\hat{\mu}(-n), \quad G \in \mathrm{A}(T)$$

and

$$\|\Phi_\mu\|_{\mathrm{A}^*(T)} = \sup_{n\in\mathbb{Z}} |\hat{\mu}(n)|.$$

To establish this lemma we will construct a μ for which

$$\Phi_\mu(f) > K \log^{1/12} N \quad \text{and} \quad \|\mu\|_{\mathrm{A}^*(T)} \le 1, \tag{9.10}$$

where K is a suitable absolute constant. Let δ_t denote the measure of mass one concentrated at the point t. Our measure μ is a linear combination of such measures. We note that the convolution $\mu * \nu$ of two measures μ and ν, defined for Borel sets E by

$$\mu * \nu(E) = \frac{1}{2\pi} \int_T \mu(E-\tau)\,d\nu(\tau),$$

is linear in each argument, $\delta_t * \delta_{t'} = \delta_{t+t'}$, and $\widehat{\mu * \nu} = \hat{\mu} * \hat{\nu}$.

Let the points of $(0,\gamma)$ where $|f(t)| = 1$ be the points

$$t_1 < \tau_1 < t_2 < \tau_2 < \cdots < t_N < \tau_N.$$

For any positive integer q let

$$n_q \quad \text{be the integer part of} \quad \sqrt{q},$$

and set

$$\varepsilon\langle q\rangle = e^{-2q}, \quad m\langle q\rangle = q^q, \quad \eta\langle q\rangle = (20\,q)^{q^2}.$$

Observe that

$$Q_N = \max\{\, q : \eta\langle q+1\rangle \le 2N \,\}$$

is well defined. Let $\mu_1 = \delta_{t_1}$ and note that $1 \le Q_N < \eta\langle Q_N + 1\rangle \le 2N$. Suppose for q that we have defined

$$\mu_q = \sum_{s=1}^{s_q} \alpha_s^{\langle q\rangle} \delta_{x_s^{\langle q\rangle}}$$

with

$$x_s^{\langle q\rangle} \in (-\tau_{\eta\langle q\rangle}, \tau_{\eta\langle q\rangle}), \quad 1 \le s \le s_q < m\langle q\rangle,$$

and satisfying the inductive hypotheses

$$\|\mu_q\|_{\mathrm{M}(T)} = \sum_{s=1}^{s_q} |\alpha_s^{\langle q\rangle}| < e^q, \tag{9.11}$$

$$\|\Phi_{\mu_q}\|_{\mathrm{A}^*(T)} \le 1, \tag{9.12}$$

and

$$\Phi_{\mu_q}(f) > 10^{-1}\, q^{1/4}. \tag{9.13}$$

Assuming $q \le Q_N$, we proceed inductively as follows. Write

$$t_\nu^{\langle q\rangle} = t_{\eta\langle q\rangle+\nu} \quad \text{and} \quad \tau_\nu^{\langle q\rangle} = \tau_{\eta\langle q\rangle+\nu} \quad \text{for} \quad \nu = 1, \ldots, N_q = \eta\langle q+1\rangle - \eta\langle q\rangle.$$

Then the variation of f on $(-\gamma, \tau_{\eta\langle q\rangle})$, denoted by V_q, is

$$V_q = 4\,\eta\langle q\rangle - 1$$

and

$$x_s^{\langle q\rangle} + t_1^{\langle q\rangle} - \tau_{N_q}^{\langle q\rangle} = x_s^{\langle q\rangle} + t_{\eta\langle q\rangle+1} - \tau_{\eta\langle q+1\rangle} > -\tau_{\eta\langle q\rangle} + \tau_{\eta\langle q\rangle+1} - \tau_{\eta\langle q+1\rangle} > -\tau_{\eta\langle q+1\rangle}.$$

We may apply the previous lemma for we have

$$\begin{aligned} n_q\big(4\,m\langle q\rangle\,\varepsilon\langle q\rangle^{-1}\,V_q + 1\big) &< q^{1/2}\left(16\,\eta\langle q\rangle\,e^{2q}q^q + 1\right) \\ &= 16\,e^{2q}\,20^{q^2+q+1/2} + q^{1/2} < (20\,q)^{(q+1)^2} \\ &< \big(20\,(q+1)\big)^{(q+1)^2} - (20\,q)^{q^2} = N_q. \end{aligned}$$

Thus we can find positive integers $\nu_1 < \cdots < \nu_{n_q}$ such that for each s

$$\begin{aligned} |f(x_s^{\langle q\rangle} + t_{\nu_k}^{\langle q\rangle} - t_{\nu_j}^{\langle q\rangle}) &- f(x_s^{\langle q\rangle} + t_{\nu_k}^{\langle q\rangle} - t_{\nu_j}^{\langle q\rangle}) \\ &- f(x_s^{\langle q\rangle} + \tau_{\nu_k}^{\langle q\rangle} - t_{\nu_j}^{\langle q\rangle}) + f(x_s^{\langle q\rangle} + \tau_{\nu_k}^{\langle q\rangle} - \tau_{\nu_j}^{\langle q\rangle})| < \epsilon\langle q\rangle \end{aligned}$$

whenever $1 \le k < j \le n_q$. We set

$$\begin{aligned} \mu_{q+1} = \big(1 - 2\,n_{q+1}^{-2}\big)\,\mu_q &+ n_q^{-5/2}\sum_{k=1}^{n_q}\frac{1}{2}\left(\delta_{t_{\nu_k}^{\langle q\rangle}} - \delta_{\tau_{\nu_k}^{\langle q\rangle}}\right) \\ &- n_q^{-3}\mu_q * \sum_{1\le k<j\le n_q}\frac{1}{2}\left(\delta_{t_{\nu_k}^{\langle q\rangle}} - \delta_{\tau_{\nu_k}^{\langle q\rangle}}\right) * \frac{1}{2}\left(\delta_{-t_{\nu_j}^{\langle q\rangle}} - \delta_{-\tau_{\nu_j}^{\langle q\rangle}}\right). \end{aligned}$$

Thus μ_{q+1} is a linear combination of point masses concentrated at the points

$$x_s^{\langle q\rangle},\ t_{\nu_k}^{\langle q\rangle},\ \tau_{\nu_k}^{\langle q\rangle},\ x_s^{\langle q\rangle} + t_{\nu_k}^{\langle q\rangle} - t_{\nu_j}^{\langle q\rangle},\ x_s^{\langle q\rangle} + t_{\nu_k}^{\langle q\rangle} - \tau_{\nu_j}^{\langle q\rangle},\ x_s^{\langle q\rangle} + \tau_{\nu_k}^{\langle q\rangle} - t_{\nu_j}^{\langle q\rangle},\ x_s^{\langle q\rangle} + \tau_{\nu_k}^{\langle q\rangle} - \tau_{\nu_j}^{\langle q\rangle}$$

which are contained on the interval $\big(-\tau_{\eta\langle q+1\rangle}, \tau_{\eta\langle q+1\rangle}\big)$ and for whose cardinality, s_{q+1}, we have the estimate

$$\begin{aligned} s_{q+1} &\le s_q + 2\,N_q + 2\,s_q\,(N_q - 1)(N_q - 2) \\ &< q^q\,(1 + 2\,q) + 2\,q^{1/2} < (q+1)^{q+1} = m\langle q+1\rangle. \end{aligned}$$

Then utilizing Lemma 9.12 and $|\hat{\mu}_q(n)| \le 1$, which is (9.12), we have

$$\begin{aligned} |\,\hat{\mu}_{q+1}(n)| = \Big|(1 - 2\,n_q^{-2})\hat{\mu}_q(n) &+ n_q^{-5/2}\sum_{k=1}^{n_q}\frac{1}{2}\big(e^{int_{\nu_k}^{\langle q\rangle}} - e^{in\tau_{\nu_k}^{\langle q\rangle}}\big) \\ &- n_q^{-3}\,\hat{\mu}_q(n)\sum_{1\le k<j\le n_q}\frac{1}{2}\big(e^{int_{\nu_k}^{\langle q\rangle}} - e^{in\tau_{\nu_k}^{\langle q\rangle}}\big)\cdot\frac{1}{2}\big(e^{-int_{\nu_j}^{\langle q\rangle}} - e^{-in\tau_{\nu_j}^{\langle q\rangle}}\big)\Big| < 1. \end{aligned}$$

From the definition of μ_q and the inductive hypothesis (9.11), we have

$$\begin{aligned} \|\mu_{q+1}\|_{\mathrm{M}(T)} &\le \big(1 - 2\,n_q^{-2}\big)\,e^q + n_q^{-3/2} + n_q^{-1}\,e^q/2 \\ &< e^q\big(1 + n_q^{-1/2}\big) + n_q^{-3/2} < e^{q+1}. \end{aligned}$$

Checking the remaining inductive hypothesis,

$$
\begin{aligned}
\Phi_{\mu_{q+1}}(f) &= \int_T f\,d\mu_{q+1} \\
&= \left(1 - 2\,n_q^{-2}\right) \int_T f\,d\mu_q + n_q^{-3/2} \\
&\quad - 4^{-1}\,n_q^{-3} \sum_{s=1}^{s_q} \alpha_s^{\langle q\rangle} \sum_{k<j} \Big[f(x_s^{\langle q\rangle} + t_{\nu_k}^{\langle q\rangle} - t_{\nu_j}^{\langle q\rangle}) + f(x_s^{\langle q\rangle} + \tau_{\nu_k}^{\langle q\rangle} - \tau_{\nu_j}^{\langle q\rangle}) \\
&\quad - f(x_s^{\langle q\rangle} + t_{\nu_k}^{\langle q\rangle} - \tau_{\nu_j}^{\langle q\rangle}) - f(x_s^{\langle q\rangle} + \tau_{\nu_k}^{\langle q\rangle} - t_{\nu_j}^{\langle q\rangle}) \Big] \\
&\geq 10^{-1}\left(1 - 2\,n_q^{-2}\right) q^{1/4} + n_q^{-3/2} - \frac{\epsilon\langle q\rangle}{8\,n_q} \sum_{s=1}^{s_q} \alpha_s^{\langle q\rangle} \\
&> 10^{-1}\,q^{1/4} + q^{-3/4} - 8^{-1}\,e^{-q} \\
&> 10^{-1}\,(q+1)^{1/4}.
\end{aligned}
$$

We have shown that the hypotheses (9.11), (9.12), and (9.13) are valid with q replaced by $q+1$ subject to the restriction $\eta\langle q+1\rangle \leq 2N$. Now, as $\eta\langle Q_N+2\rangle > 2N$, it is easily seen that

$$Q_N > K_1\,(\log N)^{1/3} > 0,$$

where K_1 is an absolute constant. Thus (9.10) holds with $\mu = \mu_{Q_N}$ and the first assertion of the lemma is proven.

Now suppose that I is an interval containing the support of μ_{Q_N}. Let h be in $\mathrm{C}(T)$ with $\|\chi_I\,h\|_\infty < e^{-N}$ and let λ be a constant. From (9.11), the variation of μ_{Q_N} is

$$\|\mu_{Q_N}\|_{\mathrm{M}(T)} = \int_T |d\mu_{Q_N}| < e^{Q_N} < e^N.$$

Thus

$$
\begin{aligned}
\|\lambda f + h\|_{\mathrm{A}(T)} &= \sup\left\{ \frac{1}{\|\Phi\|_{\mathrm{A}^*(T)}}\,\Phi(\lambda\,f + h) : \|\Phi\|_{\mathrm{A}^*(T)} \leq 1 \right\} \\
&\geq \left| \int_T (\lambda f + h)\,d\mu_{Q_N} \right| \\
&\geq |\lambda| \left| \int_T f\,d\mu_{Q_N} \right| - \left| \int_T h\,d\mu_{Q_N} \right| \\
&\geq K\,|\lambda| \log^{1/12} N - \|\chi_I\,h\|_\infty \int_T |d\mu_{Q_N}| \\
&\geq K\,|\lambda| \log^{1/12} N - 1,
\end{aligned}
$$

which is the last assertion of Lemma 9.14.

We proceed now to the proof of the theorem.

Proof of Olevskii's Theorem. Let I_k denote the interval $[\pi/(k+1), \pi/k]$ and let ψ_k denote the characteristic function χ_{I_k} of I_k. Consider the continuous function

$$f(x) = \sum_{k=1}^{\infty} (\log^{1/24} N_k)\,\psi_k(x)\,\sin k(k+1)N_k x\,,$$

where N_k, $k = 1, 2, \ldots$, is an increasing sequence of integers such that

$$\log^{1/24} N_k > e^{N_{k-1}} \quad \text{for all} \quad k.$$

Let g be a self-homeomorphism of T. Without loss of generality we may suppose that g preserves order and $g(0) = 0$. We will show that $F = f \circ g \notin \mathrm{A}(T)$. Clearly, there is a positive δ such that $F(t) = 0$ whenever $t \in (-\delta, 0]$. Set

$$t_k = g^{-1}(\pi/k) \quad \text{and} \quad \gamma_k = t_k - t_{k+1}.$$

We shall apply Lemma 9.14 to the summands of F. For a fixed positive integer ν for which $\gamma_\nu < \delta$, define

$$\varphi_\nu(t) = \nu(\nu+1)g(t + t_{\nu+1}) - \nu\pi, \quad t \in [0, \gamma_\nu].$$

Then

$$F_\nu(t) = F(t + t_{\nu+1}) = (\log^{-1/24} N_\nu) F_\nu^{(1)} + F_\nu^{(2)}$$

where

$$F_\nu^{(1)}(t) = \begin{cases} 0 & \text{for } \gamma_\nu \le t \le \pi, \\ (-1)^{\nu N_\nu} \sin N_\nu \varphi_\nu(t) & \text{for } 0 \le t \le \gamma_\nu, \\ 0 & \text{for } -\pi \le t < 0. \end{cases}$$

Let μ_ν be the corresponding Radon measure provided by Lemma 9.14. Clearly

$$\mathrm{support}(\mu_\nu) \subset [-\gamma_\nu, \gamma_\nu].$$

Now suppose $f \circ g \in \mathrm{A}(T)$. Then

$$\|f \circ g\|_{\mathrm{A}(T)} \ge 2\pi \left| \int_T f \circ g \, d\mu \right|$$

for every μ in $\mathrm{M}(T)$ with $\|\mu\|_{\mathrm{A}^*(T)} \le 1$. In particular, this will be true for μ_ν. Observing

$$\|\chi_{[-\gamma_\nu, \gamma_\nu]} F_\nu^{(2)}\|_\infty < e^{-N_\nu},$$

we have from Lemma 9.14 that

$$\|f \circ g\|_{\mathrm{A}(T)} = \|F\|_{\mathrm{A}(T)} \ge \left| \int_T F_\nu \, d\mu_\nu \right| > K \log^{1/24} N_\nu - 1.$$

We have arrived at the contradiction $K \log^{1/24} N_\nu - 1 < \|f \circ g\|_{\mathrm{A}(T)} < +\infty$ for every ν. The theorem is proved.

CHAPTER 10

Preservation of Convergence of Fourier Series

If $\mathcal{F}$ is a class of functions on a domain D, we may say that $\mathcal{F}$ is invariant under change of variable if $f \in \mathcal{F}$ implies $f \circ g \in \mathcal{F}$ for every homeomorphism $g : D \to D$. The principal ways in which these classes arise are:

(i) A class may be defined by an intrinsic property of the functions which is invariant under change of variable, for example a condition on the corresponding interval functions ($f(I) = f(b) - f(a)$, where $I = (a, b)$) which is independent of the length of the intervals, or by invariant properties of the functions' level sets, such as a condition on the Banach indicatrix.

Those that readily come to mind are the functions of bounded variation as well as the class of functions f such that $\varphi\big(\mathrm{N}(f, y)\big) \in \mathrm{L}^p$, where φ is a nonnegative function on $\mathbb{R}^+$ and $\mathrm{N}(f, y)$ is the Banach indicatrix.

(ii) Given a property P, we may consider the class of functions f such that $f \circ g$ has property P for all changes of variable g. Examples of the second type include the classes of functions f in $\mathrm{C}(T)$ whose Fourier series $S(f \circ g)$ converges everywhere for each g, or converges uniformly, or the class of functions whose Fourier coefficients are of a given order of magnitude after any change of variable.

This and the succeeding chapter will consider the problem: *Given a class of functions with a characterization of the second type, find a characterization of the first type.* The properties of convergence and uniform convergence of $S(f \circ g)$ are considered first.

We begin in Section 10.1 with a discussion of a condition of Salem for uniform convergence which inspired this line of inquiry. We prove a somewhat more general result which provides a condition for pointwise convergence as well. Section 10.2 is devoted to establishing the necessary and sufficient condition for the convergence of $S(f \circ g)$ for every change of variable g when f is continuous or regulated. In Section 10.3 we consider the analogous problem for uniform convergence.

A natural next topic would be the consideration of those integrable f which have the property that $S(f \circ g)$ is everywhere convergent for any change of variable g. This topic requires certain measure-theoretic considerations which are also pertinent to the property of preservation of order of magnitude of the Fourier coefficients, so we shall defer this discussion to the next chapter.

10.1. Tests for pointwise and uniform convergence

The Salem test for uniform convergence of $S(f)$ seems quite different from the well-known standard tests. Bary expresses the opinion that it appears "*at first glance to be hardly suitable for application*" [**8**, pages 305–310], and then proceeds to show that it implies the Dini-Lipschitz test and the Dirichlet-Jordan theorem.

Kahane and Zygmund in the introduction of Salem's **Oeuvre Mathématiques** [**120**] refer to it as "curieuse", but then say:

> *Il nous paraît qu'il ya ici une idée digne d'intérêt et qui pourrait donner des résultats intéressants.*

This idea has, in fact, been the starting point of many investigations into the convergence of Fourier series and into various classes of functions of generalized bounded variation.

For a real or complex-valued function f on T and odd integers n, let

$$T_n(x,t) = \frac{f(x+t/n) - f(x+(t+\pi)/n)}{1} + \frac{f(x+(t+2\pi)/n) - f(x+(t+3\pi)/n)}{3} + \cdots + \frac{f(x+(t+(n-1)\pi)/n) - f(x+(t+n\pi)/n)}{n}$$

and let $Q_n(x,t)$ be obtained from $T_n(x,t)$ by substituting $-t$ and $-\pi$ for t and π respectively.

The Salem test [**117**] can then be stated as follows:

THEOREM 10.1 (SALEM). *If $f \in \mathrm{C}(T)$ and if $T_n(x,0)$ and $Q_n(x,0)$ converge uniformly to 0, then the Fourier series of f converges uniformly.*

This result is contained in the following theorem of Waterman [**136**] which is a test for convergence at the Lebesgue points of a function. (See Remark 10.5 below.)

THEOREM 10.2 (WATERMAN). *Let $f \in \mathrm{L}^1(T)$.*

1. *If x is such that*

$$(*) \qquad \frac{1}{h}\int_0^h |f(x+t) + f(x-t) - 2f(x)|\, dt = o(1) \quad \text{as} \quad h \to 0$$

and

$$(**) \qquad \int_\pi^{2\pi} |T_n(x,t) + Q_n(x,t)|\, dt = o(1) \quad \text{as} \quad n \to \infty,$$

then $S(x;f)$ converges to $f(x)$.

2. *If E is a set on which $(*)$ and $(**)$ hold uniformly and on which f is bounded, then $S(x;f)$ converges uniformly to $f(x)$ on E.*

In the statements 1 and 2, condition $(**)$ may be replaced by

$$(***) \qquad \int_\pi^{2\pi} (T_n(x,t) + Q_n(x,t)) \sin t\, dt = o(1) \quad \text{as} \quad n \to \infty.$$

3. *If x is such that $(*)$ holds, then $(***)$ is a necessary and sufficient condition for $S(x;f)$ to converge to $f(x)$.*

We make three remarks before proceeding to the proof.

REMARK 10.3. Condition $(*)$ states that x is a (symmetric-) Lebesgue point of f. It would be of considerable interest to determine if condition $(*)$ can be replaced by the symmetric differentiability of $\int f$ at x, i.e. by

$$(*') \qquad \frac{1}{h}\int_0^h (f(x+t) + f(x-t) - 2f(x))\, dt = o(1) \quad \text{as} \quad h \to 0.$$

It is elementary that either $(*)$ or $(*')$ uniformly on T is equivalent to continuity.

REMARK 10.4. It is easily seen that the odd integer denominators $2k-1$ that appear in the definitions of T_n and Q_n may be replaced by k. To see this, let

$$a_j(t) = f\big(x + \big(t + 2(j-1)\pi\big)/n\big) - f\big(x + \big(t + (2j-1)\pi\big)/n\big)$$

for $n = 2k-1$ and $j = 1, \dots, k$. Then

$$\sum_{j=1}^{k} \frac{a_j(t)}{2j-1} - \frac{1}{2}\sum_{j=1}^{k} \frac{a_j(t)}{j} = \sum_{j=1}^{k} \frac{a_j(t)}{(2j-1)2j},$$

and the L^1-norm of this sum is $O(\omega_1(\pi/n; f))$, where ω_1 is the integral modulus of continuity (see page 115).

REMARK 10.5. We shall show that Theorem 10.2 is, in fact, a generalization of Salem's test. Let us suppose that $f \in C(T)$ holds and the Salem condition is satisfied. For odd integers $n = 2k-1$, by the integral mean value theorem,

$$\int_{\pi}^{2\pi} |\,T_n(x,t)|\,dt =$$
$$\int_{\pi}^{2\pi} \Big|\sum_{j=1}^{k} \frac{f\big(x + \big(t + 2(j-1)\pi\big)/n\big) - f\big(x + \big(t + (2j-1)\pi\big)/n\big)}{2j-1}\Big|\,dt$$
$$= \pi\,|\,T_n(x + \theta_n, 0)|$$

for some θ_n in $(\pi/n, 2\pi/n)$. Salem's condition implies that this is $o(1)$ as $n \to \infty$ uniformly in x, which, with the corresponding result for Q_n, implies that condition (**) of Theorem 10.2 holds uniformly.

PROOF OF THE THEOREM. From the preliminary section of Chapter 9 we have

$$S_n(x; f) - f(x) = \frac{1}{\pi}\int_0^{\pi} \varphi_x(t)\,\frac{\sin nt}{t}\,dt + o(1) \quad \text{as} \quad n \to \infty,$$

for every x and uniformly on sets on which f is bounded, where $S_n(x; f)$ is the n-th partial sum of the Fourier series and

$$\varphi_x(t) = f(x+t) + f(x-t) - 2f(x).$$

Integrating by parts, we have

$$\int_0^{\pi/n} \varphi_x(t)\,\frac{\sin nt}{t}\,dt =$$
$$-\,n\int_0^{\pi/n}\Big(\int_0^t \varphi_x(u)\,du\Big)\frac{\cos nt}{t}\,dt + \int_0^{\pi/n}\Big(\int_0^t \varphi_x(u)\,du\Big)\frac{\sin nt}{t^2}\,dt.$$

On assuming (*), the first term is

$$n\int_0^{\pi/n} o(1)\,dt = o(1)$$

and the second is

$$\int_0^{\pi/n} o(t)\,O\Big(\frac{n}{t}\Big)\,dt = o(1).$$

Hence

$$\int_0^{\pi/n} \varphi_x(t)\,\frac{\sin nt}{t}\,dt = o(1)$$

holds at x whenever $(*)$ holds at x, and holds uniformly whenever $(*)$ holds uniformly. We note that $(*')$ would suffice here.

In estimating $S_n(x;f) - f(x)$, there is no loss in generality in assuming that n is odd. Letting $\sum^e$ denote summation over even indices,

$$\begin{aligned}
&\int_{\pi/n}^{\pi} \varphi_x(t)\,\frac{\sin nt}{t}\,dt \\
&\qquad = \sum_{k=0}^{n-2} \int_{\pi/n}^{2\pi/n} \varphi_x(t)\,(t + k\pi/n)\,\frac{(-1)^k \sin nt}{t + k\pi/n}\,dt \\
&\qquad = \sum_{k=0}^{n-2}{}^{e} \int_{\pi/n}^{2\pi/n} \Big(\varphi_x(t + k\pi/n)\,\frac{1}{t + k\pi/n} \\
&\qquad\qquad\qquad - \varphi_x(t + (k+1)\pi/n)\,\frac{1}{t + (k+1)\pi/n}\Big) \sin nt\,dt \\
&\qquad = \int_{\pi/n}^{2\pi/n} \sum_{k=0}^{n-2}{}^{e} \Big(\varphi_x(t + k\pi/n) - \varphi_x(t + (k+1)\pi/n)\Big)\,\frac{\sin nt}{t + k\pi/n}\,dt \\
&\qquad\qquad + \int_{\pi/n}^{2\pi/n} \sum_{k=0}^{n-2}{}^{e} \varphi_x(t + (k+1)\pi/n) \\
&\qquad\qquad\qquad \cdot \Big(\frac{1}{t + k\pi/n} - \frac{1}{t + (k+1)\pi/n}\Big) \sin nt\,dt \\
&\qquad = \int_{\pi/n}^{2\pi/n} I_x(t)\,dt + \int_{\pi/n}^{2\pi/n} J_x(t)\,dt.
\end{aligned}$$

Let

$$\Phi(t) = \Phi_x(t) = \int_0^t |\varphi_x(u)|\,du.$$

Then

$$\begin{aligned}
\Big|\int_{\pi/n}^{2\pi/n} J_x(t)\,dt\Big| &= \Big|\int_{\pi/n}^{2\pi/n} \sum_{k=0}^{n-2}{}^{e} \varphi_x\big(t + (k+1)\pi/n\big) \\
&\qquad \cdot \Big(\frac{1}{t + k\pi/n} - \frac{1}{t + (k+1)\pi/n}\Big) \sin nt\,dt\Big| \\
&\le \frac{n}{\pi} \sum_{k=0}^{n-2}{}^{e} \int_{\pi/n}^{2\pi/n} \big|\varphi_x\big(t + (k+1)\pi/n\big)\big|\,\frac{1}{(k+1)(k+2)}\,dt \\
&\le \frac{n}{\pi} \sum_{k=0}^{n-2}{}^{e} \big(\Phi\big((k+3)\pi/n\big) - \Phi\big((k+2)\pi/n\big)\big)\frac{1}{(k+1)^2} \\
&\le \frac{n}{\pi} \sum_{k=0}^{n-3}{}^{e} \big(\Phi\big((k+3)\pi/n\big) - \Phi\big((k+2)\pi/n\big)\big)\frac{1}{(k+1)^2} \\
&\le \frac{n}{\pi}\Big\{ \sum_{k=1}^{n-3}{}^{e} \Phi\big((k+2)\pi/n\big)\Big(\frac{1}{k^2} - \frac{1}{(k+1)^2}\Big) + \Phi(\pi)/(n-2)^2 \Big\} \\
&= \frac{n}{\pi} \sum_{k=1}^{N}{}^{e} (\cdots) + \frac{n}{\pi}\Big\{ \sum_{N+1}^{n-3}{}^{e} (\cdots) + \Phi(\pi)/(n-2)^2 \Big\}.
\end{aligned}$$

In the last line, on assuming $N/n \to 0$ when $n \to \infty$, the first term satisfies

$$n \sum_{k=1}^{N} {}^{e}\, o\Big(\frac{k}{n}\Big) \frac{1}{k^3} = o(1) \quad \text{as} \quad n \to \infty.$$

if $(*)$ is satisfied. The other term does not exceed

$$Cn\big(|f(x)| + \|f\|_1\big)\big(1/N^2 + 1/n^2\big),$$

where C is an appropriate constant. Choosing $N \approx n^{2/3}$, we see that

$$\int_{\pi/n}^{2\pi/n} J_x(t)\, dt = o(1) \quad \text{as} \quad n \to \infty$$

for every x and that this holds uniformly on any set where $f(x)$ is bounded and $\Phi(h) = \Phi_x(h) = o(h)$ uniformly.

We observe now that

$$\begin{aligned}
\int_{\pi/n}^{2\pi/n} I_x(t)\, dt &- \frac{1}{\pi}\int_{\pi}^{2\pi} \big(T_n(x,u) + Q_n(x,u)\big) \sin u\, du \\
&= \int_{\pi/n}^{2\pi/n} \sum_{k=0}^{n-2} {}^{e} \big(\varphi_x(t + k\pi/n) - \varphi_x(t + (k+1)\pi/n)\big) \\
&\qquad \cdot \Big(\frac{1}{t + k\pi/n} - \frac{1}{(k+1)\pi/n}\Big) \sin nt\, dt \\
&\quad - \int_{\pi/n}^{2\pi/n} \big(f(x + t + (n-1)\pi/n) - f(x + t + \pi) \\
&\qquad + f(x - t - (n-1)\pi/n) - f(x - t - \pi)\big) \sin nt\, dt
\end{aligned}$$

and this last term is dominated by $\omega_1(\pi/n; f)$. Clearly the remaining term is $o(1)$ uniformly in x if f is continuous. Otherwise it is dominated by

$$\begin{aligned}
n \sum_{k=0}^{n-2} {}^{e} &\{\Phi((k+2)\pi/n) - \Phi((k+1)\pi/n)\}/(k+1)^2 \\
&+ n \sum_{k=0}^{n-2} {}^{e} \{\Phi((k+3)\pi/n) - \Phi((k+2)\pi/n)\}/(k+1)^2
\end{aligned}$$

and these sums are estimated just as was done for $\int_{\pi/n}^{2\pi/n} J_x(t)\, dt$. Thus $(***)$ is equivalent to

$$\int_{\pi/n}^{2\pi/n} I_x(t)\, dt = o(1) \quad \text{as} \quad n \to \infty \quad \text{whenever} \quad (*) \quad \text{holds.}$$

Hence, under condition $(*)$, condition $(***)$ is necessary and sufficient for the convergence of $S(x; f)$ to $f(x)$. The estimate for the difference between $\int_{\pi/n}^{2\pi/n} I_x(t)\, dt$ and $\frac{1}{\pi}\int_{\pi}^{2\pi} \big(T_n(x,u) + Q_n(x,u)\big) \sin u\, du$ is uniform on any set on which $f(x)$ is bounded.

From the estimates made in the proof of this theorem we may derive a proposition which will be of use in the study of preservation of uniform convergence.

Let us now set, for positive odd integers n,

$$P_n^R(x;f) = \int_0^\pi T_n(x, t+\pi) \sin t \, dt \quad \text{and} \quad P_n^L(x;f) = \int_0^\pi Q_n(x, t+\pi) \sin t \, dt.$$

(The reference to f in the above formulas will often be dropped.) In the argument for the above theorem we estimated

$$S_n(x;f) - f(x) = \frac{1}{\pi} \int_0^\pi \varphi_x(t) \frac{\sin nt}{t} + o(1)$$

and the estimate we obtained may be expressed as

PROPOSITION 10.6. *Let $f \in \mathrm{L}^1(T)$. For each x,*

$$(\mathrm{E_U}) \qquad f(x) - S_n(x;f) = \frac{1}{\pi^2}\big(P_n^R(x;f) + P_n^L(x;f)\big) + o(1) \quad \text{as} \quad n \to \infty$$

whenever

$$(*) \qquad \frac{1}{h} \int_0^h |f(x+t) + f(x-t) - 2f(x)| \, dt = o(1) \quad \text{as} \quad h \to 0.$$

The estimate $(\mathrm{E_U})$ *is uniform on every set on which f is bounded and* $(*)$ *holds uniformly, in particular, on every closed bounded set of points of continuity.*

The following proposition applies the idea of the Salem theorem to the question of the convergence of the Fourier series at points where the one-sided limits exist, as they do at each point for regulated functions. (See page 113 for regulated functions.)

If f is a regulated function on T and if $\delta > 0$, we have

$$S_n(x;f) - f(x) = \frac{1}{\pi} \int_0^\delta \big(f(x+t) - f(x+)\big) \frac{\sin nt}{t} \, dt + \frac{1}{\pi} \int_0^\delta \big(f(x-t) - f(x-)\big) \frac{\sin nt}{t} \, dt + o(1)$$

as $n \to \infty$ uniformly in x.

PROPOSITION 10.7. *If $f \in \mathrm{R}(T)$, and if $\varepsilon > 0$ and $\pi \geq \delta > 0$, then there is an $n(\varepsilon, \delta)$ such that $n > n(\varepsilon, \delta)$ implies*

$$\left| \int_0^\delta \big(f(x+t) - f(x+)\big) \frac{\sin nt}{t} \, dt \right| < \varepsilon + \frac{1}{\pi} \int_0^\pi \left| \sum_{k=1}^{N}{}^{o} \, k^{-1} \big(f(x + (t + k\pi)/n) - f(x + (t + (k+1)\pi)/n)\big) \right| dt,$$

where $\sum^o$ denotes summation over odd indices and N is the greatest odd number such that $N + 1 < n\delta/\pi$. An analogous estimate holds for

$$\left| \int_0^\delta \big(f(x-t) - f(x-)\big) \frac{\sin nt}{t} \, dt \right|.$$

These estimates are uniform on any closed interval of points of continuity.

PROOF. Letting

$$h(t) = h_x(t) = f(x+t) - f(x+),$$

we see that

$$\int_0^{\pi/n} h(t)\,\frac{\sin nt}{t}\,dt < \pi \sup_{0<t<\pi/n} |h(t)| = o(1)$$

for each x and uniformly on a closed interval of points of continuity. Now

$$\begin{aligned}\int_{\pi/n}^{\delta} h(t)\,\frac{\sin nt}{t}\,dt &= \sum_{k=1}^{N} \int_{k\pi/n}^{(k+1)\pi/n} h(t)\,\frac{\sin nt}{t}\,dt + \int_{(N+1)\pi/n}^{\delta} h(t)\,\frac{\sin nt}{t}\,dt \\ &= I_1 + I_2,\end{aligned}$$

where $N+1$ is the integer part of $n\delta/\pi$. Clearly $I_2 = o(1)$ uniformly in x and

$$I_1 = \int_0^{\pi} \sum_{k=1}^{N} h\Big(\frac{t+k\pi}{n}\Big)(-1)^k\,\frac{\sin t}{t+k\pi}\,dt.$$

When N is even, the absolute value of the integrand is dominated by

$$\Big|\sum_{k=1}^{N-1}{}^{o}\Big\{h\Big(\frac{t+k\pi}{n}\Big)\,\frac{1}{t+k\pi} - h\Big(\frac{t+(k+1)\pi}{n}\Big)\,\frac{1}{t+(k+1)\pi}\Big\}\Big|$$

which we shall denote as

$$\Big|\sum_{k=1}^{N-1}{}^{o}\Big(\frac{\alpha_k(t)}{t+k\pi} - \frac{\beta_k(t)}{t+(k+1)\pi}\Big)\Big|.$$

If N is odd, then

$$\int_{N\pi/n}^{(N+1)\pi/n} h(t)\,\frac{\sin nt}{t}\,dt = o(1),$$

just as was I_2, and hence, by removing this term, we reduce the problem to one in which the sum has an even number of terms. We shall, therefore, assume N to be even. Note that

$$\begin{aligned}\frac{\alpha_k(t)}{t+k\pi} - \frac{\beta_k(t)}{t+(k+1)\pi} & \\ &= \frac{\alpha_k(t)-\beta_k(t)}{k\pi} + \alpha_k(t)\Big(\frac{1}{t+k\pi} - \frac{1}{k\pi}\Big) \\ &\qquad - \beta_k(t)\Big(\frac{1}{t+(k+1)\pi} - \frac{1}{k\pi}\Big).\end{aligned}$$

Given $\eta > 0$ and choosing N_0 so that $\sum_{k=N_0+1}^{\infty} 1/k^2 < \eta$, for $0 \le t \le \pi$, the second and third terms on the right are dominated by

$$\Big|\sum_{k=1}^{N-1}{}^{o}\,\alpha_k(t)\,\frac{t}{(t+k\pi)k\pi}\Big| \le \sum_{k=1}^{N-1} |\alpha_k(t)|/k^2 = \sum_{k=1}^{N_0}{}^{o}(\cdots) + \sum_{k=N_0+1}^{N-1}{}^{o}(\cdots)$$

and

$$\Big|\sum_{k=1}^{N-1}{}^{o}\,\beta_k(t)\,\frac{t}{(k\pi)(t+(k+1)\pi)}\Big| \le \sum_{k=1}^{N-1} |\beta_k(t)|/k^2 = \sum_{k=1}^{N_0}{}^{o}(\cdots) + \sum_{k=N_0+1}^{N-1}{}^{o}(\cdots),$$

where, for each x, the second sums on the right are bounded by $2\eta \sup|f(x)|$ and the first sums are bounded by

$$\sup_{0<u\leq(N_0+2)\pi/n} |f(x+u)-f(x+)| \cdot \sum_{k=1}^{N_0}{}^{o}\, 1/k^2 = o(1)$$

as $n \to \infty$, and is $o(1)$ uniformly in any closed interval of points of continuity. Given ε and δ, we see that we can choose η and N_0 and, finally, $n(\varepsilon, \delta)$ to obtain

$$I_1 \leq \frac{\varepsilon}{2} + \frac{1}{\pi}\int_0^{\pi}\Big|\sum_{k=1}^{N}{}^{o}\, k^{-1}\big(f(x+(t+k\pi)/n) - f(x+(t+(k+1)\pi)/n)\Big|\, dt,$$

and the desired result will follow.

10.2. Fourier series of regulated functions

For each positive integer n, let I_{ni}, $i = 1, \dots, k_n$, be disjoint closed intervals in T where the indexing is from left to right, i.e., $I_{n,i-1}$ is to the left of I_{ni}. Let x be point of T such that for every positive number δ there is an N such that $I_{ni} \subset (x, x+\delta)$ for every i whenever $n > N$. Then

$$\mathcal{I} = \{\, I_{ni} : i = 1, 2, \dots, k_n,\ n = 1, 2, \dots \}$$

is called a **right system of intervals at** x. A **left system** is defined similarly with $I_{ni} \subset (x-\delta, x)$ and the indexing by i is from right to left for each n.

As in the introduction to the chapter, for a real-valued functions f and an intervals $I = [a, b]$, we define the interval function

$$f(I) = f(b) - f(a), \quad I = [a, b].$$

The reader should be aware of the notational distinction we have made between the interval function $f(I)$ and the image $f[I]$ of I under f.

Goffman and Waterman [**67**] considered the condition GW.

DEFINITION 10.8. A function f is said to satisfy **condition GW** if, for every x,

$$\text{(h)} \qquad \lim_{n\to\infty} \sum_{i=1}^{k_n} i^{-1} f(I_{ni}) = 0$$

for every right and every left system at x.

The definition is invariant under self-homeomorphisms of T.

PROPOSITION 10.9. *The following statements are equivalent for functions f.*

1. *f satisfies condition GW.*
2. *$f \circ g$ satisfies condition GW for some change of variable g.*
3. *$f \circ g$ satisfies condition GW for every change of variable g.*

Observe that g is uniformly continuous. The proof is left to the reader.

There is the following theorem.

THEOREM 10.10 (GOFFMAN-WATERMAN). *For f in* C(T), *the Fourier series $S(f \circ g)$ converges everywhere for every change of variable g if and only if f satisfies condition* GW.

At various times it has been said that $\mathrm{C}(T)$ may be replaced by $\mathrm{R}(T)$ in the above result [**67**] without change of proof, but careful consideration has shown that the analogous result for regulated functions requires further arguments. The principal goal of this section is to demonstrate that result [**109**].

THEOREM 10.11 (PIERCE-WATERMAN). *For $f \in \mathrm{R}(T)$, condition* GW *is necessary and sufficient for the convergence of $S(f \circ g)$ for every change of variable g.*

By $\mathrm{R}(T)$ we shall mean, as before (page 113), the regulated functions on T, those with one-sided limits at each point and value equal to the arithmetic mean of those limits. It is clear that if a function satisfies condition GW, then it has right limits and left limits at each point and so equals a regulated function except on an at most countable set. Thus we may formulate Theorem 10.11 in a somewhat more general manner.

THEOREM 10.12. *If f satisfies condition* GW*, then $S(f \circ g)$ converges everywhere for every change of variable g. If $f \in \mathrm{R}(T)$, then condition* GW *is necessary for the everywhere convergence of $S(f \circ g)$ for every change of variable g.*

We shall note that Goffman and Waterman defined a system of intervals somewhat differently. They assumed the restriction

$$k_n \to \infty \quad \text{and} \quad k_n/n \to 0.$$

The restricted (h) condition is equivalent to the condition (h) whenever f is measurable [**5**]. Clearly, condition (h) implies the restricted (h). To see the reverse implication, assume that the restricted (h) holds. Let

$$\mathcal{I} = \{\, I_{nm} : m = 1, 2, \ldots, m_n,\ n = 1, 2, \ldots \}$$

be any system. Choose an increasing sequence h_n, $n = 1, 2, \ldots$, of positive integers such that $m_n \le \sqrt{h_n}$. For each integer p in $[h_n, h_{n+1})$, we define k_p to be the integer part of $\sqrt{h_n}$. We may define $J_{pk} = I_{nk}$ for $k \le m_n$ and, for any remaining indices k, $m_n < k \le k_p$, select suitably placed intervals J_{pk} such that

$$\sum_{m_n < k \le k_p} k^{-1} |f(J_{pk})| < p^{-1}. \tag{10.1}$$

This is easily achieved because the points of approximate continuity of f are dense. Then

$$\mathcal{J} = \{\, J_{pk} : k = 1, 2, \ldots, k_p,\ p = 1, 2, \ldots \}$$

is a restricted system. Indeed, we have $k_p \ge \sqrt{h_n} - 1$ and $k_p/p \le \sqrt{h_n}/h_n$ for $p \in [h_n, h_{n+1})$. Now the restricted (h) applied to the system together with (10.1) implies the GW condition holds for f. Note that we need not assume measurability if we permit the J_{pk} in (10.1) to be degenerate intervals, that is, intervals consisting of a single point.

The notion of a system can be further simplified.

PROPOSITION 10.13. *In the* GW *condition, the requirement that the I_{ni} be disjoint can be relaxed to be nonoverlapping.*

PROOF. The proof is immediate upon observing that the interval function $f(I)$ is additive. That is,

$$f([a, c]) = f([a, b]) + f([b, c]).$$

In our next result we show that a seemingly weaker condition is equivalent to the GW condition.

PROPOSITION 10.14. *A function f satisfies condition* GW *if and only if for each positive number ε and each x in T there is a positive number δ such that*

$$\Big|\sum_{j=1}^{k} j^{-1} f(I_j)\Big| < \varepsilon$$

whenever I_j, $j = 1, 2, \ldots, k$, is a finite sequence of nonoverlapping intervals indexed from left to right with $\bigcup_{j=1}^{k} I_j \subset (x, x+\delta)$, or indexed from right to left with $\bigcup_{j=1}^{k} I_j \subset (x-\delta, x)$.

PROOF. If the GW condition fails, then at some x for some right or left system $\mathcal{I} = \{\, I_{nj} : j = 1, 2, \ldots, j_n,\ n = 1, 2, \ldots \}$, we have

$$\limsup_{n\to\infty} \Big|\sum_{j=1}^{j_n} j^{-1} f(I_{nj})\Big| > 0.$$

When the system is a right one, $\bigcup_{j=1}^{j_n} I_{nj} \subset (x, x+\delta)$ as $n \to \infty$ and the condition of the theorem does not hold. The analogous statement holds when the system is a left one.

On the other hand, suppose for some x and some positive number ε that there is for each positive integer n a finite collection of nonoverlapping intervals I_{nj}, $j = 1, \ldots, j_n$, indexed from left to right such that $\bigcup_{j=1}^{j_n} I_{nj} \subset (x, x+\frac{1}{n})$ (or, from right to left such that $\bigcup_{j=1}^{j_n} I_{nj} \subset (x-\frac{1}{n}, x)$) and $\big|\sum_{j=1}^{j_n} j^{-1} f(I_{nj})\big| \geq \varepsilon$. Then the "relaxed" system $\mathcal{I} = \{\, I_{nj} : j = 1, 2, \ldots, j_n,\ n = 1, 2, \ldots \}$ (as in Proposition 10.13) is one for which the condition (h) fails. In view of Proposition 10.13, the GW condition fails.

We first consider the sufficiency of condition GW.

PROOF OF SUFFICIENCY. Let f satisfy condition GW. Since f equals a regulated function except on an at most countable set, it is enough to prove that the Fourier series of a regulated function f converges to itself everywhere whenever it satisfies condition GW. As regulated functions are bounded, we have observed in Section 9.1 that, for $\delta > 0$, the function f satisfies

$$\begin{aligned} S_n(x; f) - f(x) &= \frac{1}{\pi}\int_0^{\delta} \big(f(x+t) - f(x+)\big)\frac{\sin nt}{t}\,dt \\ &\quad + \frac{1}{\pi}\int_0^{\delta} \big(f(x-t) - f(x-)\big)\frac{\sin nt}{t}\,dt + o(1) \end{aligned} \tag{10.2}$$

as $n \to \infty$ uniformly in x. Consider the first integral. By Proposition 10.7, given $\varepsilon > 0$ and $\delta > 0$ we have

$$\begin{aligned} &\Big|\int_0^{\delta} \big(f(x+t) - f(x+)\big)\frac{\sin nt}{t}\,dt\Big| < \varepsilon \\ &\quad + \frac{1}{\pi}\int_0^{\pi} \Big|\sum_{k=1}^{N}{}^{o}\, k^{-1}\big(f(x+(t+k\pi)/n) - f(x+(t+(k+1)\pi)/n)\big)\Big|\,dt, \end{aligned} \tag{10.3}$$

where $\sum^o$ denotes summation over odd indices and N is the greatest odd number such that $N + 1 < n\delta/\pi$.

As f satisfies the seemingly weaker form of the GW condition given in Proposition 10.14, the integrand on the right in (10.3) can be made small uniformly in t and n by choosing δ small because

$$\frac{k\pi}{n} \leq \frac{N\pi}{n} < \delta \quad \text{whenever} \quad k = 1, 2, \ldots, N.$$

Choose δ so that this integral and the corresponding integral obtained from the other term on the right in (10.2) are each less than ε, a choice which depends on x. Once this choice of δ has been made, we may then choose $n(\varepsilon)$ so that the $o(1)$ term of (10.2) will be so small that

$$\left|S_n(x; f) - f(x)\right| < 5\varepsilon/\pi \quad \text{whenever} \quad n > n(\varepsilon),$$

thereby proving the sufficiency.

To proceed with the proof of the necessity of the GW condition, we will need some additional tools.

An elementary but useful mean value theorem for integrals asserts that if G is a nonnegative integrable function on $[a, b]$ and H is continuous, then there is a θ in (a, b) such that

$$H(\theta) \int_a^b G(t)\, dt = \int_a^b H(t)\, G(t)\, dt.$$

It has recently been observed [**138**] that this result has an equally useful extension to regulated functions H. Let

$$\overline{H}(t) = \max\{H(t+), H(t-)\} \quad \text{and} \quad \underline{H}(t) = \min\{H(t+), H(t-)\}.$$

The result is

PROPOSITION 10.15 (An Integral Mean Value Theorem). *If H is a regulated function defined on $[a, b]$ and G is a nonnegative, integrable function on $[a, b]$, then there is a θ in (a, b) such that*

$$c(H, \theta) \int_a^b G(t)\, dt = \int_a^b H(t)\, G(t)\, dt,$$

for some choice $c(H, \theta)$ in $[\underline{H}(\theta), \overline{H}(\theta)]$.

The construction which will establish necessity requires the following lemma.

LEMMA 10.16. *Let k_n, $k = 1, 2 \ldots$, be a sequence of integers with the properties*

$$\lim_{n\to\infty} k_n = \infty \quad \textit{and} \quad \lim_{n\to\infty} \frac{k_n}{n} = 0.$$

Then for each n there is a number ε_n with $0 < \varepsilon_n < \pi/n$ such that for every regulated function h with $h(0+) = 0$, there exists a θ_n in $(0, \pi/n - \varepsilon_n)$ such that the sequences

$$\int_0^{(2k_n+1)\pi/n} h(t)\, \frac{\sin nt}{t}\, dt \quad \textit{and} \quad \frac{1}{\pi} c(H_n, \theta_n) \quad \textit{are equiconvergent,}$$

where $c(H_n, \theta)$ is a choice for the regulated function

$$H_n(t) = \sum_{i=1}^{k_n} \frac{1}{i} \left\{ h\left(t + \frac{2i\pi}{n}\right) - h\left(t + \frac{(2i-1)\pi}{n}\right) \right\}$$

that satisfies the mean value condition

$$c(H_n,\theta)\int_0^{\pi/n-\varepsilon_n}\sin nt\,dt=\int_0^{\pi/n-\varepsilon_n}H_n(t)\,\sin nt\,dt$$

with $0<\theta<\pi/n-\varepsilon_n$.

PROOF. We write

$$\begin{aligned}\int_0^{(2k_n+1)\pi/n}h(t)\,\frac{\sin nt}{t}\,dt&=\int_0^{\pi/n}h(t)\,\frac{\sin nt}{t}\,dt+\sum_{i=1}^{2k_n}\int_{i\pi/n}^{(i+1)\pi/n}h(t)\,\frac{\sin nt}{t}\,dt\\&=P+Q.\end{aligned}$$

Now

$$\begin{aligned}|P|=\Big|\int_0^{\pi/n}h(t)\,\frac{\sin nt}{t}\,dt\Big|&\le\sup_{t\in(0,\pi/n)}|h(t)|\,n\,\frac{\pi}{n}\\&=\pi\sup_{t\in(0,\pi/n)}|h(t)|=o(1)\quad\text{as}\quad n\to\infty.\end{aligned}$$

We also have

$$\begin{aligned}Q=\sum_{i=1}^{2k_n}\int_{i\pi/n}^{(i+1)\pi/n}h(t)\,\frac{\sin nt}{t}\,dt&=\sum_{i=1}^{2k_n}\int_0^{\pi}h\Big(\frac{t+i\pi}{n}\Big)\,\frac{\sin(t+i\pi)}{t+i\pi}\,dt\\&=\int_0^{\pi}\sin t\sum_{i=1}^{2k_n}\frac{(-1)^i}{t+i\pi}\,h\Big(\frac{t+i\pi}{n}\Big)\,dt\\&=\int_0^{\pi-\eta_n}(\cdots)\,dt+\int_{\pi-\eta_n}^{\pi}(\cdots)\,dt\\&=Q'+R,\end{aligned}$$

where $\eta_n\in(0,\pi)$ will be chosen presently. For convenience, let

$$h_n=\sup\{\,|h(t)|:t\in(0,(2k_n+1)\pi/n)\,\}.$$

We observe that

$$\begin{aligned}|R|&=\Big|\int_{\pi-\eta_n}^{\pi}\sin t\sum_{i=1}^{2k_n}\frac{(-1)^i}{t+i\pi}\,h\Big(\frac{t+i\pi}{n}\Big)\,dt\Big|\\&\le\eta_n^2\,\frac{1}{\pi}\sum_{i=1}^{2k_n}\frac{1}{i}\,\sup\{\,|h(t)|:t\in(0,(2k_n+1)\pi/n)\,\}\\&\le\eta_n^2\,\frac{1}{\pi}\,\big(1+\log(2k_n)\big)\,h_n.\end{aligned}$$

Now by choosing $\eta_n=O\big((\log k_n)^{-\frac12}\big)$, we have

$$|R|\le C\,h_n=o(1)\quad\text{as}\quad n\to\infty,$$

which yields

$$\int_0^{(2k_n+1)\pi/n}h(t)\,\frac{\sin nt}{t}\,dt=Q'+o(1).$$

We may write

$$Q'=\int_0^{\pi-\eta_n}\sin t\sum_{i=1}^{k_n}\Big\{\frac{1}{t+2i\pi}\,h\Big(\frac{t+2i\pi}{n}\Big)-\frac{1}{t+(2i-1)\pi}\,h\Big(\frac{t+(2i-1)\pi}{n}\Big)\Big\}\,dt$$

and the sum in the integrand may be written as

$$\sum_{i=1}^{k_n} \frac{1}{t+2i\pi} \left\{ h\Big(\frac{t}{n} + \frac{2i\pi}{n}\Big) - h\Big(\frac{t}{n} + \frac{(2i-1)\pi}{n}\Big) \right\}$$
$$- \sum_{i=1}^{k_n} \frac{\pi}{(t+2i\pi)(t+(2i-1)\pi)} h\Big(\frac{t}{n} + \frac{(2i-1)\pi}{n}\Big)$$
$$= S_1 + S_2\,.$$

For an estimate of S_2, we have

$$|S_2| \le \frac{1}{\pi} \sum_{i=1}^{k_n} \frac{1}{(2i)(2i-1)}\, h_n \le C\, h_n = o(1) \quad \text{as} \quad n \to \infty.$$

Turning to S_1, we observe that

$$\left| \frac{1}{t+2i\pi} - \frac{1}{2i\pi} \right| = \left| \frac{t}{(t+2i\pi)(2i\pi)} \right| \le \left| \frac{t}{4\,i^2\pi^2} \right| \le \frac{1}{4\,i^2\pi}$$

since $0 \le t \le \pi$. So if we replace $t + 2i\pi$ by $2i\pi$ in the denominators of the terms of S_1, the error is less than

$$C\, h_n = o(1) \quad \text{as} \quad n \to \infty.$$

We now have

$$Q' = \frac{1}{2\pi} \int_0^{(\pi-\eta_n)/n} \left(\sum_{i=1}^{k_n} \frac{1}{i} \left\{ h\Big(t + \frac{2i\pi}{n}\Big) - h\Big(t + \frac{(2i-1)\pi}{n}\Big) \right\} \right) n \sin nt\, dt + o(1).$$

We see that the sum in the parenthesis is the regulated function $H_n(t)$ in the statement of the lemma. From the integral mean value theorem we obtain

$$Q' = (1 + \cos\eta_n)\, \frac{1}{2\pi}\, c(H_n, \theta_n) + o(1),$$

where θ_n satisfies $0 < \theta_n < (\pi - \eta_n)/n$. If we now replace $1 + \cos\eta_n$ by 2, we introduce an error bounded by

$$\frac{1}{2\pi}\, |(1-\cos\eta_n)\, c(H_n, \theta_n)| \le C\, \eta_n^2\, (\log k_n)\, h_n.$$

This last quantity is $o(1)$ since $\eta_n = O\big((\log k_n)^{-\frac{1}{2}}\big)$. We now have

$$\int_0^{(2k_n+1)\pi/n} h(t)\, \frac{\sin nt}{t}\, dt = \frac{1}{\pi}\, c(H_n, \theta_n) + o(1).$$

Finally

$$\varepsilon_n = \frac{\eta_n}{n} = O\big(n^{-1} (\log k_n)^{-\frac{1}{2}}\big)$$

gives us the desired result.

We are now able to continue with the proof of Theorem 10.12.

Proof of Necessity. Let f be a regulated function. The proof will be by contraposition. We assume that there is a restricted right or a restricted left system at a point x for which the limit (h) of condition GW fails (see page 138). We may assume that $x = 0$ and $f(0) = 0$ and that it is a right system

$$\mathcal{I} = \{\, I_{ni} : i = 1, 2, \ldots, k_n,\ n = 1, 2, \ldots \}$$

such that, for some positive number α,

$$\limsup_{n\to\infty} \sum_{i=1}^{k_n} i^{-1} f(I_{ni}) > \alpha \neq 0.$$

We shall construct a change of variable g such that $S(f \circ g)$ does not converge to $f \circ g$ when $x = 0$. As the proof is quite long, we shall give an outline of it. From equation (10.2), for each change of variable g, we have that

$$\begin{aligned} S_n(x; f \circ g) - f \circ g(x) &= \frac{1}{\pi} \int_0^\pi \big(f \circ g(x+t) - f \circ g(x+)\big) \frac{\sin nt}{t}\, dt \\ &\quad + \frac{1}{\pi} \int_0^\pi \big(f \circ g(x-t) - f \circ g(x-)\big) \frac{\sin nt}{t}\, dt + o(1) \end{aligned}$$

as $n \to \infty$. The required change of variable g is constructed in two parts. First, it is constructed to be a self-homeomorphism of $[0, \pi]$ with

$$\liminf_{n\to\infty} \frac{1}{\pi} \int_0^\pi \big(f \circ g(t) - f \circ g(0+)\big) \frac{\sin m_n t}{t}\, dt > \frac{\alpha}{\pi^2}$$

for some increasing sequence m_n, $n = 1, 2, \ldots$. Then, for any preassigned subsequence m_n, $n = 1, 2, \ldots$, a self-homeomorphism g is constructed on $[-\pi, 0]$ in such a manner that

$$\limsup_{n\to\infty} \frac{1}{\pi} \int_0^\pi \big(f \circ g(-t) - f \circ g(0-)\big) \frac{\sin m_n t}{t}\, dt \geq 0.$$

We infer from these inequalities that

$$\limsup_{n\to\infty} S_n(0; f \circ g) - f \circ g(0) > \frac{\alpha}{\pi^2}$$

and hence the Fourier series of $f \circ g$ will not converge to $f \circ g$ at $x = 0$.

We turn to the construction of g on $[0, \pi]$. For the above restricted right system $\mathcal{I}$ at 0, let its intervals be denoted by

$$I_{ni} = [\tau_{n,i}, \tau'_{n,i}].$$

As the system is restricted, we have $k_n \to \infty$ and $k_n/n \to 0$. Since f is continuous except for a set that is at most countable and $f(x) = \frac{1}{2}\big(f(x+) + f(x-)\big)$ at each x, without loss of generality, we may assume that the end points of the intervals I_{ni} are points of continuity of f because the intervals I_{ni}, $i = 1, 2, \ldots, k_n$, are mutually disjoint. For convenience, we shall let α_n be defined by

$$\alpha_n = \sum_{i=1}^{k_n} i^{-1} f(I_{ni}) = \sum_{i=1}^{k_n} i^{-1} \big(f(\tau'_{n,i}) - f(\tau_{n,i})\big).$$

The required change of variable g will be constructed inductively. We begin the first step of the induction by choosing a positive integer m_1 and a number b_1 with $\tau'_{m_1,k_{m_1}} < b_1 < \pi$ such that

(i$_1$) $$\alpha_{m_1} > \frac{\alpha}{2}$$

and

(ii$_1$) $$\left| \int_{\beta_1}^\pi f \circ g(t) \frac{\sin m_1 t}{t}\, dt \right| < \frac{1}{2}, \quad \text{where} \quad \beta_1 = (2k_{m_1} + 1)\pi/m_1,$$

and where g is the increasing linear function which maps $\big[\beta_1, \pi\big]$ onto $[b_1, \pi]$. To see that such a choice of m_1 and b_1 is possible we note that

$$\begin{aligned}
&\Big|\int_{\beta_1}^{\pi} f\circ g(t)\,\frac{\sin m_1 t}{t}\,dt\Big| \\
&\qquad \le |f(\pi-)|\,\Big|\int_{\beta_1}^{\pi}\frac{\sin m_1 t}{t}\,dt\Big| + \int_{\beta_1}^{\pi}|f(\pi-)-(f\circ g)(t)|\,\Big|\frac{\sin m_1 t}{t}\Big|\,dt \\
&\qquad \le |f(\pi-)|\,\Big|\int_{\beta_1}^{\pi}\frac{\sin m_1 t}{t}\,dt\Big| + \sup_{t\in(b_1,\pi)}|f(\pi-)-f(t)|\int_{\beta_1}^{\pi}\frac{1}{t}\,dt \\
&\qquad \le |f(\pi-)|\,\Big|\int_{\beta_1}^{\pi}\frac{\sin m_1 t}{t}\,dt\Big| + \sup_{t\in(b_1,\pi)}|f(\pi-)-f(t)|\,C\,\log m_1.
\end{aligned}$$

In the last line above, the first term can be made small by choosing m_1 large and, for any choice of m_1, the second term will be small if b_1 is sufficiently close to π. We finish the first step of the inductive construction of g by defining it on the interval $[\pi/m_1, \beta_1]$. Let ε_n, $n = 1, 2, \ldots$, be the sequence determined in Lemma 10.16. For each j, $j = 1, 2, \ldots, 2\,k_{m_1}$, let

$$J_j = \Big[\frac{j\pi}{m_1}, \frac{(j+1)\pi}{m_1} - \varepsilon_{m_1}\Big] \quad\text{and}\quad J_j' = \Big[\frac{(j+1)\pi}{m_1} - \varepsilon_{m_1}, \frac{(j+1)\pi}{m_1}\Big].$$

Define, for $i = 1, 2, \ldots, k_{m_1}$,

$$g\Big(\frac{(2i-1)\pi}{m_1}\Big) = \tau_{m_1,i} \quad\text{and}\quad g\Big(\frac{2\,i\pi}{m_1}\Big) = \tau_{m_1,i}'.$$

Recalling that we have arranged f to be continuous at $\tau_{m_i,i}$ and $\tau_{m_i,i}'$, we now define g to be linear on each of the intervals J_j and J_j', making the slope of g on each J_j so small that

$$\text{(iii}_1\text{)}\qquad \Big|\alpha_{m_1} - \sum_{i=1}^{k_{m_1}}\frac{1}{i}\Big\{f\circ g\Big(\frac{2i\pi}{m_1}+\theta\Big) - f\circ g\Big(\frac{(2i-1)\pi}{m_1}+\theta\Big)\Big\}\Big| < \frac{1}{2}$$
$$\text{whenever}\quad 0<\theta<\frac{\pi}{m_1}-\varepsilon_{m_1}.$$

The construction of g on $[\pi/m_1, \pi]$ is now completed.

Suppose for $n \le r$ that we have defined m_n with $m_1 < m_2 < \cdots < m_r$ and a homeomorphism g from $[\pi/m_r, \pi]$ onto $[\tau_{m_r,1}, \pi]$ with $g(\pi) = \pi$ such that, for $n = 1, 2, \ldots, r$, we have

$$\text{(i}_n\text{)}\qquad \alpha_{m_n} > \alpha\Big(1-\frac{1}{2^n}\Big),$$

$$\text{(ii}_n\text{)}\qquad \Big|\int_{\beta_n}^{\pi} f\circ g(t)\,\frac{\sin m_n t}{t}\,dt\Big| < \frac{1}{2^n}, \quad\text{where}\quad \beta_n = (2k_{m_n}+1)\pi/m_n,$$

$$\text{(iii}_n\text{)}\qquad \Big|\alpha_{m_n} - \sum_{i=1}^{k_{m_n}}\frac{1}{i}\Big\{f\circ g\Big(\frac{2i\pi}{m_n}+\theta\Big) - f\circ g\Big(\frac{(2i-1)\pi}{m_n}+\theta\Big)\Big\}\Big| < \frac{1}{2^n}$$
$$\text{whenever}\quad 0<\theta<\frac{\pi}{m_n}-\varepsilon_{m_n}.$$

With

$$h(t) = f\circ g(t) - f(0+)$$

we shall exhibit equivalent forms of (ii_n) and (iii_n). Indeed,

$$\left| \left| \int_{\beta_n}^{\pi} h(t) \frac{\sin m_n t}{t} \, dt \right| - \left| \int_{\beta_n}^{\pi} f \circ g(t) \frac{\sin m_n t}{t} \, dt \right| \right| < |f(0+)| \left| \int_{\beta_n}^{\pi} \frac{\sin m_n t}{t} \, dt \right|,$$

and the right side can be made arbitrarily small by selecting m_n sufficiently large. So (ii_n) and (ii'_n) below will be equivalent for a suitably chosen m_n.

$$(\text{ii}'_n) \qquad \left| \int_{\beta_n}^{\pi} h(t) \frac{\sin m_n t}{t} \, dt \right| < \frac{1}{2^n}.$$

Clearly (iii_n) is equivalent to

$$(\text{iii}'_n) \qquad \left| \alpha_{m_n} - \sum_{i=1}^{k_{m_n}} \frac{1}{i} \left\{ h\Big(\frac{2i\pi}{m_n} + \theta\Big) - h\Big(\frac{(2i-1)\pi}{m_n} + \theta\Big) \right\} \right| < \frac{1}{2^n}$$

$$\text{whenever} \quad 0 < \theta < \frac{\pi}{m_n} - \varepsilon_{m_n}.$$

Now choose m_{r+1} so that

$$\beta_{r+1} < \pi/m_r, \quad \tau'_{m_{r+1}, k_{m_{r+1}}} < \tau_{m_r,1},$$

and (i_{r+1}) is satisfied. Choose it so large that (ii_{r+1}) and (ii'_{r+1}) are satisfied when g is extended as the increasing linear mapping of $[\beta_{r+1}, \pi/m_r]$ onto $[b_{r+1}, \tau_{m_r,1}]$, where b_{r+1} is to the left of and sufficiently close to $\tau_{m_r,1}$. Next, extend g as an increasing, continuous, piecewise linear map of $[\pi/m_{r+1}, \beta_{r+1}]$ onto $[\tau_{m_{r+1}}, b_{r+1}]$ in a manner analogous to the definition of g on $[\pi/m_1, \beta_1]$. In particular, it is extended in such a manner that (iii_{r+1}) and hence (iii'_{r+1}) holds.

Thus we have defined an increasing sequence of integers m_n, $n = 1, 2, \ldots$, with properties (i_n), (ii'_n) and (iii'_n) for all n. By setting $g(0) = 0$, we have a self-homeomorphism of $[0, \pi]$. We claim that

$$\liminf_{n \to \infty} \int_0^{\pi} h(t) \frac{\sin m_n t}{t} \, dt = \liminf_{n \to \infty} \int_0^{\pi} (f \circ g(t) - f(0+)) \frac{\sin m_n t}{t} \, dt \geq \frac{\alpha}{\pi}.$$

To see this we consider

$$\int_0^{\pi} h(t) \frac{\sin m_n t}{t} \, dt = \int_0^{\beta_n} h(t) \frac{\sin m_n t}{t} \, dt + \int_{\beta_n}^{\pi} h(t) \frac{\sin m_n t}{t} \, dt$$
$$= A_n + B_n.$$

By Lemma 10.16, for each n, there exists a θ_{m_n} with

$$0 < \theta_{m_n} < \frac{\pi}{m_n} - \varepsilon_{m_n},$$

such that A_n is equiconvergent with $\pi^{-1} c(H_{m_n}, \theta_{m_n})$. As H_{m_n} is a regulated function, there is a point $\theta^*_{m_n}$ at which H_{m_n} is continuous such that

$$0 < \theta^*_{m_n} < \frac{\pi}{m_n} - \varepsilon_{m_n} \quad \text{and} \quad c(H_{m_n}, \theta^*_{m_n}) - \frac{1}{2^n} < c(H_{m_n}, \theta_{m_n}).$$

Thus

$$\begin{aligned} c(H_{m_n}, \theta_{m_n}) &> c(H_{m_n}, \theta^*_{m_n}) - \frac{1}{2^n} \\ &> \alpha_{m_n} - \frac{1}{2^n} - \frac{1}{2^n} && \text{by } (\mathrm{iii}'_{m_n}) \\ &> \alpha\left(1 - \frac{1}{2^n}\right) - \frac{1}{2^{n-1}} && \text{by } (\mathrm{i}_{m_n}). \end{aligned}$$

Therefore

$$A_n = \frac{1}{\pi}\, c(H_{m_n}, \theta_{m_n}) + o(1) > \frac{\alpha}{\pi} + o(1).$$

Since $B_n < \frac{1}{2^{n-1}}$ (by property (ii'_{m_n})), we have

$$A_n + B_n > \frac{\alpha}{\pi} + o(1),$$

implying

$$\liminf_{n\to\infty} \int_0^{\pi} (f \circ g(t) - f(0+))\, \frac{\sin m_n t}{t}\, dt \geq \frac{\alpha}{\pi}.$$

The required self-homeomorphism g of $[0, \pi]$ has now been constructed.

Finally, we turn to the self-homeomorphism of $[-\pi, 0]$. We will show that for any regulated function p satisfying $p(0-) = 0$ there is a homeomorphism g of $[-\pi, 0]$ onto itself with the property that

$$\limsup_{n\to\infty} \int_{-\pi}^{0} p \circ g(t)\, \frac{\sin m_n t}{t}\, dt \geq 0,$$

where m_n, $n = 1, 2, \ldots$, is any increasing sequence of positive integers. Later we will set $p(t) = f(t) - f(0-)$. The construction of g will be made inductively on the intervals $[-\pi/r, -\pi/(r+1)]$, $r = 1, 2, \ldots$. We shall give the initial step and then next step. The inductive step will be evident from these two.

Step 1. We claim that we may choose ε'_1 in $(0, \pi/2)$, b_1 in $(-\pi, 0)$, m_{n_1}, and δ_1 in $(-\pi, 0)$ such that if

$$g(-\pi) = -\pi, \quad g\left(-\frac{\pi}{2} - \varepsilon'_1\right) = b_1, \quad g\left(-\frac{\pi}{2}\right) = \delta_1,$$

and g is linear on the intervals

$$\left[-\pi, -\frac{\pi}{2} - \varepsilon'_1\right] \quad \text{and} \quad \left[-\frac{\pi}{2} - \varepsilon'_1, -\frac{\pi}{2}\right],$$

then, no matter how g is defined on $[-\pi/2, 0]$,

$$\left| \int_{-\pi}^{0} p \circ g(t)\, \frac{\sin m_{n_1} t}{t}\, dt \right| < 1.$$

To see this we first write

$$\begin{aligned} \int_{-\pi}^{-\pi/2} p \circ g(t)\, \frac{\sin m_{n_1} t}{t}\, dt &= \int_{-\pi}^{-\pi/2 - \varepsilon'_1} p \circ g(t)\, \frac{\sin m_{n_1} t}{t}\, dt \\ &\qquad + \int_{-\pi/2 - \varepsilon'_1}^{-\pi/2} p \circ g(t)\, \frac{\sin m_{n_1} t}{t}\, dt \\ &= A'_1 + B'_1. \end{aligned}$$

Now

$$\begin{aligned}|A_1'| &= \Big| \int_{-\pi}^{-\pi/2-\varepsilon_1'} \big(p\circ g(t) - p(-\pi+) + p(-\pi+)\big) \frac{\sin m_{n_1} t}{t}\, dt \Big| \\ &\le C \sup_{-\pi\le t\le b_1} |p(t) - p(-\pi+)| + \frac{2\,|p(-\pi+)|}{m_{n_1}}.\end{aligned}$$

The first term of the last line can be made small by choosing b_1 close enough to $-\pi$. The second is small for large m_{n_1}. And B_1' can be made small by choosing ε_1' small, for

$$|B_1'| = \Big| \int_{-\pi/2-\varepsilon_1'}^{-\pi/2} p\circ g(t) \frac{\sin m_{n_1} t}{t}\, dt \Big| \le \varepsilon_1' \frac{2}{\pi} \sup_{t\in(-\pi,0)} |p(t)|.$$

For the integral over $[-\pi/2, 0]$ that remains, we have

$$\Big| \int_{-\pi/2}^{0} p\circ g(t) \frac{\sin m_{n_2} t}{t}\, dt \Big| \le \sup_{t\in(\delta_1,0)} |p(t)| \frac{\pi}{2} m_{n_1}$$

no matter what g is on $[-\pi/2, 0]$. For a fixed m_{n_1} this quantity is small whenever $|\delta_1|$ is small. Thus we can find ε_1', δ_1, b_1 and m_{n_1} so that

$$\Big| \int_{-\pi}^{0} p\circ g(t) \frac{\sin m_{n_1} t}{t}\, dt \Big| < 1$$

regardless of the definition of g on $[-\pi/2, 0]$.

Step 2. We now choose ε_2', b_2, m_{n_2} and δ_2 and define g on $[-\pi/2, -\pi/4]$ so that

$$\Big| \int_{-\pi}^{0} p\circ g(t) \frac{\sin m_{n_2} t}{t}\, dt \Big| < \frac{1}{2}$$

independent of the definition of g on $[-\pi/4, 0]$.

For ε_2' in $(0, \pi/4)$, and for b_2 and δ_2 such that

$$\delta_1 < b_2 < \delta_2 < 0,$$

set

$$g\Big(-\frac{\pi}{4} - \varepsilon_2'\Big) = b_2 \quad \text{and} \quad g\Big(-\frac{\pi}{4}\Big) = \delta_2$$

and extend g to be linear on the intervals

$$\Big[-\frac{\pi}{2}, -\frac{\pi}{4} - \varepsilon_2'\Big] \quad \text{and} \quad \Big[-\frac{\pi}{4} - \varepsilon_2', -\frac{\pi}{4}\Big].$$

Let

$$\begin{aligned}\int_{-\pi/2}^{-\pi/4} p\circ g(t) \frac{\sin m_{n_2 t}}{t}\, dt &= \int_{-\pi/2}^{-\pi/4-\varepsilon_2'} (\cdots)\, dt + \int_{-\pi/4-\varepsilon_2'}^{-\pi/4} (\cdots)\, dt \\ &= A_2' + B_2'.\end{aligned}$$

Now

$$\begin{aligned}|A_2'| &= \Big| \int_{-\pi/2}^{-\pi/4-\varepsilon_2'} \big(p\circ g(t) - p(\delta_1+) + p(\delta_1+)\big) \frac{\sin m_{n_2} t}{t}\, dt \Big| \\ &\le C \sup_{t\in(\delta_1,b_2)} |p(t) - p(\delta_1+)| + \frac{C'\,|p(\delta_1+)|}{m_{n_2}}.\end{aligned}$$

The first term can be made small if b_2 is close enough to δ_1. The second can be made small if m_{n_2} is large enough. Now

$$|B_2'| = \left| \int_{-\pi/4-\varepsilon_2'}^{-\pi/4} p \circ g(t) \, \frac{\sin m_{n_2} t}{t} \, dt \right| \le \varepsilon_2' \, \frac{4}{\pi} \sup_{t \in (-\pi,0)} |p(t)| \,,$$

so B_2' can be made small if ε_2' is small. Also

$$\left| \int_{-\pi/4}^{0} p \circ g(t) \, \frac{\sin m_{n_2} t}{t} \, dt \right| \le \sup_{t \in (\delta_2,0)} |p(t)| \, m_{n_2} \,,$$

and for a fixed m_{n_2} this can be made small if $|\delta_2|$ is small. These values may be chosen so that

$$\left| \int_{-\pi}^{0} p \circ g(t) \, \frac{\sin m_{n_2} t}{t} \, dt \right| < \frac{1}{2} \,,$$

and the second step is completed.

We may continue this process so that at the r-th step we have

$$\left| \int_{-\pi}^{0} p \circ g(t) \, \frac{\sin m_{n_r} t}{t} \, dt \right| < \frac{1}{2^{r-1}} \,.$$

Hence

$$\limsup_{n \to \infty} \int_{0}^{\pi} \left(f \circ g(-t) - f \circ g(0-) \right) \frac{\sin m_n t}{t} \, dt \ge 0$$

and the theorem now follows from the description of our proof at the beginning.

10.3. Uniform convergence of Fourier series

We have seen in the last section that the GW condition of Section 10.2 is necessary and sufficient for the preservation of the pointwise convergence of $S(f \circ g)$ for every change of variable g. We now turn our attention to the analogous problem of the preservation of uniform convergence.

First note that if a function f is the uniform sum of its Fourier series, then $f \in \mathrm{C}(T)$. Note also that the continuous functions of bounded variation have uniformly convergent Fourier series [**147**, volume I, pages 57–59] and the variation of a function is invariant under change of variable. As we shall see in Section 11.2, there are other, more general, definitions of bounded variation which suffice to produce uniform convergence.

We define here the UGW condition which will be shown to be necessary and sufficient for the uniform convergence of the Fourier series $S(f \circ g)$ for every change of variable g whenever f is continuous. We remind the reader that the definition of right and left systems of intervals at x is given on page 138.

DEFINITION 10.17. A function f satisfies the **UGW condition** if, for every x, each right and each left system $\mathcal{I} = \{ I_{ni} : i = 1, 2, \ldots, k_n, \ n = 1, 2, \ldots \}$ at x satisfies

$$\text{(h)} \qquad \lim_{n \to \infty} \sum_{i=1}^{k_n} i^{-1} f(I_{ni}) = 0$$

and

$$\text{(hh)} \qquad \lim_{n\to\infty} \sum_{i=1}^{k_n} (k_n + 1 - i)^{-1} f(I_{ni}) = 0.$$

We note again that we can relax the "disjoint" requirement to nonoverlapping in the definition of systems. As the the UGW condition implies the GW condition, if f satisfies condition UGW then f is equal almost everywhere to a regulated function. Note that step functions satisfy both equations (h) and (hh). Hence any characterization that uses condition UGW must also include continuity.

Our principal result is the theorem of Baernstein and Waterman [**5**].

THEOREM 10.18 (BAERNSTEIN-WATERMAN). *For $f \in \mathrm{C}(T)$, condition* UGW *is necessary and sufficient for the uniform convergence of $S(f \circ g)$ for every change of variable g.*

We shall also show that there are functions in $\mathrm{C}(T)$ which satisfy condition GW but not UGW.

In Proposition 10.14 we gave a condition on finite sequences of intervals which was equivalent to GW. Here we will give two conditions each of which is equivalent to UGW and continuity. Indeed, the next proposition shows that the UGW condition with continuity is really "uniform GW".

PROPOSITION 10.19. *For functions f defined on T, the condition that f be continuous and satisfy condition* UGW *is equivalent to each of the following conditions*:

(A) *Given $\varepsilon > 0$ there is a positive number δ such that, for every x, each finite sequence I_j, $j = 1, 2, \ldots, k$, of nonoverlapping intervals indexed from left to right with $\bigcup_{j=1}^{k} I_j \subset (x, x+\delta)$ or indexed from right to left with $\bigcup_{j=1}^{k} I_j \subset (x-\delta, x)$ satisfies*

$$\left|\sum_{j=1}^{k} j^{-1} f(I_j)\right| < \varepsilon.$$

(B) *Given $\varepsilon > 0$ there is a positive number δ such that*

$$\left|\sum_{j=1}^{k} j^{-1} f(I_j)\right| < \varepsilon$$

whenever I_j, $j = 1, 2, \ldots, k$, is a finite sequence of nonoverlapping intervals indexed from right to left or left to right with

$$\operatorname{diam} \bigcup_{j=1}^{k} I_j < \delta.$$

PROOF. Let us show that (A) and (B) are equivalent. It is obvious that (B) implies (A). Now suppose that (A) holds with a pair ε and δ and let I_j, $j = 1, 2, \ldots, k$, be a finite sequence of intervals as in (B) with $\operatorname{diam} \bigcup_{j=1}^{k} I_j < \delta$. If the sequence is indexed from left to right, there is an x such that $\bigcup_{j=1}^{k} I_j \subset (x, x+\delta)$. The analogous interval $(x-\delta, x)$ exists when the sequence is indexed form right to left. Hence (B) follows.

As condition (B) is stronger than uniform continuity, we obviously have that (B) implies continuity.

Now suppose that

$$\mathcal{I} = \{ I_{ni} : i = 1, 2, \dots, k_n,\ n = 1, 2, \dots \}$$

is a left or a right system at x for which (h) or (hh) of the UGW condition fails. Then, for some positive number ε,

$$\limsup_{n\to\infty} \Big| \sum_{i=1}^{k_n} i^{-1} f(I_{ni}) \Big| > \varepsilon$$

or

$$\limsup_{n\to\infty} \Big| \sum_{i=1}^{k_n} (k_n + 1 - i)^{-1} f(I_{ni}) \Big| > \varepsilon,$$

respectively. Clearly, condition (B) fails because diam $\bigcup_{i=1}^{k_n} I_{ni} \to 0$ as $n \to \infty$. Hence (B) implies continuity and UGW.

In the other direction, assume that (B) fails for some continuous f. Then for some positive number ε and for each n there is a finite sequence of nonoverlapping intervals I_j^n with diam $\bigcup_j I_j^n \to 0$ as $n \to \infty$ and

$$\Big| \sum_j j^{-1} f(I_j^n) \Big| > \varepsilon \quad \text{for} \quad n = 1, 2, \dots.$$

Without loss of generality we may assume that, for each n, the finite sequence of intervals I_j^n is indexed from left to right. Now

$$\Big| \sum_j j^{-1} f(I_j^n) \Big| \le \sum_{\{ j : f(I_j^n) > 0 \}} j^{-1} f(I_j^n) + \Big| \sum_{\{ j : f(I_j^n) < 0 \}} j^{-1} f(I_j^n) \Big| .$$

Consider one of the terms on the right which exceeds $\varepsilon/2$. By shifting the indices if necessary, replacing the j which appear with $j = 1, 2, \dots$, in the same order, we will obtain a collection of intervals I_j^n such that

$$\Big| \sum_j j^{-1} f(I_j^n) \Big| > \varepsilon/2 \tag{10.4}$$

and the terms in the sum are different from 0 and of the same sign, say positive.

For each n, let a_n be the left end point of I_1^n. Since f is continuous, we can move the a_n's very slightly as necessary so that the sequence a_n, $n = 1, 2, \dots$, is one-to-one and condition (10.4) is still satisfied. By passing to a subsequence if necessary, we see that we may assume that the sequence is monotone and converges to some x. Thus we have either $a_n < x$ for all n or $a_n > x$ for all n.

In the first case, (h) fails for the right system $I_{nj} = I_j^n$, $j = 1, 2, \dots, k_n$, at x. Let us consider the second case. Let j_n be the greatest j such that the left end point of I_j^n is at most x. The continuity of f implies $f(I_{j_n}^n) \to 0$ as $n \to \infty$. Thus for large n we have either

$$\sum_{j > j_n} j^{-1} f(I_j^n) > \varepsilon/4 \tag{10.5}$$

or

$$\sum_{j < j_n} j^{-1} f(I_j^n) > \varepsilon/4 . \tag{10.6}$$

If (10.5) holds for infinitely many n, then for those n the collection of intervals $I_{nj} = I_{j_n + j}$, $j = 1, 2, \dots, k_n - j_n$, form a right system at x for which (h) fails. If (10.6) holds for infinitely many n, then for those n the collection $I_{nj} = I_{j_n}$, $j = 1, 2, \ldots, j_n - 1$, form a left system at x for which (hh) fails.

The proposition is now proven.

We turn now to the proof of Theorem 10.18. The argument presented here follows the lines of the paper of Baernstein and Waterman. Another argument, which is in the spirit of the proof offered for Theorem 10.12, was given in [**133**] using approximation results found in [**137**].

Proof of sufficiency of UGW condition. Suppose that f satisfies the condition (B) of Proposition 10.19.

Since $f \in C(T)$, Proposition 10.6 asserts that

$$f(x) - S_n(x; f) = \frac{1}{\pi^2}\left(P_n^R(x; f) + P_n^L(x; f)\right) + o(1) \quad \text{as} \quad n \to \infty$$

uniformly in x, where the n's are odd. We will show that

$$P_n^R(x; f) + P_n^L(x; f) \to 0 \quad \text{uniformly as} \quad n \to \infty.$$

Consider $P_n^R(x; f)$. It will be useful to put it in the interval function notation $f(I)$. To this end, let n be odd. For any odd k with $0 \le k \le n$, let

$$I_i^n(x, t) = \left[x + \frac{t + k\pi}{n}, x + \frac{t + (k+1)\pi}{n}\right]$$

and, for any even k with $0 \le k \le n$, let

$$I_i^n(x, t) = \left[x + \frac{t + k\pi}{n}, x + \frac{t + k\pi}{n}\right] \quad \text{(a degenerate interval).}$$

Then $P_n^R(x; f)$ becomes

$$P_n^R(x; f) = \int_0^\pi \sum_{k=1}^n \frac{1}{k} f\left(I_i^n(x, t)\right) \sin t \, dt.$$

Given $\varepsilon > 0$, choose δ as in Proposition 10.6 (B) and let m_n be the integer such that

$$\pi^{-1} n\delta - 1 \le m_n < \pi^{-1} n\delta + 1.$$

Since $\operatorname{diam} \bigcup_{i=1}^{m_n - 1} I_i(t) = (m_n - 1)\pi/n < \delta$ for every t and every x, we have

$$\begin{aligned} |P_n^R(x; f)| &\le \int_0^\pi \left| \sum_{i=1}^{m_n - 1} \frac{1}{i} f\left(I_i^n(x, t)\right)\right| \sin t \, dt \\ &\qquad + \int_0^\pi \left| \sum_{i=m_n}^{n} \frac{1}{i} f\left(I_i^n(x, t)\right)\right| \sin t \, dt \\ &\le \pi\varepsilon + \pi\,\omega(\pi/n; f) \sum_{i=m_n}^{n} \frac{1}{i}. \end{aligned}$$

Since $m_n > (2\pi)^{-1} n\delta$,

$$\omega(\pi/n; f) \sum_{i=m_n}^{n} \frac{1}{i} = o(1)O(1) \quad \text{as} \quad n \to \infty.$$

Thus, for all n large enough and all x,

$$|P_n^R(x; f)| < 2\pi\varepsilon.$$

Since $P_n^L(x;f)$ can be treated analogously, $S(f)$ converges uniformly. Clearly the UGW condition and the equivalent conditions of Proposition 10.19 are preserved by change of variable. Hence $S(f \circ g)$ converges uniformly for every change of variable g. The sufficiency of the UGW condition for $f \in C(T)$ has now been proven.

PROOF OF THE NECESSITY OF THE UGW CONDITION. We have seen in Section 10.2 that if (h) is false for some system, then a change of variable g may be constructed so that $S(f \circ g)$ diverges at some point. Thus we need only consider the case when (hh) fails. We shall show that this case leads to the existence of a change of variable g such that $P_n^R(x; f \circ g) + P_n^L(x; f \circ g)$ fails to converge uniformly to 0.

Assuming that (hh) fails for a right system at $x = 0$, there exists a positive number α and there exist positive even integers k_p and points $x_{p,i}$ for each i with $0 \le i \le k_p$ such that $x_{p,0}$ decreases to 0 as $p \to \infty$ and such that

$$0 < x_{p+1,0} < x_{p,k_p} \le x_{p,i+1} < x_{p,i} < \pi,$$

$$\sum_{1}^{k_p-1}{}^{o} \frac{1}{i}\big(f(x_{p,i+1}) - f(x_{p,i})\big) > \alpha, \tag{10.7}$$

and

$$|f(x) - f(y)| < \alpha\, 4^{-p} \quad \text{whenever} \quad 0 \le x \le y \le x_{p,0}. \tag{10.8}$$

Once again $\sum^o$ denotes summation over odd indices. Let n_p, $p = 1, 2, \ldots$, be an increasing sequence of odd integers such that

$$n_p > 2^{p+2} k_p \quad \text{and} \quad \lim_{p\to\infty} \omega(\pi/n_p; f)p = 0. \tag{10.9}$$

Let $v_{p,i}$ and $y_{p,i}$, $i = 0, 1, \ldots, k_p$, be points satisfying

$$\frac{\pi}{2^p} - \frac{(i+1)\pi}{n_p} < v_{p,i} < \frac{\pi}{2^p} - \frac{i\pi}{n_p} \quad \text{and} \quad x_{p,i+1} < y_{p,i} < x_{p,i}$$

whenever $0 \le i \le k_p - 1$, and

$$\frac{\pi}{2^{p+1}} < v_{p,k_p} \le \frac{\pi}{2^{p+1}} + \frac{\pi}{n_{p+1}} < \frac{\pi}{2^p} - \frac{k_p\pi}{n_p} \quad \text{and} \quad x_{p+1,0} < y_{p,k_p} < x_{p,k_p}.$$

Let $x = g(u)$ be a strictly increasing continuous function on $[-\pi, \pi]$ such that

$$g(u) = u \quad \text{for} \quad -\pi \le u \le 0,$$
$$g(\pi) = \pi,$$

and, for $i = 0, 1, \ldots, k_p$, $p = 1, 2, \ldots$,

$$g\Big(\frac{\pi}{2^p} - \frac{\pi i}{n_p}\Big) = x_{p,i},$$
$$g(v_{p,i}) = y_{p,i}.$$

If we choose $v_{p,i}$ close enough to $2^{-p}\pi - n_p^{-1}(i+1)\pi$ and $y_{p,i}$ close enough to $x_{p,i}$, $i = 1, 2, \ldots, k_p - 1$, we have

$$\Big|\int_0^\pi f \circ g\Big(\frac{\pi}{2^p} - \frac{t + i\pi}{n_p}\Big) \sin t\, dt - 2f(x_{p,i})\Big| < \frac{1}{p\, 2^{i+1}}. \tag{10.10}$$

Let r_p and q_p, for $p \geq 2$, be the unique odd integers such that

$$r_p < 2^{-p+1} n_p \leq r_p + 2, \tag{10.11}$$

and

$$q_p < 2^{-p} n_p - k_p \leq q_p + 2. \tag{10.12}$$

We choose y_{p,k_p} so close to x_{p,k_p} that (10.10) holds for $i = k_p$ and, for $0 \leq t \leq \pi$,

$$\left| f \circ g\Big(\frac{\pi}{2^p} - \frac{t + i\pi}{n_p}\Big) - f \circ g\Big(\frac{\pi}{2^p} - \frac{t + (i+1)\pi}{n_p}\Big) \right| < \frac{1}{p\, r_p} \tag{10.13}$$

whenever $k_p \leq i \leq r_p - 2$, and

$$\left| f \circ g\Big(\frac{\pi}{2^{p+1}} + \frac{t + i\pi}{n_{p+1}}\Big) - f \circ g\Big(\frac{\pi}{2^{p+1}} + \frac{t + (i+1)\pi}{n_{p+1}}\Big) \right| < \frac{1}{p\, q_{p+1}} \tag{10.14}$$

whenever $1 \leq i \leq q_{p+1} - 2$.

We will show that

$$\left| P^L_{n_p}\Big(\frac{\pi}{2^p}; f \circ g\Big) \right| \geq 2\alpha + o(1) \quad \text{as} \quad p \to \infty \tag{10.15}$$

and

$$\left| P^R_{n_p}\Big(\frac{\pi}{2^p}; f \circ g\Big) \right| \leq \alpha + o(1) \quad \text{as} \quad p \to \infty. \tag{10.16}$$

In view of Proposition 10.6 it will follow that $S(f \circ g)$ does not converge uniformly and the proof will be complete.

We split the sum defining $P^L_{n_p}(2^{-p}\pi; f \circ g)$ into the five parts,

$$\begin{aligned} P^L_{n_p}\Big(\frac{\pi}{2^p}; f \circ g\Big) &= \Big\{ \sum_{i=1}^{k_p - 1}{}^{o} + \sum_{i=k_p+1}^{r_p - 2}{}^{o} + \sum_{i=r_p}^{r_p^* - 2}{}^{o} + \sum_{i=r_p^*}^{r_p^*}{}^{o} + \sum_{i=r_p^*+2}^{n_p - 2}{}^{o} \Big\} \frac{1}{i} \int_0^\pi \Big\{ f \circ g\Big(\frac{\pi}{2^p} - \frac{t + i\pi}{n_p}\Big) \\ &\qquad - f \circ g\Big(\frac{\pi}{2^p} - \frac{t + (i+1)\pi}{n_p}\Big)\Big\} \sin t \, dt = \mathrm{I} + \mathrm{II} + \mathrm{III} + \mathrm{IV} + \mathrm{V}, \end{aligned}$$

where r_p is as in (10.11) and r_p^* is the unique odd integer such that

$$r_p^* < 2^p n_p \leq r_p^* + 2.$$

From (10.7) and (10.10) we have

$$\begin{aligned} |\,\mathrm{I}\,| &= \Big| \sum_{i=1}^{k_p - 1}{}^{o} \frac{1}{i} \int_0^\pi \Big\{ f \circ g\Big(\frac{\pi}{2^p} - \frac{t + i\pi}{n_p}\Big) \\ &\qquad - f \circ g\Big(\frac{\pi}{2^p} - \frac{t + (i+1)\pi}{n_p}\Big) \Big\} \sin t \, dt \Big| \\ &> 2 \Big| \sum_{i=1}^{k_p - 1}{}^{o} \frac{1}{i} \left(f(x_{p,i}) - f(x_{p,i+1}) \right) \Big| - \sum_{i=1}^{k_p} \frac{1}{p 2^{i+1}} > 2\alpha - \frac{1}{p}. \end{aligned}$$

In the sum II there are less than r_p terms, so (10.13) implies $|\,\mathrm{II}\,| < p^{-1}$. From (10.8) we see that the absolute value of each integral in the sum III is smaller than $2\,\alpha\, 4^{-p}$. Since $(r_p^* - 2)/r_p < 3$, we have $\mathrm{III} = O(4^{-p})$ as $p \to \infty$. The sum IV is $o(1)$ as $p \to \infty$ since it consists of a single term, $f \circ g$ is continuous, and $n_p \to \infty$.

Finally, since $2^{-p} > (2+r_p^*)/n_p$ and $g(u) = u$ on $[-\pi, 0]$, we can replace $f \circ g$ by f in each integral of the sum V. Using the second condition in (10.9) we have

$$\mathrm{V} = O\big(\omega(\pi/n_p; f)\log p\big) = o(1) \quad \text{as} \quad p \to \infty$$

and we have established (10.15).

To establish (10.16), fix p and, for $j = 1, \ldots, p-1$, let β_j be the odd integers such that $\beta_j < 2^{-p}n_p(2^{p-j}-1) \le \beta_j + 2$. Decompose $P_{n_p}^R(2^{-p}\pi; f \circ g)$:

$$\begin{aligned} P_{n_p}^R\Big(\frac{\pi}{2^p}; f \circ g\Big) &= \Big\{\sum_{i=1}^{q_p-1}{}^{o} + \sum_{i=q_p}^{\beta_{p-1}-2}{}^{o} + \sum_{j=2}^{p-1}\sum_{i=\beta_{p-j+1}}^{\beta_{p-j}-2}{}^{o} + \sum_{i=\beta_1}^{n_p-2}{}^{o}\Big\}\frac{1}{i}\int_0^{\pi}\Big\{f \circ g\Big(\frac{\pi}{2^p} + \frac{t+i\pi}{n_p}\Big) \\ &\qquad - f \circ g\Big(\frac{\pi}{2^p} + \frac{t+(i+1)\pi}{n_p}\Big)\Big\}\sin t\,dt \\ &= Z_0 + Z_1 + \sum_{j=2}^{p-1} Z_j + Z_p, \end{aligned}$$

where q_r is as in (10.12).

Applying (10.14) with p in place of $p+1$, we have $Z_0 = o(1)$ as $p \to \infty$. From (10.8) we have

$$|\,Z_j\,| < 2\alpha\, 4^{-(p-j)} \sum_{i=\beta_{p-j+1}}^{\beta_{p-j}-2}{}^{o}\ \frac{1}{i} \quad \text{whenever} \quad j = 1, \ldots, p-1. \tag{10.17}$$

For $j \ge 2$, a straight forward computation shows that

$$(\beta_{p-j} - 2)/\beta_{p-j+1} < 4.$$

We also have

$$(\beta_{p-1} - 2)q_p < 3$$

from inequality (10.9) and the fact that $n_p > n_{p-1}$. Thus each sum in (10.17) is less than $\log 4$ and we have

$$\sum_{j=1}^{p-1} |\,Z_j\,| < 2\alpha(\log 4)\sum_{j=1}^{p-1} 4^{-p-j} < \frac{2}{3}\alpha \log 4 < \alpha.$$

Each integral in the sum Z_p has absolute value less than $2\,\omega(\pi/n_p; f \circ g)$. Since $(\beta_1 + 2)/n_p \ge 3/8$ for $p \ge 3$ we have

$$Z_p = O\big(\omega(\pi/n_p; f \circ g)\big) = o(1) \quad \text{as} \quad p \to \infty.$$

Thus (10.16) has been established and the theorem has been proven.

We close the chapter by giving an example of a continuous function f which satisfies the GW condition but not the UGW condition. Let a_n, $n = 1, 2, \ldots$, and b_n, $n = 1, 2, \ldots$, be sequences in $(0, \pi)$ with $a_n \to 0$ as $n \to \infty$ and

$$a_{n+1} < b_{n+1} < a_n < b_n.$$

Let λ_n, $n = 1, 2, \ldots$, be a decreasing sequence of positive numbers such that

$$\sum_{n=1}^{\infty} \lambda_n < \infty.$$

Choose a sequence of integers k_n, $n = 1, 2, \ldots$, such that

$$\lambda_n \sum_{i=1}^{k_n} \big(i \log(i+1)\big)^{-1} \geq 1. \tag{10.18}$$

For each n let

$$x_{n,i} = a_n + (b_n - a_n)i/(2k_n) \quad \text{for} \quad 0 \leq i \leq 2k_n.$$

Define f at each $x_{n,i}$ to be

$$f(x_{n,i}) = \begin{cases} \lambda_n / \log\big(k_n - (i-3)/2\big) & \text{for odd } i \text{ in } [0, 2k_n], \\ 0 & \text{for even } i \text{ in } [0, 2k_n], \end{cases}$$

and define f to be linear on each interval $[\,x_{n,i},\, x_{n,i+1}\,]$ and to be 0 elsewhere in T. Then $f \in \mathrm{C}(T)$.

Note that

$$f([\,x_{n,2j},\, x_{n,2j+1}\,]) = f(x_{n,2j+1}) = \frac{\lambda_n}{\log(k_n + 1 - j)}. \tag{10.19}$$

With the aid of (10.18) and (10.19) above, the finite systems

$$\{\,[\,x_{n,2j},\, x_{n,2j+1}\,] : j = 0, 1, \ldots, k_n - 1\}, \quad n = 1, 2, \ldots,$$

can be used to show that f fails to satisfy the necessary and sufficient condition (B) of Proposition 10.19 for continuous functions to satisfy the UGW condition.

Let us turn to the GW condition. We use the characterization given in Proposition 10.14. Let $x \neq 0$ and $\varepsilon > 0$. There is a positive number δ such that the total variation of f satisfies $\mathrm{V}(f, [x - \delta, x + \delta]) < \varepsilon$. Hence the condition in the characterization holds for $x \neq 0$. To handle $x = 0$, we start by giving names to the intervals constructed above. For each n, let

$$J_{n,i} = [\,x_{n,i},\, x_{n,i+1}\,] \quad \text{for} \quad 0 \leq i \leq 2k_n - 1.$$

The collection

$$\mathcal{J} = \{\, J_{n,2j} : j = 0, 1, \ldots, k_n - 1,\ n = 1, 2, \ldots\}$$

will play an important role in the estimates computed below. It is enough to compute upper estimates for the sums

$$\sum_{p=1}^{P} \frac{1}{p}\, |f(I_p)|$$

where

$$\{\, I_p : p = 1, 2, \ldots, P\}$$

are finite collections of nonoverlapping intervals indexed from left to right with

$$\bigcup\nolimits_{p=1}^{P} I_p \subset (0, b_m).$$

Clearly we may assume $|f(I_p)| > 0$ for every p. Let us show that, for each p, we may assume that there is a $J_{n,i}$ with

$$I_p \subset J_{n,i}.$$

Indeed, suppose that $I_p = [c, d]$ is such that $f(c) > f(d)$, $c \in J_{n_1,i_1}$ and $d \in J_{n_2,i_2}$. If, on the one hand, $J_{n_1,i_1} = J_{n_2,i_2}$, then $I_p \subset J_{n_1,i_1}$. And, on the other hand, if

$J_{n_1,i_1} \neq J_{n_2,i_2}$, then J_{n_1,i_1} is to the left of J_{n_2,i_2}. Hence $I'_p = I_p \cap J_{n_1,i_1}$ is a closed interval with

$$|f(I_p)| < f(c) = |f(I'_p)|.$$

So we may replace I_p with I'_p in our computations for bounds. Of course, the case $f(c) < f(d)$ is treated in an analogous manner.

For each n and each j, let

$$\mathcal{P}_{n,j} = \{ p : I_p \subset J_{n,2j} \cup J_{n,2j+1} \}.$$

When $\mathcal{P}_{n,j} \neq \emptyset$, a simple calculation will show that

$$\sum_{p \in \mathcal{P}_{n,j}} \frac{1}{p} |f(I_p)| \leq \frac{2}{p_{n,j}} |f(J_{n,2j})|,$$

where

$$p_{n,j} = \min \mathcal{P}_{n,j}.$$

Let

$$\{ j_{n,r} : r = 0, 1, \ldots, r_n \} = \{ j : \mathcal{P}_{n,j} \neq \emptyset \}.$$

Then

$$\begin{aligned} \sum_{p=1}^{P} \frac{1}{p} |f(I_p)| &\leq 2 {\sum_{n=m}^{\infty}}^{*} \sum_{r=0}^{r_n} \frac{1}{p_{n,j_{nr}}} |f(J_{n,2j_{nr}})| \\ &\leq 2 {\sum_{n=m}^{\infty}}^{*} \sum_{r=0}^{r_n} \frac{1}{r+1} |f(J_{n,2j_{nr}})|, \end{aligned}$$

where $\sum^*$ denotes summation over all n's for which $\mathcal{P}_{nj} \neq \emptyset$ for some j. As

$$f(J_{n,2j}) = \frac{\lambda_n}{\log(k_n + 1 - j)},$$
$$j_{n,0} < j_{n,1} < \cdots < j_{n,r_n} \leq k_n - 1 \quad \text{and} \quad r_n \leq k_n - 1,$$

we easily see that

$$f(J_{n,2j_{nr}}) \leq \frac{\lambda_n}{\log(r_n + 2 - r)} \quad \text{for} \quad r = 0, 1, \ldots, r_n.$$

Hence

$$\sum_{p=1}^{P} \frac{1}{p} |f(I_p)| \leq 2 {\sum_{n=m}^{\infty}}^{*} \lambda_n \sum_{r=0}^{r_n} \frac{1}{r+1} \frac{1}{\log(r_n + 2 - r)}.$$

Let us prove

$$\sum_{r=0}^{r_n} \frac{1}{(r+1)\log(r_n + 2 - r)} \leq 3.$$

Let q_n be the integer defined by

$$\frac{1}{2} r_n \leq q_n \leq \frac{1}{2} (r_n + 1).$$

Then

$$\begin{aligned}
\sum_{r=0}^{r_n}\big((r+1)\log(r_n+2-i)\big)^{-1} &= \sum_{r=0}^{q_n}(\cdots)+\sum_{i=q_n+1}^{r_n}(\cdots)\\
&\le \big(\log(r_n+2-q_n)\big)^{-1}\sum_{r=0}^{q_n}(r+1)^{-1}\\
&\qquad\qquad + (\log 2)^{-1}\sum_{r=q_n+1}^{r_n}(r+1)^{-1}\\
&\le \big(\log(q_n+1)\big)^{-1}(1+\log q_n)+(\log 2)^{-1}\log\frac{r_n}{q_n}\\
&< 2+1.
\end{aligned}$$

This establishes

$$\sum_{p=1}^{P}\frac{1}{p}\,|f(I_p)|\le 6\sum_{n=m}^{\infty}\lambda_n,$$

and the necessary and sufficient condition of Proposition 10.14 has been verified. Hence the continuous function f satisfies condition GW and fails to satisfy condition UGW.

CHAPTER 11

Fourier Series of Integrable Functions

In the previous chapter we exhibited the necessary and sufficient conditions for the convergence of $S(f \circ g)$ for every change of variable g under the assumption that f was regulated. Here we will drop that assumption, but it will be seen that it reappears in an essential way. In fact, in order that convergence or the order of magnitude of the coefficients be preserved, it is necessary that f be **λ-absolutely equivalent** to a regulated function, i.e., f must equal a regulated function except on a set all of whose homeomorphic images have λ-measure zero. (Recall that λ is Lebesgue measure.)

In Section 11.1 we will discuss λ-absolute equivalence, λ-absolute essential boundedness and integrability, and functions that are λ-absolutely equivalent to regulated functions. In Section 11.2 we will show that for an integrable function f, the Fourier series $S(f \circ g)$ converges everywhere for every change of variable g if and only if there is a function satisfying the GW condition (hence regulated) which is λ-absolutely equivalent to f. Various classes of functions of generalized bounded variation will be considered in Section 11.3. These classes are invariant under change of variable. Convergence and summability properties of Fourier series of function in these classes will be discussed and the order of magnitude of the Fourier coefficients will be computed. An application to localization theory of double Fourier series will be made. In the last section, Section 11.4, we show that the necessary and sufficient condition that the Fourier coefficients of $f \circ g$ have a fixed order of magnitude for all changes of variable g is that f be λ-absolutely equivalent to a function of generalized bounded variation.

11.1. Absolutely measurable functions

In the following sections of this chapter we will be dealing with the Fourier series of the compositions of a function f with every change of variable g. In order for this to be meaningful, we must have $f \circ g$ measurable[1] and integrable for every such g. Functions f with the property that $f \circ g$ remains measurable for each change of variable g have already been encountered in Section 2.2 of Chapter 2. We follow the lead of that section and call such functions **absolutely measurable**.

Associated with absolutely measurable functions are the **absolutely measurable sets**. Among these are those that have absolute measure 0. A set E has **absolute[2] measure zero** (AMZ) if the images $g[E]$ satisfy

$$\lambda(g[E]) = 0 \quad \text{for every change of variable} \quad g\,.$$

We refer to these sets as **AMZ-sets**. Obviously countable sets are AMZ-sets.

[1] We will suppress the symbol "λ" since only one-dimensional Lebesgue measure will be used.

[2] In the papers to which we refer in this chapter, "universal" was used rather than "absolute", so these sets are called UMZ-sets there.

PROPOSITION 11.1. *There exist uncountable* AMZ*-sets.*

PROOF. With the aid of the continuum hypothesis, an uncountable subset E of T whose intersection with every nowhere dense set in T is at most countable may be constructed [**54**, pages 146–147]. This property is preserved by any change of variable g. Given $\varepsilon > 0$, there is a nowhere dense perfect set P in T whose measure exceeds $2\pi - \varepsilon$. Thus $P \cap g[E]$ is at most countable, implying that the outer measure of $g[E]$ is at most ε, so that $g[E]$ must have measure zero, as was to be proved.

The construction we referred to in the above proof uses the continuum hypothesis. We have already made reference in Chapter 2 to the nice discussion of the above proposition in Oxtoby's book [**106**, pages 81 and 99]. Oxtoby points out that, without the aid of the continuum hypothesis, Sierpiński and Szpilrajn [**123**] have shown that the cardinality of the sets in the proposition is at least $\aleph_1$.

A function f is said to be **absolutely essentially bounded** if there is a real number M such that $\{ x : |f(x)| > M \}$ is an AMZ-set. The **absolute essential supremum** of f may be defined in the obvious way to be the infimum of all M such that $\{ x : f(x) > M \}$ is an AMZ-set, denoted $\operatorname{ab.ess.sup} f$. As sets E that are not AMZ-sets have an image $g[E]$ with positive outer measure for some change of variable g, it is clear that absolute essential upper bounds are invariant under change of variable.

If we are to study the Fourier series of $f \circ g$ for every change of variable g, we must have $f \circ g \in \mathrm{L}^1(T)$ for every change of variable g, i.e., f must be **absolutely integrable**. We shall assume that *the functions in this chapter are absolutely measurable.*

Goffman and Waterman gave a characterization of such functions in [**69**] without proof. Here we will remedy that. We begin by showing that absolutely essentially bounded functions may be described in other ways.

THEOREM 11.2. *The following are equivalent:*

(i) *f is absolutely essentially bounded.*
(ii) *There is a positive number M such that $|f \circ g(t)| \le M$ a.e. for every change of variable g.*
(iii) *$f \circ g$ is essentially bounded for every change of variable g.*

PROOF. It is clear that (i) $\Rightarrow$ (ii) $\Rightarrow$ (iii). It is easily seen that (ii) $\Rightarrow$ (i), for if M is as in (ii) and g is any change of variable, then $g[A] = \{ t : |f \circ g^{-1}(t)| > M \}$, which has measure 0, where $A = \{ t : |f(t)| > M \}$. Thus A is an AMZ-set.

We shall show that (iii) $\Rightarrow$ (ii). If f were such that the collection of all essential suprema of $|f \circ g|$, where g is any change of variable, is bounded above, we would have (ii); so let us assume for each k that there is an M_k and a corresponding self-homeomorphisms g_k of I with $\operatorname{ess.sup}|f \circ g_k| = M_k$ so that M_k, $k = 0, 1, 2, \ldots$, increases to ∞ and $\operatorname{ess.sup}|f| = M_0$. Let

$$H_k = \{ t : |f(t)| \in (M_{k-1}, M_k] \}$$

Then H_k, $k = 1, 2, \ldots$, are pairwise disjoint and $\lambda(H_k) = 0$ for each k.

Let

$$E_k = \{ x : |f \circ g_k(x)| \in (M_{k-1}, M_k] \} = g_k^{-1}[H_k].$$

Note that $\lambda(E_k) > 0$ for each k.

For $k = 1, 2, \dots$, choose t_k in H_k so that E_k has metric density one at $x_k = g_k^{-1}(t_k)$ and $x_k \neq x_{k'}$ whenever $k' < k$. By passing to a subsequence if necessary, we may assume that t_k, $k = 1, 2, \dots$, is monotonic, say increasing to $\bar{t}$. We can choose pairwise disjoint closed intervals I_k, $k = 1, 2, \dots$, such that t_k is an interior point of I_k for each k. Now $J_k = g_k^{-1}[I_k]$ is an interval such that $\lambda(E_k \cap J_k) > 0$. So, for each k, define h_k be the linear map of J_k onto I_k. Then $g_k \circ h_k^{-1}$ is a self-homeomorphism of I_k that maps $h_k[E_k \cap J_k]$ onto $H_k \cap I_k$. We now define g to be $g_k \circ h_k^{-1}$ on each I_k and to be the identity on the remainder of I. As g is an order-preserving, one-to-one, onto map, it is a self-homeomorphism. Clearly, for each k, we have $\{\, x : |f \circ g(x)| \geq M_{k-1} \,\} \supset h_k[E_k \cap J_k]$ and $\lambda\big(h_k[E_k \cap J_k]\big) > 0$. Hence $\operatorname{ess.sup}|f \circ g| = \infty$ and our proof is complete.

In passing we remark that the above proof can be modified so as to achieve

$$\operatorname{ab.ess.sup}|f| = \max\{\, \operatorname{ess.sup}|f \circ g| : g \text{ is a change of variable} \,\}.$$

We are now in a position to characterize absolute integrability.

THEOREM 11.3. *A function is absolutely integrable if and only if it is absolutely essentially bounded.*

PROOF. This is obvious in one direction, so let us assume that f is not absolutely essentially bounded. Then $f \circ g$ is not essentially bounded for some g. It should cause no confusion to replace $f \circ g$ by f, so we will assume that f is not essentially bounded on I. Recall that we have assumed f to be absolutely measurable. We will show that there is a change of variable g such that $f \circ g \notin \mathrm{L}^1(I)$.

The Lebesgue points of f are a set of full measure. Hence we can find a sequence of Lebesgue points t_n for which $|f(t_n)| > n$. Since this sequence has a monotone subsequence, we can assume without loss of generality that $I = [a, b]$, $t_n \to a$ as $n \to \infty$ and $|f(t_n)| > n$ for every n.

Let I_n, $n = 1, 2, \dots$, be a sequence of pairwise disjoint closed intervals in I with I_n centered at t_n for each n. Let $0 < \gamma < 1$. As $I_n \cap \{\, t : |f(t)| > n \,\}$ contains a perfect set of positive measure, there is a self-homeomorphism h_n of I_n such that $\lambda\big(I_n \cap \{\, t : |f \circ h_n(t)| > n \,\}\big) > \gamma\, \lambda(I_n)$. Define h to be the self-homeomorphism of I given by the mappings h_n in each interval I_n and the identity on the remainder of I. Hence h has the property that

$$\lambda\big(I_n \cap \{\, t : |f \circ h(t)| > n \,\}\big) > \gamma\, \lambda(I_n). \tag{11.1}$$

Choose α_n, $n = 1, 2, \dots$, to be a decreasing sequence with

$$\sum_{n=1}^{\infty} \alpha_n\, n = \infty \quad \text{and} \quad \sum_{n=1}^{\infty} \alpha_n < b - a.$$

Let J_n, $n = 1, 2, \dots$, be a sequence of pairwise disjoint closed intervals in (a, b) with $\lambda(J_n) = \alpha_n$ for each n such that the end points of J_n converge downward to a. Let H be a self-homeomorphism of I such that J_n is mapped linearly onto I_n for each n. With $g = h \circ H$ let

$$E_n = \{\, x : |f \circ g(x)| > n \,\}$$

for each n. Then

$$E_n \cap J_n = H^{-1}\big[\{\, t : |f \circ h(t)| > n \,\} \cap I_n\big].$$

As H is linear on J_n, we have from (11.1) that

$$\begin{aligned}\lambda(E_n \cap J_n) &= \lambda(\{t : |f \circ h(t)| > n\}) \frac{\lambda(J_n)}{\lambda(I_n)} \\ &> (\gamma\, \lambda(I_n)) \frac{\lambda(J_n)}{\lambda(I_n)} \\ &= \gamma\, \alpha_n.\end{aligned}$$

Thus

$$\int_T |f \circ g(t)|\, dt \ge \sum_{n=1}^{\infty} \int_{E_n \cap J_n} |f \circ g(t)|\, dt \ge \gamma \sum_{n=1}^{\infty} n\, \alpha_n = \infty,$$

that is, $f \circ g \notin \mathrm{L}^1(T)$, and the result is established.

We shall require one other change-of-variable invariant class of functions, the **absolutely regulated** functions, i.e., those f which are absolutely equivalent to a regulated function (see page 113 for the definition). A pair of functions are said to be **absolutely equivalent** if they are equal except on an AMZ-set.

Let $\operatorname{osc}(f; E)$ be the oscillation of the function f over the set E. That is,

$$\operatorname{osc}(f; E) = \operatorname{diam} f[E].$$

As we noted in the introduction to this chapter, the regulated functions play an essential role in the solution of the problems we are considering here. We show

THEOREM 11.4. *A function f is absolutely regulated if and only if one of the following conditions holds.*

(A) *There is an AMZ-set H such that for each positive number ε and each x, there is a positive number $\delta = \delta(\varepsilon, x)$ such that*

$$\operatorname{osc}\big(f; (x, x+\delta) \setminus H\big) < \varepsilon \quad \textit{and} \quad \operatorname{osc}\big(f; (x-\delta, x) \setminus H\big) < \varepsilon.$$

(B) *For each positive number ε and each x, there is an AMZ-set H_x^ε and a positive number $\delta = \delta(\varepsilon, x)$ such that*

$$\operatorname{osc}\big(f; (x, x+\delta) \setminus H_x^\varepsilon\big) < \varepsilon \quad \textit{and} \quad \operatorname{osc}\big(f; (x-\delta, x) \setminus H_x^\varepsilon\big) < \varepsilon.$$

PROOF. We begin by showing that (A) and (B) are equivalent. Clearly (A) implies (B), so we suppose that (B) is satisfied. For each n, the collection of intervals $(x-\delta, x+\delta)$, $x \in I$, $\delta = \delta(1/n, x)$ covers I. Let I_{ni}, $i = 1, 2, \ldots, N_n$, be a finite subcovering with respective centers x_{ni} and AMZ-sets $H_{x_{ni}}^{1/n}$. Let us show that

$$H = \bigcup_{n=1}^{\infty} \bigcup_{i=1}^{N_n} H_{x_{ni}}^{1/n}$$

is the set required in (A). Given $\varepsilon > 0$ and x, choose n_0 so that $1/n_0 < \varepsilon$. If $x = x_{n_0 i}$ for some i, then the conclusion of (A) holds with $\delta = \lambda(I_{n_0 i})/2$. Otherwise, $I_{n_0 i} \setminus \{x_{n_0 i}\}$ is an open set containing x for some i. Let δ be chosen to be a radius of a neighborhood of x that is contained in this open set. Thus we have shown that (A) and (B) are equivalent statements.

If f is absolutely regulated then (A) clearly holds. Assuming now that (A) holds, we must show that there is a regulated function h such that f is absolutely equivalent to h. We shall use the following convex hull operator K in the construction:

$$\mathrm{K}(E) = \bigcap \{\, J : J \text{ is a closed interval containing } E \,\},$$

where E is a subset of $\mathbb{R}$. With H as in (A) and x in I, consider the closed intervals, possibly unbounded,

$$\mathrm{K}\big(f[(x, x+\delta) \setminus H]\big) \quad \text{for} \quad \delta > 0.$$

From the oscillation condition in (A), the Cantor nested interval theorem yields that the intersection of all these intervals consists of exactly one point. Moreover, for each positive number ε, there is a δ such that $\mathrm{K}\big(f[(x, x+\delta) \setminus H]\big)$ is contained in the ε-neighborhood of this point. Let $h^{(+)}(x)$ be the unique number determined by the above intersection. Using the intervals $(x-\delta, x)$, we can define $h^{(-)}(x)$ in the analogous way. Let us show that the function $h^{(+)}$ defined on I has limits from the right and limits from the left at every x. Given x and $\varepsilon > 0$, let δ_0 be a positive number such that

$$\{\, y : |h^{(+)}(x) - y| < \varepsilon \,\} \supset \mathrm{K}\big(f[(x, x+\delta_0) \setminus H]\big).$$

For each t in $(x, x+\delta_0)$ there is a δ such that

$$(t, t+\delta) \subset (x, x+\delta_0).$$

Hence

$$h^{(+)}(t) \in \mathrm{K}\big(f[(x, x+\delta_0) \setminus H]\big),$$

and the limit from the right is $h^{(+)}(x)$. For the limit from the left of $h^{(+)}$ at x, a similar argument will show that the limit is $h^{(-)}(x)$. Hence we find the following limits:

$$\lim_{t \to x+} h^{(+)}(t) = h^{(+)}(x), \qquad \lim_{t \to x-} h^{(+)}(t) = h^{(-)}(x),$$
$$\lim_{t \to x+} h^{(-)}(t) = h^{(+)}(x), \qquad \lim_{t \to x-} h^{(-)}(t) = h^{(-)}(x).$$

Clearly the function

$$h = \tfrac{1}{2}\left(h^{(+)} + h^{(-)}\right)$$

is regulated.

To see that h is absolutely equivalent to f we need only consider the x's in $I \setminus H$ at which $f(x) \neq h(x)$. The collection of x's at which $h(x) \neq h^{(+)}(x)$ is countable. And the collection of x's at which $f(x) \neq h(x)$ and $h^{(+)}(x) = h^{(-)}(x)$ is countable because such a point x is a removable point of discontinuity of the restricted function $f|(I \setminus H)$.

We have finally a result which characterizes those absolutely essentially bounded functions which are absolutely regulated.

THEOREM 11.5. *An absolutely essentially bounded function f is absolutely regulated if and only if for each pair r and s with $r > s$ and for each x there is a positive number δ such that at least one of the sets*

$$\{\, t : f(t) > r \,\} \cap (x, x+\delta) \quad \text{and} \quad \{\, t : f(t) < s \,\} \cap (x, x+\delta)$$

and at least one of the sets

$$\{\, t : f(t) > r \,\} \cap (x-\delta, x) \quad \text{and} \quad \{\, t : f(t) < s \,\} \cap (x-\delta, x)$$

are AMZ-*sets.*

PROOF. If f is regulated then condition (B) of the previous result clearly implies the condition of this theorem. Let us assume therefore that f is absolutely essentially bounded and satisfies the condition of the theorem. For a given x,

$$r_0 = \inf\{\, r : \{\, t : f(t) > r \,\} \cap (x, x+\delta) \text{ is an AMZ-set for some positive } \delta \,\}$$

and

$$s_0 = \sup\{\, s : \{\, t : f(t) < s \,\} \cap (x, x+\delta) \text{ is an AMZ-set for some positive } \delta \,\}$$

are well-defined in view of boundedness of f on excluding an AMZ-set. Clearly $r_0 \geq s_0$. If $r_0 > s_0$, choose r and s so that $r_0 > r > s > s_0$. Then for every positive δ,

$$\text{neither} \quad \{\, t : f(t) < s \,\} \cap (x, x+\delta) \quad \text{nor} \quad \{\, t : f(t) > r \,\} \cap (x, x+\delta)$$

is an AMZ-set, contrary to the condition of the theorem. So let $y = r_0 = s_0$. Then for each positive ε there is a δ such that

$$\{\, t : f(t) > y + \varepsilon/2 \,\} \cap (x, x+\delta) \quad \text{and} \quad \{\, t : f(t) < y - \varepsilon/2 \,\} \cap (x, x+\delta)$$

are AMZ-sets. This is the part of condition (B) of Theorem 11.4 concerning $(x, x+\delta)$. The part concerning $(x-\delta, x)$ may be established similarly and then the result will have been proven.

11.2. Convergence of Fourier series after change of variable

We shall say that f is **absolutely GW** if f is absolutely equivalent to a function which satisfies the GW condition (see page 138).

Our result is

THEOREM 11.6. *A necessary and sufficient condition that $S(f \circ g)$ converges everywhere for every change of variable g is that f be absolutely* GW.

COROLLARY 11.7. *If $S(f \circ g)$ converges everywhere for every change of variable g, then f is absolutely regulated.*

REMARK 11.8. If we assume f to be absolutely regulated, then $f = f^*$ except on an AMZ-set for some regulated function f^*. And $S(f \circ g)$ will converge everywhere if and only if $S(f^* \circ g)$ converges everywhere, which is equivalent (by Theorem 11.6) to f^* satisfying the GW condition, i.e., f is absolutely GW. On the other hand, the GW condition implies regulated, so if f is absolutely GW, it must be absolutely regulated.

We see then that the theorem will be established if we prove its corollary, i.e., everywhere convergence of $S(f \circ g)$ for every change of variable g implies that f is absolutely regulated.

PROOF. We will assume that f is absolutely integrable but is **not** absolutely regulated. We will show that there is a change of variable g such that $S(f \circ g)$ diverges at some point. In view of Theorem 11.5 we may assume, without loss of generality, that

$$\{\, t : f(t) > 1 \,\} \cap (0, \delta) \quad \text{and} \quad \{\, t : f(t) < -1 \,\} \cap (0, \delta)$$

are not AMZ-sets for every positive δ.

Let us construct an order-preserving homeomorphism g of $[0,\pi]$ onto itself and a sequence n_k, $k=1,2,\ldots$, so that

$$\lim_{k\to\infty}\int_0^\pi f\circ g(t)\,\frac{\sin n_k t}{t}\,dt=\infty.$$

There is a decreasing sequence a_n, $n=0,1,2,\ldots$, with $a_0=\pi$ and $a_n\to 0$ as $n\to\infty$ such that

$$\{\,t:(-1)^n f(t)>1\,\}\cap(a_{n+1},a_n)$$

is not an AMZ-set for every n. For a given odd integer n_1 greater than 1 we can construct a continuous, strictly increasing map g of $[\,\pi-(i+1)\pi/n_1,\ \pi-i\pi/n_1\,]$ onto $[a_{i+1},a_i]$ for $i=0,1,\ldots,n_i-2$ so that, for each i, the relative measure of $\{\,t:(-1)^i f\circ g(t)>1\,\}$ in $[\,\pi-(i+1)\pi/n_1,\ \pi-i\pi/n_1\,]$ is as close to one as we wish. Since

$$\operatorname{ess.sup}|f\circ g(t)|\le \operatorname{ab.ess.sup}|f(t)|=M<\infty,$$

we may choose g so that

$$\int_{\pi/n_1}^{\pi} f\circ g(t)\,\frac{\sin n_1 t}{t}\,dt>\int_{\pi/n_1}^{\pi}\frac{|\sin n_1 t|}{t}\,dt-1.$$

We will choose n_1 so that the right side here is positive.

Suppose that, for some k, we have defined a continuous strictly increasing g on $[\pi/n_k,\pi]$ mapping it onto $[a_{i_k},\pi]$ so that n_k is odd and

$$\int_{\pi/n_k}^{\pi} f\circ g(t)\,\frac{\sin n_k t}{t}\,dt>k-1. \tag{11.2}$$

We have done this for $k=1$ with $i_1=n_1-1$. Choose an odd n_{k+1} larger than n_k so that

$$\left|\int_{\pi/n_k}^{\pi} f\circ g(t)\,\frac{\sin n_{k+1} t}{t}\,dt\right|<\frac{1}{3}$$

holds and so that the greatest odd integer m that satisfies

$$\frac{m}{n_{k+1}}<\frac{1}{n_k}$$

will also satisfy

$$m>6M\quad\text{and}\quad\int_{\pi/n_{k+1}}^{m\pi/n_{k+1}}\frac{|\sin n_{k+1}t|}{t}\,dt>k+1.$$

We extend g to a continuous, strictly increasing function mapping

$$\left[\frac{m-1}{n_{k+1}}\pi,\frac{\pi}{n_k}\right]\quad\text{onto}\quad[a_{i_k+1},a_{i_k}],$$

$$\left[\frac{m-2}{n_{k+1}}\pi,\frac{m-1}{n_{k+1}}\pi\right]\quad\text{onto}\quad[a_{i_k+2},a_{i_k+1}],$$

and

$$\left[\frac{m-3}{n_{k+1}}\pi,\frac{m-2}{n_{k+1}}\pi\right]\quad\text{onto}\quad[a_{i_k+3},a_{i_k+2}],$$

and continue in this manner until we have defined g on $[\pi/n_{k+1}, \pi]$ onto $[a_{i_{k+1}}, \pi]$ with $i_{k+1} = i_k + m - 1$. We choose g so that the relative measure of the set $\{t : (-1)^i f \circ g(t) > 1\}$ in each of the intervals

$$\left[\frac{m-(i+1)}{n_{k+1}}\pi, \frac{m-i}{n_{k+1}}\pi\right] \quad \text{for} \quad i = 0, 1, 2, \ldots, m-2,$$

is so close to one that

$$\int_{\pi/n_{k+1}}^{m\pi/n_{k+1}} f \circ g(t) \frac{\sin n_{k+1}t}{t}\, dt > \int_{\pi/n_{k+1}}^{m\pi/n_{k+1}} \frac{|\sin n_{k+1}t|}{t}\, dt - \frac{1}{3}.$$

Now

$$\begin{aligned}\int_{m\pi/n_{k+1}}^{\pi/n_k} \left| f \circ g(t) \frac{\sin n_{k+1}t}{t} \right| dt &\le \left(\frac{\pi}{n_k} - \frac{m\pi}{n_{k+1}}\right) M \frac{n_{k+1}}{m\pi} \\ &\le \frac{2\pi}{n_{k+1}} M \frac{n_{k+1}}{m\pi} < \frac{1}{3},\end{aligned}$$

since

$$\frac{n_{k+1}}{n_k} - m \le 2.$$

Thus

$$\int_{\pi/n_{k+1}}^{\pi} f \circ g(t) \frac{\sin n_{k+1}t}{t}\, dt > \int_{\pi/n_{k+1}}^{m\pi/n_{k+1}} \frac{|\sin n_{k+1}t|}{t}\, dt - 1 > (k+1) - 1,$$

and we have shown that we may define g so that (11.2) holds for $k = 1, 2, \ldots$. Since a_{i_k} decreases to 0 as $k \to \infty$, setting $g(0) = 0$ and observing that

$$\int_0^{\pi/n_k} f \circ g(t) \frac{\sin n_k t}{t}\, dt = O(1) \quad \text{as} \quad k \to \infty,$$

we see that we have defined a self-homeomorphism g of $[0, \pi]$ so that

$$\int_0^{\pi} f \circ g(t) \frac{\sin n_k t}{t}\, dt \to \infty \quad \text{as} \quad k \to \infty.$$

Turning now to the interval $[-\pi, 0]$, define g on $[-\pi, -\pi/n_1]$ to be the increasing linear mapping of that interval onto itself. Choose $k(1)$ with $n_{k(1)} > n_1$ such that

$$\left| \int_{-\pi}^{-\pi/n_1} f \circ g(t) \frac{\sin n_{k(1)}t}{t}\, dt \right| < \frac{1}{2}.$$

In $[-\pi/n_1, -\pi/n_{k(1)}]$ there is a set E of positive measure such that

$$\operatorname{osc}(f; E) < 1/\log n_{k(1)}.$$

There is a continuous strictly increasing extension of g to $[-\pi/n_1, -\pi/n_{k(1)}]$, mapping this interval onto itself, so that the relative measure of $g^{-1}[E]$ is so close to one that

$$\begin{aligned}&\left| \int_{-\pi/n_1}^{-\pi/n_{k(1)}} f \circ g(t) \frac{\sin n_{k(1)}t}{t}\, dt \right| \\ &\qquad < \frac{1}{\log n_{k(1)}} \int_{-\pi/n_1}^{-\pi/n_{k(1)}} \left| \frac{\sin n_{k(1)}t}{t} \right| dt + M \left| \int_{\pi/n_1}^{\pi/n_{k(1)}} \frac{\sin n_{k(1)}t}{t}\, dt \right| + 1.\end{aligned}$$

Continuing in this way, we can define inductively a sequence $k(j)$, $j = 0, 1, 2, \dots$, with $k(0) = 1$ and an increasing homeomorphism g of $[-\pi, 0]$ onto itself such that

$$\left|\int_{-\pi/n_{k(j-1)}}^{-\pi/n_{k(j)}} f \circ g(t) \frac{\sin n_{k(j)}t}{t}\, dt\right| < \frac{1}{\log n_{k(j)}} \int_{-\pi/n_{k(j-1)}}^{-\pi/n_{k(j)}} \left|\frac{\sin n_{k(j)}t}{t}\right| dt + O(1) = O(1)$$

as $j \to \infty$ and, for each j,

$$\left|\int_{-\pi}^{-\pi/n_{k(j-1)}} f \circ g(t) \frac{\sin n_{k(j)}t}{t}\, dt\right| < \frac{1}{2^j}.$$

Since

$$\int_{-\pi/n_{k(j)}}^{0} f \circ g(t) \frac{\sin n_{k(j)}t}{t}\, dt = O(1),$$

we have

$$\int_{-\pi}^{0} f \circ g(t) \frac{\sin n_{k(j)}t}{t}\, dt = O(1).$$

As

$$S_n(0; f \circ g) = \frac{1}{\pi} \int_{-\pi}^{\pi} f \circ g(t) \frac{\sin nt}{t}\, dt + o(1) \quad \text{as} \quad n \to \infty,$$

we see that the sequence of $n_{k(j)}$-th partial sums of the Fourier series of $f \circ g$ diverges at $x = 0$ as $j \to \infty$.

11.3. Functions of generalized bounded variation

Various classes of functions of generalized bounded variation have proven to be of interest in the study of Fourier series. It will be evident from the definitions that these classes, like the functions of ordinary (Jordan) bounded variation are classes of regulated functions and are invariant under change of variable.

We will discuss the convergence of Fourier series of functions in these classes and the order of magnitude of the Fourier coefficients. An excellent survey of this area has been made by Avdispahić [**3**].

11.3.1. Λ-bounded variation. Let $\Lambda = \{\, \lambda_n : n = 1, 2, \dots \}$ be an increasing sequence of positive real numbers such that

$$\sum_{n=1}^{\infty} \frac{1}{\lambda_n} = \infty.$$

We say that f is a function of Λ-**bounded variation** if, for every finite or infinite collection I_n, $n = 1, 2, \dots$, of nonoverlapping intervals contained in I,

$$\sum_{n=1}^{\infty} \frac{|f(I_n)|}{\lambda_n} < \infty.$$

Equivalently we can say that there is a finite M such that, for every such collection I_n, $n = 1, 2, \dots$,

$$\sum_{n=1}^{\infty} \frac{|f(I_n)|}{\lambda_n} < M.$$

The infimum of such M is called the **Λ-variation of** f, denoted $\mathrm{V}_\Lambda(f, I)$. These classes were first considered by Waterman in [**130, 132**], where the equivalence is established.

The class of functions of Λ-bounded variation on a domain D is denoted by $\Lambda\mathrm{BV}(D)$ or simply by $\Lambda\mathrm{BV}$ when the domain is understood.

It is easily seen that $\Lambda\mathrm{BV} = \mathrm{BV}$ if and only if Λ is a bounded sequence.

The class $\Lambda\mathrm{BV}$ with $\Lambda = \{1/n : n = 1, 2, \dots\}$ is called the functions of **harmonic bounded variation** and is denoted HBV. The class HBV was anticipated by Goffman in [**58**].

11.3.2. Properties of functions of ΛBV. We shall not delve deeply into the properties of functions of ΛBV. In [**130, 132**] it is shown that:

(1) The Helly selection theorem holds for ΛBV functions.
(2) The function $\mathrm{V}_\Lambda(f, [a, x])$ has continuity properties which mimic those of f, i.e., $\mathrm{V}_\Lambda(f, [a, x])$ is right (left) continuous at a point if and only if f is right (left) continuous at that point.
(3) $\Lambda\mathrm{BV}([a, b])$ is a Banach space with norm

$$\|f\|_\Lambda = |f(a)| + \mathrm{V}_\Lambda(f, [a, b]).$$

The following proposition [**132**] will be useful.

PROPOSITION 11.9. *If f is in ΛBV on an interval containing $[a, b]$, then*

$$\mathrm{V}_\Lambda\big(f, (a, b)\big) \to 0 \quad \textit{as} \quad b \to a+ \quad \textit{and} \quad \textit{as} \quad a \to b- .$$

If f is in $\Lambda\mathrm{BV}(I)$ and is continuous at each point of a closed interval J in the interior of I, then

$$\mathrm{V}_\Lambda\big(f, (x - \delta, x)\big) \to 0 \quad \textit{and} \quad \mathrm{V}_\Lambda\big(f, (x, x + \delta)\big) \to 0 \quad \textit{as} \quad \delta \to 0+$$

uniformly in x on J.

PROOF. We examine only the second assertion since it contains the first.

The proof is by way of a contradiction. Suppose for an f in ΛBV that the conclusion is false. Then there is a positive ε and sequences δ_k and x_k, $k = 1, 2, \dots$, such that δ_k decreases to 0 and, for each k,

$$x_n \in J \quad \text{and} \quad \mathrm{V}_\Lambda\big(f, (x_n, x_n + \delta_n)\big) > \varepsilon.$$

By choosing a subsequence if necessary, we may assume that x_k is monotone, say decreasing, with limit x_0. The other case is handled in an analogous manner. Then in $(x_1, x_1 + \delta_1)$ there is a collection of nonoverlapping intervals I_k, $k = 1, \dots, k_1$, such that

$$\sum_{k=1}^{k_1} \frac{|f(I_k)|}{\lambda_k} > \varepsilon.$$

Since f obviously has limits from the right at every point, we can choose n_2 so that $x_{n_2} + \delta_{n_2} < x_1$ and

$$\operatorname{osc}\big(f; (x_n, x_n + \delta_n)\big) < \frac{\varepsilon}{\sum_{k=1}^{k_1} {\lambda_k}^{-1}} \qquad \text{whenever} \quad n > n_2.$$

Then there exist nonoverlapping intervals I_k, $k = k_1 + 1, \dots, k_{n_2}$ contained in $(x_{n_2}, x_{n_2} + \delta_{n_2})$ so that

$$\sum_{k_1 < k \leq k_{n_2}} \frac{|f(I_k)|}{\lambda_k} > \frac{\varepsilon}{2}.$$

Continuing in this manner to select a sequence $(x_{n_i}, x_{n_i} + \delta_{n_i})$, $i = 1, 2, \dots$, of disjoint intervals so that in each there is a collection of nonoverlapping intervals I_k, $k = k_{i-1} + 1, \dots, k_n$, with

$$\sum_{k_{i-1} < k \leq k_i} \frac{|f(I_k)|}{\lambda_k} > \frac{\varepsilon}{2},$$

we have a collection I_k, $k = 1, 2, \dots$, such that

$$\sum_{k=1}^{\infty} \frac{|f(I_k)|}{\lambda_k} = \infty$$

and so $f \notin \Lambda\mathrm{BV}$, a contradiction.

11.3.3. GW and HBV. The definitions of GW and HBV arise from our interest in the convergence of Fourier series. It is natural to expect them to be related.

THEOREM 11.10. GW $\supset$ HBV *and* UGW $\supset$ HBV $\cap$ C.

If we were to give a proof of the next Theorem 11.11 which did not utilize the classes GW and UGW, as in [**130, 136**], then this result would be a consequence of Theorem 11.11. Here we prove it directly from the definitions of the classes following an argument used by Goffman [**58**].

PROOF. Suppose $f \in \mathrm{R}(T) \setminus \mathrm{GW}$. Then there is a t_0 in T and a system $\mathcal{I} = \{ I_{ni} : i = 1, 2, \dots, k_n,\ n = 1, 2, \dots \}$ at t_0 (say a right system) so that, for some positive α,

$$\limsup_{n \to \infty} \left| \sum_{i=1}^{k_n} i^{-1} f(I_{ni}) \right| > \alpha.$$

Choose n_1 so that

$$\sum_{i=1}^{k_{n_1}} i^{-1} |f(I_{n_1 i})| > \alpha.$$

Next choose a positive δ so that

$$\operatorname{osc}\bigl(f, (t_0, t_0 + \delta)\bigr) < \alpha/(2k_{n_1}) \quad \text{and} \quad (t_0, t_0 + \delta) \cap I_{n_1 1} = \emptyset.$$

Then there is a large n_2 such that

$$\textstyle\bigcup_{i=1}^{k_{n_2}} I_{n_2 i} \subset (t_0, t_0 + \delta) \quad \text{and} \quad \left| \sum_{i=1}^{k_2} i^{-1} f(I_{n_2 i}) \right| > \alpha,$$

whence

$$\sum_{i=k_1+1}^{k_{n_2}} \frac{|f(I_{n_2 i})|}{i} > \frac{\alpha}{2}.$$

Continuing in this manner and renumbering the intervals, we obtain a sequence of pairwise disjoint intervals I_n such that

$$\sum_{n=1}^{\infty} \frac{|f(I_n)|}{n} = \infty.$$

Now suppose $f \in C(T) \setminus \mathrm{UGW}$. The condition (h) of UGW is simply GW, so if f does not satisfy the condition (h), the previous argument shows $f \notin \mathrm{HBV}$. Let us assume that f does not satisfy the condition (hh) of UGW. Then there is a t_0 and a system $\mathcal{I}$ at t_0 (say a right one) and a positive α such that

$$\limsup_{n\to\infty} \left|\sum_{i=1}^{k_n} (k_n + 1 - i)^{-1} f(I_{ni})\right| > \alpha.$$

Using an argument very similar to that employed in the first part of this proof, we can construct a sequence of pairwise disjoint intervals I_n such that

$$\sum_{n=1}^{\infty} \frac{|f(I_n)|}{n} = \infty$$

implying $f \notin \mathrm{HBV}$.

11.3.4. Convergence of Fourier series of functions in HBV. The principal result on Fourier series of functions of generalized bounded variation is the generalized Dirichlet-Jordan theorem of Waterman [**130**]. (See also Goffman [**58**].)

THEOREM 11.11 (WATERMAN). *If f is in* HBV(T), *then $S(f)$ converges to* $\frac{1}{2}(f(x+) + f(x-))$ *for every x and convergence is uniform on any closed interval of points of continuity.*

PROOF. Since GW $\supset$ HBV (Theorem 11.10), the convergence portion of the theorem is immediate.

Let I be the closed interval of points of continuity. The continuous function

$$h(t) = \begin{cases} f(t) & \text{for } t \in I, \\ \text{linear} & \text{for } t \in T \setminus I. \end{cases}$$

is in HBV and, therefore, in UGW by Theorem 11.10. Its Fourier series converges uniformly and, using a standard equiconvergence theorem (see [**147**, volume I, page 53] for example), we have that $S(f)$ converges uniformly on any interval I' interior to I. To obtain uniform convergence on I, a different argument is needed. From Proposition 10.7 we have, given $\varepsilon > 0$ and $\delta > 0$, the existence of an $n(\varepsilon, \delta)$ such that, whenever $n > n(\varepsilon, \delta)$,

$$\left|\int_0^\delta \big(f(x+t) - f(x)\big)\frac{\sin nt}{t}\,dt\right| < \varepsilon$$
$$+ \int_0^\delta \left|\sum_{k=1}^{N}{}^{o} \frac{1}{k}\Big(f\big(x + (t + k\pi)/n\big) - f\big(x + (t + (k+1)\pi/n)\big)\Big)\right| dt$$

uniformly in I, where $\sum^o$ denotes summation over odd indices and N is the greatest odd number such that $N + 1 < n\delta/\pi$. The integrand in this estimate is bounded above by $\mathrm{V_H}\big(f, [x, x+2\delta]\big)$ which tends to zero uniformly in view of Proposition 11.9. A similar estimate holds for

$$\left|\int_0^{\delta} \big(f(x-t) - f(x)\big) \frac{\sin nt}{t}\, dt\right|.$$

Thus

$$\pi\,|S_n(x; f) - f(x)| \le 2\varepsilon + \delta\left(\mathrm{V_H}(f, [x, x+2\delta]) + \mathrm{V_H}(f, [x-2\delta, x])\right) + o(1)$$

uniformly in I as $n \to \infty$, implying that $S(f)$ converges uniformly on I.

It is important to note that HBV is the largest ΛBV class for which the conclusions of the Dirichlet-Jordan theorem hold [**130**].

THEOREM 11.12 (WATERMAN). *If* $\Lambda\mathrm{BV}\setminus\mathrm{HBV} \ne \emptyset$, *then there is an* f *in* ΛBV *such that* $S(f)$ *diverges at some point*; f *may be chosen to be continuous.*

PROOF. Let $f^* \in \Lambda\mathrm{BV} \setminus \mathrm{HBV}$. Then there is a collection of nonoverlapping intervals I_n, $n = 1, 2, \dots$, such that

$$\sum_{n=1}^{\infty} \frac{|f^*(I_n)|}{n} = \infty \quad \text{and} \quad \sum_{n=1}^{\infty} \frac{|f^*(I_n)|}{\lambda_n} < \infty.$$

For the sake of simplicity, let $a_n = |f^*(I_n)|$ for each n. We may assume that a_n decreases to 0 as $n \to \infty$. Let

$$B = \sum_{i=1}^{\infty} \frac{a_i}{\lambda_i}.$$

For each n define f_n on T by

$$f_n(x) = \begin{cases} a_i & \text{for } (2i-2)\pi < (n+\frac{1}{2})x < (2i-1)\pi,\ i = 1, \dots, n+1, \\ 0 & \text{elsewhere.} \end{cases}$$

Then the ΛBV norm of f_n satisfies

$$\|f_n\|_{\Lambda} \le 2B. \tag{11.3}$$

Now

$$\begin{aligned} S_n(0; f_n) &= \frac{1}{\pi}\int_0^{2\pi} f_n(t)\, \frac{\sin(n+\frac{1}{2})t}{2\sin\frac{1}{2}t}\, dt \\ &\ge \frac{1}{\pi}\int_0^{2\pi} f_n(t) \frac{\sin(n+\frac{1}{2})t}{t}\, dt \\ &\ge \frac{1}{\pi}\sum_{i=1}^{n+1} a_i \frac{(n+\frac{1}{2})}{(2i-1)\pi} \int_{(2i-2)\pi/(n+\frac{1}{2})}^{(2i-1)\pi/(n+\frac{1}{2})} \sin(n+\tfrac{1}{2})t\, dt \\ &= \frac{2}{\pi^2}\sum_{i=1}^{n+1} \frac{a_i}{2i-1} > \frac{1}{\pi^2}\sum_{i=1}^{n+1} \frac{a_i}{i}. \end{aligned} \tag{11.4}$$

So the dual norm of the linear functional L_n on the Banach space ΛBV given by

$$L_n(f) = S_n(0; f)$$

satisfies

$$\|L_n\|_{\Lambda^*} \ge \frac{S_n(0; f_n)}{\|f_n\|_{\Lambda}} > \frac{\sum_{i=1}^{n+1} a_i/i}{2\pi^2 B}.$$

Hence

$$\|L_n\|_{\Lambda^*} \to \infty \quad \text{as} \quad n \to \infty.$$

We infer from the Banach-Steinhaus theorem the existence of an f in ΛBV with $\limsup_{n\to\infty} L_n(f) = \infty$. Thus $S_n(0, f)$, $n = 1, 2, \ldots$, diverges, that is, $S(f)$ diverges at $x = 0$.

Note that $\Lambda\mathrm{BV} \cap \mathrm{C}$ is itself a Banach space, since convergence in ΛBV implies uniform convergence. The above f_n can be modified so as to be continuous without much effect on the computations that show (11.3) and (11.4). The Banach space argument is exactly the same for $\Lambda\mathrm{BV} \cap \mathrm{C}(T)$.

11.3.5. Order of magnitude of Fourier coefficients of ΛBV functions. The estimation of the size of $\hat{f}(n)$ for f in ΛBV was made by Avdispahić [**1**], Schramm and Waterman [**121**], and Wang [**129**] almost simultaneously.

THEOREM 11.13. *If $f \in \Lambda\mathrm{BV}$, then*

$$|\hat{f}(n)| \leq \mathrm{V}_\Lambda(f)\left(2\sum_{j=1}^{n}\frac{1}{\lambda_j}\right)^{-1}.$$

PROOF. We begin much as in Section 9.1 by observing that, for $j = 1, 2, \ldots, n$,

$$\begin{aligned}\hat{f}(n) &= \frac{1}{2\pi}\int_T f\left(t + \frac{(2j-2)\pi}{n}\right)e^{-int}\,dt \\ &= -\frac{1}{2\pi}\int_T f\left(t + \frac{(2j-1)\pi}{n}\right)e^{-int}\,dt.\end{aligned}$$

Hence

$$\begin{aligned}|\hat{f}(n)|\sum_{j=1}^{n}\frac{1}{\lambda_j} &\leq \frac{1}{4\pi}\int_T \sum_{j=1}^{n}\frac{1}{\lambda_j}\left|f\left(t + \frac{(2j-2)\pi}{n}\right) - f\left(t + \frac{(2j-1)\pi}{n}\right)\right|dt \\ &\leq \frac{1}{2}\,\mathrm{V}_\Lambda(f).\end{aligned}$$

11.3.6. φ-bounded variation. Let φ be a strictly increasing function defined on $[0, \infty)$. The notion of φ-bounded variation (φ-BV) was introduced by L. C. Young [**143**] as a generalization of the p-bounded variation of Wiener, who considered the case $\varphi(t) = t^p$ where $p > 1$.

A function f is of φ-bounded variation if there is a positive M such that

$$\sum_{i=1}^{n}\varphi(|f(I_i)|) < M$$

for all finite collections of nonoverlapping intervals I_i, $i = 1, 2, \ldots, n$. In order for this notion to be a generalization of BV, it is clearly required that $\varphi(t)/t \to 0$ as $t \to 0+$. It is usual to consider φ to be convex and even to satisfy the Δ_2-condition:

$$\varphi(2x)/\varphi(x) \leq d \quad \text{for all} \quad x,$$

where d is a constant with $d \geq 2$. With these requirements, the class V_φ of functions of φ-bounded variation is a linear space.

We have the following result which was demonstrated for continuous functions by Salem [**118**] and is, in the general case, a consequence of the convergence theorem

(Theorem 11.11) of Waterman [**130**]. In this result ψ is the function **complementary** to φ in the sense of W. H. Young given by

$$\psi(t) = \max\{\, ut - \varphi(u) : u \geq 0\,\}.$$

THEOREM 11.14 (WATERMAN). *If f is of φ-bounded variation on T and*

$$\sum_{n=1}^{\infty} \psi\Big(\frac{1}{n}\Big) < \infty,$$

then $S(f)$ converges everywhere and converges uniformly on closed intervals of points of continuity.

This is an immediate consequence of Theorem 11.11 and the following

PROPOSITION 11.15. $\mathrm{HBV} \supset \mathrm{V}_\varphi$ *if*

$$\sum_{n=1}^{\infty} \psi\Big(\frac{1}{n}\Big) < \infty.$$

PROOF. For each sequence I_n, $n = 1, 2, \ldots$, of nonoverlapping intervals we have

$$\frac{|f(I_n)|}{n} \leq \varphi(|f(I_n)|) + \psi\Big(\frac{1}{n}\Big) \quad \text{for each} \quad n$$

by W. H. Young's inequality. Hence

$$\mathrm{V}_\Lambda(f) \leq \mathrm{V}_\varphi(f) + \sum_{n=1}^{\infty} \psi\Big(\frac{1}{n}\Big).$$

See Goffman [**58**] for related theorems expressed in terms of Köthe spaces.

The following result shows that Theorem 11.14 is sharp. It was shown by Baernstein [**4**] and also by Oskolkov [**105**].

THEOREM 11.16. *If*

$$\sum_{n=1}^{\infty} \psi\Big(\frac{1}{n}\Big) = \infty,$$

then there is a continuous f in V_φ such that $S(f)$ diverges.

The subject of φ-bounded variation was explored in detail by Lesniewicz, Musielak, and Orlicz [**93, 99**].

11.3.7. The modulus of variation and the class $\mathrm{V}[h]$. If $\{\, I_i \,\}$ denotes an arbitrary collection of nonoverlapping intervals in the domain of a function f, then

$$\nu(n; f) = \sup\Big\{ \sum_{i=1}^{n} |f(I_i)| : \{\, I_i \,\} \Big\}$$

was called the **modulus of variation** of f by Chanturiya [**32**] and he showed that $\nu(n; f)$ is positive and concave downward. For any function $h(n)$ with these properties, he defined a class of functions of generalized bounded variation, $\mathrm{V}[h]$, to be the functions f for which

$$\nu(n; f) = O\big(h(n)\big).$$

Clearly the class $\mathrm{V}[h]$ is well-defined without the concavity condition. Avdispahić [**2**] pointed out the following inclusion relation for $\Lambda\mathrm{BV}$ and $\mathrm{V}[h]$.

THEOREM 11.17 (AVDISPAHIĆ). $\Lambda BV \subset V\left[n\left(\sum_{k=1}^n 1/\lambda_k\right)^{-1}\right]$.

PROOF. We have for any finite collection I_i, $i = 1, 2, \ldots, n$, of nonoverlapping intervals,

$$\frac{|f(I_1)|}{\lambda_1} + \frac{|f(I_2)|}{\lambda_2} + \cdots + \frac{|f(I_n)|}{\lambda_n} \le V_\Lambda(f).$$

Consider the n such expressions obtained by cyclic permutation of the denominators, e.g.,

$$\frac{|f(I_1)|}{\lambda_n} + \frac{|f(I_2)|}{\lambda_1} + \cdots + \frac{|f(I_n)|}{\lambda_{n-1}} \le V_\Lambda(f).$$

Adding all of these we have

$$\Big(\sum_{i=1}^n |f(I_i)|\Big)\Big(\sum_{i=1}^n \frac{1}{\lambda_i}\Big) \le n\, V_\Lambda(f).$$

from which the result follows.

An analogous result holds for φ-bounded variation as well [**32**].

We have the following inclusion result for $V[h]$ and V_φ.

THEOREM 11.18. $V_\varphi \subset V[n\varphi^{-1}(1/n)]$.

Conditions on h which allow one to prove a result of Dirichlet-Jordan type yield classes smaller than HBV.

We have the following result of Chanturiya [**33**] on the order of magnitude of $\hat{f}(n)$.

THEOREM 11.19 (CHANTURIYA). *If* $f \in V[h]$, *then*

$$|\hat{f}(n)| \le \frac{1}{2}\frac{\nu(n; f)}{n}.$$

PROOF. As in the case of ΛBV, for each

$$I_i(t) = \left[t + \frac{(2i-2)\pi}{n}, t + \frac{(2i-1)\pi}{n}\right]$$

we have

$$|\hat{f}(n)| \le \frac{1}{4\pi}\int_T |f(I_i(t))|\, dt,$$

whence

$$n|\hat{f}(n)| \le \frac{1}{2}\nu(n; f).$$

11.3.8. The Banach indicatrix. If f is a continuous function on an interval I, let G be the graph of f, that is, the subset $\mathrm{graph}(f) = \{(x, y) : f(x) = y\}$ of $\mathbb{R}^2$. The Banach indicatrix $N(f, I, y)$ (see page 193) may be thought of as the cardinality of the intersection of the horizontal line at height y with G. The function $N(f, I, y)$ thus defined for a continuous f is easily generalized to regulated functions.

One way of doing this is to introduce the generalized graph G^* of a function f that has left and right limits at every point. Let

$$\overline{f}(x) = \max\{f(x), f(x+), f(x-)\} \quad \text{and} \quad \underline{f}(x) = \min\{f(x), f(x+), f(x-)\}.$$

If for each x we adjoin to G the line segment connecting $(x, \underline{f}(x))$ and $(x, \overline{f}(x))$, we obtain the generalized graph G^*. For regulated functions f, the indicatrix $\mathrm{N}(f, I, y)$ is then defined as above with G^* replacing G.

Banach [**6**] proved the following result in the case f is continuous: If f is regulated and

$$V = \int \mathrm{N}(f, I, y)\, dy < \infty,$$

then f is of bounded variation and V is its total variation.

A natural object for study is the class G_φ of regulated functions f for which

$$\int \varphi\big(\mathrm{N}(f, I, y)\big)\, dy < \infty,$$

where φ is nonnegative and defined on $[0, +\infty]$. Clearly the class G_φ is invariant under change of variable. Garsia and Sawyer [**45**] considered this class for continuous f and $\varphi(x) = \log x$. They showed:

If f is continuous on T and is in $\mathrm{G}_{\log}$, then the Fourier series $S(f)$ converges uniformly.

This result is a consequence of the convergence theorem for Fourier series of functions in the class HBV, for HBV $\supset \mathrm{G}_{\log}$. In fact we have [**130**]:

THEOREM 11.20 (WATERMAN). *If $\varphi(n) \sim \sum_{k=1}^{n} 1/\lambda_k$ as $n \to \infty$, then*

$$\mathrm{G}_\varphi \subsetneqq \Lambda\mathrm{BV}.$$

The theorem is a consequence of the following lemma [**130**].

LEMMA 11.21. *For a sequence E_i, $i = 1, 2, \dots$, of μ-measurable sets of a measure space $(X, \mathcal{A}, \mu)$, let $S_\infty = \limsup E_i$ and, for each positive integer n, let S_n be the set of points that are members of E_i for exactly n indices i. If a_i, $i = 1, 2, \dots$, is a decreasing sequence of nonnegative real numbers, then*

$$\sum_{i=1}^{\infty} a_i\, \mu(E_i) \le \sum_{n=1}^{\infty} \mu(S_n)\Big(\sum_{i=1}^{n} a_i\Big) + \sum_{i=1}^{\infty} a_i\, \mu(S_\infty).$$

PROOF. For each i let χ_i denote the characteristic function of E_i. Since a_i, $i = 1, 2, \dots$, is decreasing, we have

$$\sum_{i=1}^{\infty} a_i\, \chi_i(x) \le \sum_{i=1}^{n} a_i \quad \text{whenever} \quad x \in S_n.$$

Hence

$$\begin{aligned}
\sum_{i=1}^{\infty} a_i\, \mu(E_i) &= \int \Big(\sum_{i=1}^{\infty} a_i\, \chi_i(x)\Big)\, d\mu(x) \\
&= \sum_{n=1}^{\infty} \int_{S_n} \Big(\sum_{i=1}^{\infty} a_i\, \chi_i(x)\Big)\, d\mu(x) + \sum_{i=1}^{\infty} a_i\, \mu(S_\infty) \\
&\le \sum_{n=1}^{\infty} \mu(S_n)\Big(\sum_{i=1}^{n} a_i\Big) + \sum_{i=1}^{\infty} a_i\, \mu(S_\infty),
\end{aligned}$$

which is the desired result.

PROOF OF THEOREM 11.20. Let φ be as in the theorem and let f be a regulated function defined on an interval I that is not in ΛBV. Then there are nonoverlapping intervals $I_n = [a_n, b_n]$, $n = 1, 2, \ldots$, such that $\sum_{n=1}^{\infty} |f(I_n)|/\lambda_n = \infty$. Let E_i be the interval with end points $f(a_i)$ and $f(b_i)$. If $y \in S_k$, then $\mathrm{N}(f, I, y) \geq k$. If S_∞ has positive Lebesgue measure, then $\mathrm{N}(f, I, y) = \infty$ on that set, implying

$$\int \varphi(\mathrm{N}(f, I, y))\, dy = \infty.$$

Otherwise there is a positive constant C such that

$$C \int \varphi(\mathrm{N}(f, I, y))\, dy \geq \sum_{n=1}^{\infty} \lambda(S_n) \sum_{i=1}^{n} 1/\lambda_i \geq \sum_{n=1}^{\infty} |f(I_n)|/\lambda_n = \infty$$

and thereby $\mathrm{G}_\varphi \subset \Lambda$BV.

Thus the class $\mathrm{G}_{\log}$, like the class V_φ with $\sum_{k=1}^{\infty} \psi(1/k) < \infty$, is contained in HBV but not as a linear subspace [**42**]. (The function ψ is the complementary function in the sense of W. H. Young, see page 173.) The relation of these classes to HBV was explicated in a remarkable result of Prus-Wiśniowski [**111**]:

The collection BV *of functions of bounded variation is dense in the Banach space* HBV.

Consequently, any subset of HBV that contains BV, for example the subsets V_φ with $\sum_{k=1}^{\infty} \psi(1/k) < \infty$ and $\mathrm{G}_{\log}$, is dense in HBV.

Finally, we note that Goffman [**58**] established the inclusion relation for HBV and $\mathrm{G}_{\log}$ before the notion of HBV was formalized in [**130**].

11.3.9. A generalization of ΛBV to higher dimensions. Here we digress from the theme of homeomorphism and change of variable to discuss ΛBV in higher dimensions. The concept of Λ-bounded variation can be extended to higher dimensions in a manner analogous to that of the extension of BV by Cesari and Tonelli [**27, 127**]. The extension by Cesari and Tonelli was touched upon in Chapter 5, where connection to bi-Lipschitzian homeomorphism was discussed. The connection between the generalization of ΛBV to higher dimensions and bi-Lipschitzian homeomorphism is yet to be investigated.

For $p \geq 1$, $\alpha \geq 1$ and an interval I^n of $\mathbb{R}^n$, the class $\mathrm{V}^p_{\Lambda,\alpha}(I^n)$ consists of those f in $\mathrm{L}^p(I^n)$ with the property that for each coordinate direction i there is a function f_i that is equal to f λ_n-almost everywhere and such that, on λ_{n-1}-almost every line l_i in the i-th coordinate direction, f_i is in ΛBV and the function $\mathrm{V}_\Lambda(f_i, l_i)$, the Λ-variation of f_i on l_i, as a function of the remaining of $n-1$ variables is in $\mathrm{L}^\alpha(I^{n-1})$. (See Chapter 5 for the corresponding situation for the class BV in the sense of Cesari.) Of course the λ_n and λ_{n-1} in the above definition refer to the Lebesgue measures on $\mathbb{R}^n$ and $\mathbb{R}^{n-1}$, respectively, and should not be confused with terms of the sequence Λ.

This definition of Goffman and Waterman [**70, 71**] contained an additional hypothesis for $n > 2$ which was removed in [**61**].

Let $V_i(f)$ denote the Λ-variation of f_i on l_i, and choose each f_i to be right continuous as a function of x_i on $T = [-\pi, \pi]$ and left continuous at $x_i = \pi$ for λ_{n-1}-almost every l_i. We have the following theorem [**71**].

THEOREM 11.22 (GOFFMAN-WATERMAN). *$V^p_{\Lambda,\alpha}(T^n)$ is a Banach space with norm*

$$\|f\|_{\Lambda,p,\alpha} = \|f\|_p + \sum_{i=1}^{n} \|V_i(f)\|_\alpha .$$

This concept of bounded variation was introduced in order to study the localization principle for double Fourier series. By double Fourier series we mean the Fourier series of functions f in $L^1(T^2)$. The notion of localization here is the natural one that extends the Riemann localization principle on T^1. The Riemann localization principle on T asserts that if an integrable function vanishes identically on an open interval, the partial sums of its Fourier series converge uniformly to zero on any compact subset of that interval. (See [**147**, volume I, pages 52–55] for the principle of localization on T.) For functions of several variables, strong additional assumptions are needed in order that this principle may hold; the assumption of continuity will not suffice [**147**, volume II, pages 304–305].

For a higher dimension m one must specify the meaning of convergence of the Fourier series $S(f)$. By "convergence" we generally mean **convergence of rectangular partial sums**

$$\begin{aligned} S_{\boldsymbol{n}}(\boldsymbol{x}; f) &= S_{(n_1,n_2,\ldots,n_m)}(x_1, x_2, \ldots, x_m; f) \\ &= \sum_{j=1}^{m} \sum_{|k_j| \le n_j} \hat{f}(\boldsymbol{k})\, e^{(\boldsymbol{k}\cdot\boldsymbol{x})i}, \end{aligned}$$

where $\boldsymbol{k} \cdot \boldsymbol{x}$ is the dot product of the m-tuples $\boldsymbol{k}$ and $\boldsymbol{x}$ and where

$$\hat{f}(\boldsymbol{k}) = \frac{1}{(2\pi)^m} \int_{T^m} f(\boldsymbol{t})\, e^{-(\boldsymbol{k}\cdot\boldsymbol{t})i}\, d\boldsymbol{t},$$

of the Fourier series of f in $L^1(T^m)$, i.e., the existence of

$$\lim_{\min\{n_i\}\to\infty} S_{\boldsymbol{n}}(\boldsymbol{x}; f).$$

Localization fails for double Fourier series even for differentiable functions [**62**].

Localization fails for continuous functions even if we consider convergence of **square partial sums** (i.e., $S_{\boldsymbol{n}}(\boldsymbol{x}; f)$ with $n_1 = n_2 = \cdots = n_m$). On the positive side, if we replace partial sums with **square $(C,1)$ means**, and the ordinary limit by the L^p-limit, then localization holds in this sense for $p \ge m-1$, but fails for $p < m-1$ [**79**].

The localization theorem of Goffman and Waterman [**70, 71**] generalizes a previous result of Cesari by replacing BV with HBV [**27**].

THEOREM 11.23 (GOFFMAN-WATERMAN). *If $f \in V^1_{H,1}(T^2)$, then the localization principle for rectangular partial sums holds for f.*

They also showed that this result is best possible in a certain sense.

THEOREM 11.24 (GOFFMAN-WATERMAN). *If $\Lambda BV(T^2)$ is not contained in $HBV(T^2)$, then the localization property for square partial sums does not hold for $V^1_{\Lambda,1}(T^2)$.*

11.4. Preservation of the order of magnitude of Fourier coefficients

The idea of considering a class of functions defined by the order of magnitude of their Fourier coefficients originated with Hadamard [**75**]. He considered functions[3] f such that the restriction of f to any interval I contained in T had the property

$$|\widehat{f\chi_I}(n)| = O(1/n).$$

These are called the functions of **bounded deviation** (écart fini) (for detailed references see Zygmund [**147**, volume I, page 229]).

In [**131**] Waterman considered the effect of change of variable on these functions and showed that $f \circ g$ has this property for every change of variable g if and only if f is absolutely equivalent to a function of bounded variation.

Let us consider this problem in greater generality. In this section we assume that *h is a positive nondecreasing function on the positive integers which is concave downward and $o(n)$*. We shall characterize the functions f on T such that

$$\widehat{f \circ g}(n) = O(h(n)/n)$$

for every self-homeomorphism g on T.

There appears to be some ambiguity in the statement of the problem. We have

$$\widehat{f \circ g}(n) < C\, h(n)/n\,,$$

but C might depend on f and g or on f alone. Our result reconciles this ambiguity [**134, 135**].

THEOREM 11.25 (WATERMAN). *Let $h(n)$ be a positive nondecreasing function on the positive integers which is concave downward and $o(n)$. Then the following are equivalent:*

(i) *$|\widehat{f \circ g}(n)| < C_f\, h(n)/n$ for every change of variable g.*
(ii) *$|\widehat{f \circ g}(n)| < C_{f,g}\, h(n)/n$ for every change of variable g.*
(iii) *f is absolutely equivalent to a function in $\mathrm{V}[h]$*

REMARK 11.26. It is implicit in (i)–(iii) that f is absolutely integrable and so, by Theorem 11.3, is absolutely essentially bounded. It is elementary that functions of class $\mathrm{V}[h]$ with $h(n)/n = o(1)$ have left and right limits at every point. Thus (iii) implies that f is absolutely regulated.

We shall require some elementary results.

PROPOSITION 11.27. *If $f \in \mathrm{V}[h]$ on $[a,b]$ and on $[b,c]$, then $f \in \mathrm{V}[h]$ on $[a,c]$.*

This follows easily from the definition using the fact that $h(n)$ is concave and nondecreasing.

PROPOSITION 11.28. *If $f \notin \mathrm{V}[h]$ on $[a,b]$, then there is a point p in $[a,b]$ such that on at least one side of p, $f \notin \mathrm{V}[h]$ on any open interval terminating at p.*

In view of the previous proposition this is obtained by using the standard repeated bisection argument and observing that if f is in $\mathrm{V}[h]$ on the interior of a closed interval I, then it is in $\mathrm{V}[h]$ on I.

The following Lemma is somewhat more subtle and is critical in the proof of our result. The analysis of the effect of change of variable on the magnitude of the

[3] In fact he considered only *continuous* functions, a restriction no longer made.

Fourier coefficients differs from our previous concerns of the convergence of Fourier series in that considerations of sign play an important role.

LEMMA 11.29. *If a regulated f is not in* $\mathrm{V}[h]$ *and if* $C > 0$, *then for arbitrarily large n we may find disjoint intervals* H_i, $i = 1, 2, \ldots, n$, *and disjoint intervals* J_i, $i = 1, 2, \ldots, n$, *so that*

$$C\,h(n) < \sum_{i=1}^{n} f(H_i) \quad \textit{and} \quad -C\,h(n) > \sum_{i=1}^{n} f(J_i)\,,$$

the end points of these intervals being points of continuity of f.

PROOF. Choosing $M > 2\left(C + 2\sup|f(x)|/h(1)\right)$, for arbitrarily large n we can find nonoverlapping intervals I_i, $i = 1, 2, \ldots, n$, so that

$$\sum_{i=1}^{n} |f(I_i)| > M\,h(n)\,.$$

Letting $\sum^+$ and $\sum^-$ indicate summation over those i such that $f(I_i) \geq 0$ and $f(I_i) < 0$ respectively,

$$\sum\nolimits^+ f(I_i) + \left|\sum\nolimits^- f(I_i)\right| > M\,h(n)$$

and therefore one of the 2 terms in the above sum must exceed $(M/2)h(n)$, say $\sum^+ f(I_i) > M\,h(n)$ holds (the other option is covered by using $-f$). It may happen that there are fewer than n intervals at this point of the argument. By appropriately partitioning the intervals in the above sum, we can produce a nonoverlapping collection A_i, $i = 1, 2, \ldots, n$, so that

$$\sum_{i=1}^{n} f(A_i) > (M/2)h(n)\,. \tag{11.5}$$

Let us first construct the collection H_i, $i = 1, 2, \ldots, n$. Suppose $A_i = [a_i, b_i]$ and the collection is ordered from left to right. We may assume that the end points of each A_i are points of continuity of f. Indeed, if A_i and A_{i+1} have $b_i = a_{i+1}$, then we may choose a point of continuity b_i' in (a_i, b_{i+1}) and use $[a_i, b_i']$ and $[b_i', b_{i+1}]$ instead. If a_i is not a point of continuity and $[a_i, b_i]$ does not meet A_j whenever $j < i$, there is a point of continuity a_i' such that $f([a_i', b_i]) \geq f(A_i)$ and $[a_i', b_i]$ does not meet A_j with $j < i$. The left end point b_i is treated analogously. Now for each i select a closed interval H_i contained in the interior of A_i so that the inequality (11.5) holds with H_i replacing A_i.

To produce the collection J_i, $i = 1, 2, \ldots, n$, let $[c, d]$ denote the smallest closed interval that contains all of the H_i's. Define J_i, $i = 1, 2, \ldots, n-1$, to be the closures of the components of $[c, d] \setminus \bigcup_{i=1}^{n-1} H_i$. Then

$$\sum_{i=1}^{n} f(H_i) = -\sum_{i=1}^{n-1} f(J_i) + f(d) - f(c)$$

and hence

$$-\sum_{i=1}^{n-1} f(J_i)/h(n) > M/2 - 2\sup|f(x)|/h(1) > C\,.$$

Now adjoin a suitable closed interval J_n to the above collection to complete the proof.

Let us extend the last lemma to absolutely essentially bounded functions.

LEMMA 11.30. *If f is absolutely essentially bounded but not absolutely regulated, then for some positive β and for some point x_0, there is a sequence H_i, $i = 1, 2, \dots$, of disjoint closed intervals and there is a sequence J_i, $i = 1, 2, \dots$, of disjoint closed intervals, each converging to x_0 and with the end points of their intervals being points of approximate continuity of f, such that*

$$n\,\beta < \sum_{i=1}^{n} f(H_i) \quad \textit{and} \quad -\,n\,\beta > \sum_{i=1}^{n} f(J_i) \quad \textit{for each} \quad n\,.$$

If, in addition, $f \notin \mathrm{V}[h]$ and $C > 0$, then

$$C\,h(n) < \sum_{i=1}^{n} f(H_i) \quad \textit{and} \quad -\,C\,h(n) > \sum_{i=1}^{n} f(J_i) \tag{11.6}$$

for arbitrarily large n.

PROOF. We infer from Theorem 11.5 the first statement. The second is obvious from the first since $h(n) = o(n)$.

PROOF OF THEOREM 11.25. Clearly, (i) $\Rightarrow$ (ii) is obvious. Theorem 11.19 yields (iii) $\Rightarrow$ (i). Hence it remains to prove (ii) $\Rightarrow$ (iii).

If f is an absolutely integrable function which is not absolutely equivalent to a function of class $\mathrm{V}[h]$, then it may or may not be absolutely regulated. By Proposition 11.28 we may assume that f is not absolutely equivalent to a function of class $\mathrm{V}[h]$ on $(0, \delta)$ for any positive δ.

For any choice of positive integers k and m with $k < m$, consider

$$P = \frac{m}{h(m)} \int_{(2k+1)\pi/m}^{2\pi} f \circ g(t)\, \sin mt\, dt$$

where g is an increasing continuous mapping of $[2k + 1\pi/m, 2\pi]$ onto $(c_1, 2\pi)$ with $0 < c_1 < 2\pi$. Let x_1 be a point of approximate continuity of f in $(c_1, 2\pi)$ and M be the absolute essential supremum of $|f|$, and choose g so that, given $\varepsilon > 0$,

$$\lambda\big(\{\,t : |f \circ g(t) - f(x_1)| > \varepsilon\,\}\big) < \big(2\pi - (2k+1)\pi/m\big)\varepsilon\,.$$

Then

$$\begin{aligned} |P| &\le \frac{m}{h(m)} \left\{ \left| \int_{(2k+1)\pi/m}^{2\pi} \big(f \circ g(t) - f(x_1)\big)\, \sin mt\, dt \right| + |f(x_1)|\, \frac{2\pi}{m} \right\} \\ &\le \frac{m}{h(m)} \left\{ 2\pi\varepsilon + 2M\varepsilon + \frac{2M\pi}{m} \right\} \\ &< \frac{2\pi M}{h(1)} + 1 = C \end{aligned}$$

for fixed m if ε is chosen small enough.

We note that an elementary computation using the concavity of $h(n)$ yields

$$h(\alpha n) \le (\alpha + 1)\, h(n)$$

for integers α and n with $n \geq \alpha \geq 1$. Thus if $m/k \leq 2$ and $k \geq 2$ then

$$\frac{h(m)}{h(k)} \leq \frac{h(2k)}{h(k)} \leq 3 .$$

We use Lemmas 11.29 and 11.30 to select disjoint intervals $I_i^{(1)} = \big[a_i^{(1)}, b_i^{(1)}\big]$, $i = 1, \dots, k_1$, (ordered from left to right) whose end points are points of approximate continuity of f such that $b_{k_1}^{(1)}$ is less than the minimum of 1 and c_1 and

$$\big(h(k_1)\big)^{-1}\Big|\sum_{i=1}^{k_1} f(I_i^{(1)})\Big| > 4 + 3C .$$

Choosing an integer m_1 so that $k_1 < m_1 \leq 2k_1$, we will have

$$\big(h(m_1)\big)^{-1}\Big|\sum_{i=1}^{k_1} f(I_i^{(1)})\Big| > \frac{4}{3} + C .$$

Let

$$g(\pi/m_1) = a_1^{(1)},\ g(2\pi/m_1) = b_1^{(1)},\ g(3\pi/m_1) = a_2^{(1)},\ \dots\ ,\ g(2k_1\pi/m_1) = b_{k_1}^{(1)},$$

and let

$$g\big((2k_1 + 1)\pi/m_1\big) = c_1.$$

For each i, exclude from the interval $[i\pi/m_1, (i+1)\pi/m_1]$ a small portion ending at $(i+1)\pi/m_1$. On the remainder of that interval let g be linear and of very small positive slope. On each excluded portion, let g be linear and increasing so that it is continuous on $[\pi/m_1, (2k_1+1)\pi/m_1]$. On $\big[(2k_1+1)\pi/m_1, 2\pi\big]$ let g be the increasing continuous function of the integral P defined above with $k = k_1$ and $m = m_1$. Now

$$\begin{aligned}
\int_{\pi/m_1}^{(2k_1+1)\pi/m_1} & f \circ g(x) \sin m_1 x\, dx \\
&= \sum_{i=0}^{2k_1-1} \int_{\pi/m_1}^{2\pi/m_1} f \circ g(x + i\pi/m_1) \sin(m_1 x + i\pi)\, dx \\
&= \sum_{i=1}^{k_1} \int_0^{\pi/m_1} \Big\{ f \circ g(x + 2i\pi/m_1) \\
&\qquad\qquad - f \circ g(x + (2i-1)\pi/m_1)\Big\} \sin m_1 x\, dx \\
&= \frac{2}{m_1}\Big\{\sum_{i=1}^{k_1} f(I_i^{(1)}) + \eta_1\Big\},
\end{aligned}$$

where $|\eta_1|$ can be made as small as we wish by making the "excluded portions" small and giving the almost horizontal segments of g sufficiently small slopes. With the choice of $I_i^{(1)}$, $i = 1, 2, \dots, k_1$, as above, we have

$$\frac{m_1}{h(m_1)}\Big|\int_{\pi/m_1}^{(2k_1+1)\pi/m_1} f \circ g(x) \sin m_1 x\, dx\Big| > 2\Big(\frac{4}{3} + C\Big) - \frac{2|\eta_1|}{h(m_1)} > 1 + C$$

with $|\eta_1|$ made small by an adjustment of g. Then

$$\frac{m_1}{h(m_1)}\Big|\int_{\pi/m_1}^{2\pi} f \circ g(x) \sin m_1 x\, dx\Big| > 1 .$$

Suppose now that k_n and m_n are defined for $n = 1, \ldots, r$ such that

$$(2k_n + 1)/m_n < 1/m_{n-1} \quad \text{for} \quad n \geq 2$$

and that g is an increasing homeomorphism from $[\pi/m_r, 2\pi]$ onto $[a_1^{(r)}, 2\pi]$ with $a_1^{(r)} < 1/r$ such that

$$\frac{m_n}{h(m_n)} \left| \int_{\pi/m_n}^{2\pi} f \circ g(x) \sin m_n x \, dx \right| > n \, . \tag{11.7}$$

By Lemmas 11.29 and 11.30, in $(0, a_1^{(r)})$ we can find collections of disjoint intervals $H_i^{(r+1)}$, $i = 1, 2, \ldots, k_{r+1}$, and $J_i^{(r+1)}$, $i = 1, 2, \ldots, k_{r+1}$, (ordered from left to right) consisting of closed intervals whose end points are points of approximate continuity of f, such that $k_{r+1} > 3m_r$, the lower end points of $H_1^{(r+1)}$ and $J_1^{(r+1)}$ are less than $1/(r+1)$ and such that

$$\begin{aligned} \big(h(k_{r+1})\big)^{-1} \Big\{ \sum_{i=1}^{k_{r+1}} f(H_i^{(r+1)}) \Big\} &> \quad (r+2)3(m_r+1)/2 \, , \\ \big(h(k_{r+1})\big)^{-1} \Big\{ \sum_{i=1}^{k_{r+1}} f(J_i^{(r+1)}) \Big\} &> -(r+2)3(m_r+1)/2 \, . \end{aligned} \tag{11.8}$$

Choose c_{r+1} between the larger of the upper end points of $H_{k_{r+1}}^{(r+1)}$ and $J_{k_{r+1}}^{(r+1)}$ and $a_1^{(r)}$. Then choose m_{r+1} so that

$$3k_{r+1}m_r > m_{r+1} > (2k_{r+1} + 1)m_r \, .$$

Extend g continuously as an increasing linear mapping of $\big[(2k_{r+1})\pi/m_{r+1}, \pi/m_r\big]$ onto $[c_{r+1}, a_1^{(r)}]$. Let

$$P_{r+1} = \frac{m_{r+1}}{h(m_{r+1})} \int_{(2k_r+1)\pi/m_{r+1}}^{2\pi} f \circ g(x) \sin m_{r+1} x \, dx \, .$$

Choose one of the above collections of intervals $H_i^{(r+1)}$, $i = 1, 2, \ldots, k_{r+1}$, and $J_i^{(r+1)}$, $i = 1, 2, \ldots, k_{r+1}$, so that the **sum in** (11.8) **has the same sign as** P_{r+1}. If $P_{r+1} = 0$, either can be chosen. The intervals of the chosen collection will be called

$$I_i^{(r+1)} = \big[a_i^{(r+1)}, b_i^{(r+1)}\big]$$

and ordered from left to right.

Now let $g(\pi/m_{r+1}) = a_1^{(r+1)}$, $g(2\pi/m_{r+1}) = b_1^{(r+1)}$, $g(3\pi/m_{r+1}) = a_2^{(r+1)}$, $\ldots$, $g(2k_{r+1}\pi/m_{r+1}) = b_{k_{r+1}}^{(r+1)}$, and let $g\big((2k_{r+1}+1)\pi/m_{r+1}\big) = c_{r+1}$. From each interval $[i\pi/m_{r+1}, (i+1)\pi/m_{r+1}]$ we exclude a small portion terminating at the right end. On the remainder of each interval, g is to be linear and of small positive slope. On each "excluded" portion let g be linear and increasing so that on the whole of $[\pi/m_{r+1}, (2k_{r+1}+1)\pi/m_{r+1}]$ it is a continuous increasing map onto $[a_1^{(r+1)}, c_{r+1}]$. Now let

$$Q_{r+1} = \int_{\pi/m_{r+1}}^{(2k_{r+1}+1)\pi/m_{r+1}} f \circ g(x) \sin m_{r+1} x \, dx.$$

Then

$$Q_{r+1} = \sum_{i=1}^{k_{r+1}} \int_0^{\pi/m_{r+1}} \Big\{ f \circ g(x + 2i\pi/m_{r+1}) - f \circ g(x + (2i-1)\pi/m_{r+1}) \Big\} \sin m_{r+1} x \, dx$$

$$= \frac{2}{m_{r+1}} \Big\{ \sum_{i=1}^{k_{r+1}} f(I_i^{(r+1)}) + \eta_{r+1} \Big\}$$

where $|\eta_{r+1}| < h(m_{r+1})/2$ can be assured by making the excluded portions small and making the almost horizontal segments of g have sufficiently small slopes. Then

$$\frac{m_{r+1}}{h(m_{r+1})} |Q_{r+1}| > \frac{2}{h(m_{r+1})} \{ h(k_{r+1})(r+2)(3m_r+1)/2 - |\eta_{r+1}| \}$$
$$\geq r + 2 - 2|\eta_{r+1}|/h(m_{r+1}) > r + 1$$

since

$$\frac{h(k_{r+1})}{h(m_{r+1})} \geq \frac{h(k_{r+1})}{h(3m_r k_{r+1})} \geq \frac{1}{3m_r + 1}.$$

Then

$$\frac{m_{r+1}}{h(m_{r+1})} \int_{\pi/m_{r+1}}^{\pi} f \circ g(x) \sin m_{r+1} x \, dx = \frac{m_{r+1}}{h(m_{r+1})} Q_{r+1} + P_{r+1}$$

exceeds $r+1$ in absolute value, which is (11.7) with $n = r+1$.

By this inductive process we can define g on $(0, 2\pi]$ and, setting $g(0) = 0$, we see that g is a change of variable such that

$$\frac{m_n}{h(m_n)} \int_{\pi/m_n}^{2\pi} f \circ g(x) \sin m_n x \, dx \neq O(1).$$

Since

$$\frac{m_n}{h(m_n)} \int_0^{\pi/m_n} f \circ g(x) \sin m_n x \, dx = O(1),$$

we have constructed a change of variable g such that

$$\widehat{f \circ g}(n) \neq O(h(n)/n)$$

and the argument is complete.

A close analysis of the preceding argument shows that somewhat more can be said about functions which are not absolutely regulated.

THEOREM 11.31. *Let f be absolutely integrable but not absolutely regulated. Then for any sequence γ_n, $n = 1, 2, \ldots$, decreasing to 0 there is a change of variable g such that*

$$\widehat{f \circ g}(n) \neq O(\gamma_n).$$

PROOF. Without loss of generality we may assume $\gamma_1 = 1$ and, as in the proof of the previous result, that in any interval $(0, \delta)$ we can find disjoint intervals I_i,

$i = 1, 2, \ldots, k$, (ordered left to right) whose end points are points of approximate continuity of f, so that

$$\left|\sum_{i=1}^{k} f(I_i)\right| > 2k \quad \text{for each} \quad k,$$

and the sign of this sum can be chosen at will. Let $M = \operatorname{ab.ess.sup}|f|$, noting that our last assumption ensures $M > 1$. Choose an integer m_1 with $m_1 > \pi M$ and set $k_1 = m_1 - 1$. Then choose any c_1 in $(0, 2\pi)$ and define g on $\big[(2k_1 + 1)\pi/m_1, 2\pi\big]$ to be the increasing linear map of that interval onto $[c_1, 2\pi]$. Select intervals $I_i^{(1)}$, $i = 1, \ldots, k_1$, as above, in the interval $(0, c_1)$ and so that

$$\sum_{i=1}^{k_1} f(I_i^{(1)}) \quad \text{and} \quad \int_{(2k_1+1)\pi/m_1}^{2\pi} f \circ g(x) \sin m_1 x \, dx$$

have the same sign. Defining g as in the last proof so that $I_i^{(1)}$ is the interval $\big[g\big((2i-1)\pi/m_1\big), g(2i\pi/m_1])\big]$, then we will have

$$\begin{aligned}\left|\int_{\pi/m_1}^{(2k_1+1)\pi/m_1} f \circ g(x) \sin m_1 x \, dx\right| &= \frac{2}{m_1}\left|\sum_{i=1}^{k_1} f(I_i^{(1)}) + \eta_1\right| \\ &> \frac{2}{m_1}(2m_1 - 2) - |\eta_1| \\ &= 4 - \frac{4 + |\eta_1|}{m_1} \\ &> 2\end{aligned}$$

if g is chosen so that $|\eta_1|$ is small enough. Thus, no matter how g may be extended to a self homeomorphism of T,

$$\begin{aligned}\left|\int_0^{2\pi} f \circ g(x) \sin m_1 x \, dx\right| &> 2 - \left|\int_0^{\pi/m_1} f \circ g(x) \sin m_1 x \, dx\right| \\ &\geq 2 - \pi M/m_1 \\ &> 1 > \gamma_{m_1}.\end{aligned}$$

Suppose we have chosen m_r, $r = 1, 2, \ldots, n$, and have defined an increasing homeomorphism g that maps each $[\pi/m_r, 2\pi]$ onto $[g(\pi/m_r), 2\pi]$ with $0 < g(\pi/m_r) < 1/r$ such that

$$\left|\int_0^{2\pi} f \circ g(x) \sin m_r x \, dx\right| > r\,\gamma_{m_r}$$

no matter how g is extended to be a self homeomorphism of T.

Choose m_{n+1} so that $m_{n+1} > 4m_n$ and

$$\frac{2}{m_n} > \frac{1}{m_{n+1}}(7 + \pi M) + (n+1)\,\gamma_{m_{n+1}}.$$

Let k_{n+1} be the integer satisfying

$$\frac{m_{n+1}}{m_n} - 3 \leq 2k_{n+1} < \frac{m_{n+1}}{m_n} - 1.$$

Choose c_{n+1} in $\big(0, g(\pi/m_n)\big)$ and define g on $\big[(2k_{n+1} + 1)\pi/m_{n+1}, \pi/m_n\big)$ to be the increasing linear function which maps that interval onto $\big[c_{n+1}, g(\pi/m_n)\big)$. As in the last proof, select $I_i^{(n+1)}$, $i = 1, 2, \ldots, k_{n+1}$, in the interval $(0, c_{n+1})$ so that

$$\sum_{i=1}^{k_{n+1}} f(I_i^{(n+1)}) \quad \text{and} \quad \int_{(2k_{n+1}+1)\pi/m_{n+1}}^{2\pi} f \circ g(x) \sin m_{n+1}x \, dx$$

have the same sign. Then if g is defined on $[\pi/m_{n+1}, (2k_{n+1}+1)\pi/m_{n+1}]$ in the same manner as above, we have

$$\begin{aligned}
\Big| \int_{\pi/m_{n+1}}^{(2k_{n+1}+1)\pi/m_{n+1}} & f \circ g(x) \sin m_{n+1}x \, dx \Big| \\
& > \frac{2}{m_{n+1}} \left\{ \Big| \sum_{i=1}^{k_{n+1}} f(I_i^{(n+1)}) \Big| - |\eta_{n+1}| \right\} \\
& \geq \frac{2}{m_{n+1}} \left\{ \frac{m_{n+1}}{m_n} - 3 - |\eta_{n+1}| \right\}.
\end{aligned}$$

Thus, no matter how g is extended as a self homeomorphism on T,

$$\Big| \int_0^{2\pi} f \circ g(x) \sin m_{n+1}x \, dx \Big| > \frac{2}{m_n} - \frac{1}{m_{n+1}} (6 + 2|\eta_{n+1}| + \pi M).$$

By an appropriate definition of g we can ensure that $|\eta_{n+1}| < \frac{1}{2}$, in which case

$$\Big| \int_0^{2\pi} f \circ g(x) \sin m_{n+1}x \, dx \Big| > (n+1)\gamma_{m_{n+1}}.$$

We see then that the g defined by this inductive procedure and $g(0) = 0$ will be the change of variable described in the statement of the theorem.

APPENDIX A

Supplementary Material

The appendix is devoted to the preliminary material needed in the development of the book. Moving this material to an appendix has enabled us to smooth the exposition. Some of the elementary facts are stated just for ease of citation since they can be found in standard texts in analysis. Other less accessible material are given more detailed discussions.

For subsets E of $\mathbb{R}^n$, the notations E^o and $\overline{E}$ are reserved for the interior and closure of E in the usual topology of $\mathbb{R}^n$, respectively.

Sets, Functions and Measures

A.1. Baire, Borel and Lebesgue

There are several fundamental facts that are associated with these names. A brief discussion of them will be be given below.

A.1.1. Baire category. Let E be a subset of a metric space X. Then E is said to be of **first category** if it is the countable union of nowhere dense sets, and it is said to be of the **second category** in the contrary case. A metric space X is said to be **completely metrizable** if there is an equivalent metric that makes X into a complete metric space. We state the famous Baire category theorem.

Theorem A.1 (Baire Category Theorem). *Let X be a completely metrizable space. If G_n, $n = 1, 2, \ldots$, is a sequence of dense open sets, then $\bigcap_{n=1}^{\infty} G_n$ is a dense set.*

The next theorem is a useful characterization of completely metrizable spaces.

Theorem A.2. *Let X be a metric space. Then the following are equivalent.*

(1) *X is completely metrizable.*
(2) *X is homeomorphic to a G_δ set contained in a complete metric space.*
(3) *X is a G_δ set in every metric space in which it is embedded.*

A topological space for which the Baire category theorem holds is called a **Baire space**, that is, if it is always true that the countable intersection of dense open sets is a dense set. Unfortunately, the name 'Baire space' has been used in general topology in at least two distinct contexts, see [**98**, page 427] and [**38**, page 326] for example. Perhaps a more descriptive but clumsy name would be 'Baire Category II space'. We have used the simpler name here.

A.1.2. Borel sets. For a metric space X, the members of the smallest σ-algebra $\mathcal{B}$ that contains the collection of all open sets are called **Borel sets** of X. Of particular interest to us are completely metrizable spaces. With the aid of the countable ordinal numbers one can describe various classes of $\mathcal{B}$ by using closure under countable unions or countable intersections. We will not find this useful for very large ordinal numbers. But the classes called the F_σ sets and the G_δ sets will prove useful. An F_σ **set** is the countable union of closed sets, and a G_δ **set** is the countable intersection of open sets.

Now let X be a separable, completely metrizable space. In this context there is another class of sets called the **analytic** or **Suslin** sets. This class consists of the continuous images of Borel sets. Clearly, every Borel set is an analytic set. These sets have been characterized to be those sets that are the continuous image of the space of irrational numbers. (For a nice development of analytic sets, see [**41**, pages 59–72].) There is the following theorem.

THEOREM A.3. *Let X be a separable, completely metrizable space. Every uncountable analytic set contains a topological copy of the Cantor ternary set.*

The collection of analytic sets is not a σ-algebra.

A real-valued function f defined on X is said to be **Borel measurable** if $f^{-1}[U]$ is a Borel set for every open set U of $\mathbb{R}$.

A.1.3. Finite Baire classes. For a metric space X, the members of the smallest collection of real-valued functions containing the continuous functions and closed under pointwise convergence of sequences are called **Baire functions**. With the aid of the countable ordinal numbers one can describe various classes of Baire functions. The **Baire class 0** is the class of all continuous functions. The **Baire class 1** is the largest class of functions that are the pointwise limits of functions from the Baire class 0. Analogously, the **Baire class 2** is the largest class of functions that are the pointwise limits of functions from the Baire class 1. We have the following theorems.

THEOREM A.4. *Let X be a separable, completely metrizable space. Then the following statements hold.*

(1) *A function $f\colon X \to \mathbb{R}$ is a Baire function if and only if it is a Borel measurable function.*
(2) *A function $f\colon X \to \mathbb{R}$ is in the Baire class 1 if and only if $f^{-1}\big[\{ y : y > \alpha \}\big]$ and $f^{-1}\big[\{ y : y < \alpha \}\big]$ are F_σ sets for every α.*

THEOREM A.5. *Let X be a separable, completely metrizable space. A function $f\colon X \to \mathbb{R}$ is in the Baire class 1 if and only if the set of points of discontinuity of $f|A$ is a set of the first category in A for each nonempty perfect subset A of X.*

THEOREM A.6. *The uniform limit of a sequence of functions in the Baire class 1 is a member of the Baire class 1.*

A.1.4. Lebesgue measure. It is assumed that the reader has familiarity with the properties of the Lebesgue measure. Totally imperfect sets are used in the book. A set E in a metric space X is said to be **totally imperfect** if the only compact perfect subset P of X that is contained in E is the empty set. There is the following theorem whose proof is a straightforward transfinite induction.

THEOREM A.7. *If X be a separable, completely metrizable space, then there are disjoint, totally imperfect sets E and F such that $X = E \cup F$. Consequently, if P is a nonempty, compact, perfect set, then both $P \cap E$ and $P \cap F$ are uncountable. And, if P is a compact set with $P \cap E = \emptyset$, then P is a countable set.*

The notion of density is used often here. This notion has many variations. The one that is used here involves n-cubes $I(x, \delta)$ in $\mathbb{R}^n$ that are centered at a point x and have diameter equal to δ. For a measurable set E in $\mathbb{R}^n$, the **density of E at x** is the limit

$$\lim_{\delta \to 0} \frac{\lambda_n(I(x,\delta) \cap E)}{\lambda_n\big(I(x,\delta)\big)}.$$

Of course, the limit need not exist. The interesting cases are the limits 0 and 1. In such cases, the point x is called a **point of dispersion** or **point of density of E**, respectively. Often it will be convenient to replace the n-cube centered at x with an n-sphere centered at x. This replacement in the above limit will result in the same density values at x when x is a point of dispersion or a point of density of the measurable set E. An equivalent way to compute the points of dispersion and density of a measurable set E is to relax the requirement that the point x be at the center of the n-cube. This follows from the fact that each n-cube whose diameter is δ and contains the point x is contained in the n-cube $I(x, 2\delta)$ and the ratios of the measures of these two n-cubes are bounded by a constant that depends only on n.

The Vitali covering theorem, or its many variants, is also used frequently. The coverings that we use will always be by n-cubes or n-spheres.

Finally, we mention the classical definition of approximate continuity of a function f at x. Here, the requirement is that there exists a measurable set E that contains x and has x as a point of density such that $f|E$ is continuous at x. We shall give an equivalent definition in the context of the density topology in a later section of the Appendix.

(Porous sets were mentioned in Chapter 1 in a very minor way. For the reader who wishes to know more of this notion, we direct them to the book [**126**] by Thomson as a beginning point.)

A.2. Lipschitzian functions

As usual we define a function $f \colon X \to Y$ from a metric space (X, ρ_X) to a metric space (Y, ρ_Y) to be **Lipschitzian with constant** M if

$$\rho_Y\big(f(x), f(x')\big) \leq M\rho_X(x, x') \quad \text{whenever} \quad x,\, x' \in X.$$

The number

$$\operatorname{Lip}(f) = \inf\{\, M : \rho_Y\big(f(x), f(x')\big) \leq M\rho_X(x,x'), \quad x,\, x' \in X\,\}$$

is called the **Lipschitz constant of** f.

A function is said to be a **bi-Lipschitzian map** if it is a one-to-one, onto Lipschitzian function that has a Lipschitzian inverse.

In general, we shall be dealing with spaces that are subspaces of Euclidean spaces. In this setting we may often suppose that X and Y are closed sets since Lipschitzian functions are uniformly continuous. The class of Lipschitzian functions has many pleasant properties, more than we can list here. We shall state a few that will prove useful.

A.2.1. Extension. It is known that a real-valued Lipschitzian function defined on a subset of $\mathbb{R}^n$ has a Lipschitzian extension to all of $\mathbb{R}^n$. There are sharper results along this line, but we will not need them (see, for example, [**41**, page 201, Kirszbraun's theorem]).

Clearly, there are bi-Lipschitzian maps of subsets of $\mathbb{R}^n$ onto subsets of $\mathbb{R}^n$ that do not have extensions to bi-Lipschitzian maps of $\mathbb{R}^n$ onto itself. Of course, the difficulty is really an algebraic topological obstruction.

A.2.2. Measure properties. When X and Y are contained in $\mathbb{R}^n$, Lipschitzian maps preserve sets of Lebesgue measure zero. Consequently, if Z is the set of points where the derivative of a Lipschitzian map is zero, then we infer from the Vitali covering theorem that the image of Z has Lebesgue measure equal to 0.

It also follows that, for Lipschitzian maps of $\mathbb{R}^n$ into itself, the image of a Lebesgue measurable set is Lebesgue measurable. Indeed, this is due to the fact that Lebesgue measurable sets are unions of F_σ sets and sets of measure 0. For more on these comments, see Theorem 5.21 on page 73.

A.2.3. Differentiability. A Lipschitzian function from $\mathbb{R}^n$ into $\mathbb{R}^m$ is differentiable λ_n-almost everywhere (see [**39**, page 81], for example).

A.2.4. Bi-Lipschitzian maps. The following is rather elementary.

PROPOSITION A.8. *If $f\colon S \to T$ is an onto Lipschitzian mapping of subsets S and T of $\mathbb{R}^n$, then*

$$f\big[S \cap B(x,r)\big] \subset T \cap B\big(f(x), \operatorname{Lip}(f)\, r\big),$$

and consequently, for measurable sets S,

$$\lambda_n(T) \leq \big(2\operatorname{Lip}(f)\big)^n \lambda_n(S).$$

Theorem 5.21 establishes that bi-Lipschitzian maps of $\mathbb{R}^n$ into itself preserve points of dispersion and points of density of measurable sets. A stronger statement concerning bi-Lipschitzian maps has been proved by Buczolich [**25**] in that the map f is required to be bi-Lipschitzian exactly on the measurable set S.

THEOREM A.9 (BUCZOLICH). *If $f\colon S \to T$ is a bi-Lipschitzian map of a measurable set S onto a measurable set T, then h maps the set of density points of S onto the set of density points of T and maps the set of dispersion points of S onto the set of dispersion points of T.*

PROOF. In view of Proposition A.8 we may assume that 0 is a point of S and that $f(0) = 0$.

With the aid of Proposition A.8, the inclusion

$$T \cap B(0,r) \subset f\big[S \cap B\big(0, \operatorname{Lip}(f^{-1})\, r\big)\big]$$

yields that if 0 is a dispersion point of S then 0 is a dispersion point of T. Clearly, the density statement is not as obvious since the notions of density and dispersion are not complementary.

The proof of the density statement will use the dilation $\Delta_r\colon \mathbb{R}^n \to \mathbb{R}^n$ defined by $\Delta_r(x) = rx$. As is well-known, the map f has a Lipschitzian extension $\widetilde{f}\colon \mathbb{R}^n \to \mathbb{R}^n$. Of course, there is no guarantee that the extension is an injection. For each positive number r define

$$g_r = \Delta_{\frac{1}{r}} \circ \widetilde{f} \circ \Delta_r.$$

A simple calculation gives

$$g_r(0) = 0 \quad \text{and} \quad \operatorname{Lip}(g_r) = \operatorname{Lip}(\widetilde{f}).$$

Let

$$S_r = \Delta_{\frac{1}{r}}[S] \quad \text{and} \quad T_r = \Delta_{\frac{1}{r}}[T].$$

Then

$$g_r[S_r] = \big(\Delta_{\frac{1}{r}} \circ \widetilde{f}\big)[S] = \big(\Delta_{\frac{1}{r}} \circ f\big)[S] = T_r,$$

and

$$g_r|S_r \colon S_r \to T_r$$

is bi-Lipschitzian with

$$\operatorname{Lip}(g_r|S_r) \le \operatorname{Lip}(f)$$

and

$$\operatorname{Lip}\big((g_r|S_r)^{-1}\big) \le \operatorname{Lip}\big(f^{-1}\big).$$

Clearly,

$$\frac{\lambda_n(S \cap B(0,r))}{\lambda_n(B(0,r))} = \frac{\lambda_n(S_r \cap B(0,1))}{\lambda_n(B(0,1))}.$$

With the above notation we have the following lemma

LEMMA A.10. *If*

$$\lim_{r\to 0} \frac{\lambda_n(S_r \cap B(0,1))}{\lambda_n(B(0,1))} = 1,$$

then for each positive number η with $\eta < 1$ there exists a positive number δ such that

$$0 < r < \delta \quad \textit{implies} \quad g_r\big[B(0,1)\big] \supset B\big(0, \tfrac{\eta}{\operatorname{Lip}(f^{-1})}\big).$$

PROOF. The proof is by contradiction. Suppose that there exists an η with $0 < \eta < 1$ such that for each positive number δ there is an r with $0 < r < \delta$ and $B\big(0, \frac{\eta}{\operatorname{Lip}(f^{-1})}\big) \setminus g_r\big[B(0,1)\big] \neq \emptyset$. Then we can find a sequence r_k, $k = 1, 2, \ldots$, decreasing to 0 such that, for each k,

$$\frac{\lambda_n(S_{r_k} \cap B(0,1))}{\lambda_n(B(0,1))} > 1 - \frac{1}{2^k}$$

and

$$B\big(0, \tfrac{\eta}{\operatorname{Lip}(f^{-1})}\big) \setminus g_{r_k}\big[B(0,1)\big] \neq \emptyset.$$

The following subsets of $B(0,1)$

$$H_m = \textstyle\bigcap_{k\ge m} S_{r_k} \cap B(0,1)$$

form an increasing sequence, and

$$H = \textstyle\bigcup_{m\ge 1} H_m$$

is a dense subset of $B(0,1)$ because $\lambda_n(B(0,1) \setminus H) = 0$. Moreover, for each m, $g_{r_k}|H_m$ has a Lipschitzian inverse with

$$\operatorname{Lip}\big((g_{r_k}|H_m)^{-1}\big) \le \operatorname{Lip}(f^{-1}) \quad \text{whenever} \quad k > m.$$

In view of Ascoli's Theorem, we may further assume that r_k, $k = 1, 2, \ldots$, is such that the sequence g_{r_k}, $k = 1, 2, \ldots$, converges uniformly on $\overline{B(0,1)}$ to a continuous map $g\colon \overline{B(0,1)} \to \mathbb{R}^n$. Clearly, since H is dense in $B(0,1)$,

$$\operatorname{Lip}(g) \leq \operatorname{Lip}(f), \quad g(0) = 0,$$

and $g|H$ has a Lipschitzian inverse. Again as H is dense in $B(0,1)$, we have that g^{-1} is a well defined map of $g\Big[\overline{B(0,1)}\Big]$ onto $\overline{B(0,1)}$ and that

$$\operatorname{Lip}(g^{-1}) \leq \operatorname{Lip}(f^{-1}).$$

Consequently, $g\colon \overline{B(0,1)} \to g\Big[\overline{B(0,1)}\Big]$ is a bi-Lipschitzian map. By the Brouwer invariance of domain theorem [**78**, page 95], we have that 0 has a neighborhood that is contained in $g\Big[\overline{B(0,1)}\Big]$ and that $B\big(0, \frac{1}{\operatorname{Lip}(f^{-1})}\big)$ can be selected to be that neighborhood because $\operatorname{Lip}(g^{-1}) \leq \operatorname{Lip}(f^{-1})$. For each k, select a point y_k in $B\big(0, \frac{\eta}{\operatorname{Lip}(f^{-1})}\big) \setminus g_{r_k}\big[B(0,1)\big]$. There is a subsequence y_{k_j}, $j = 1, 2, \ldots$, that converges to a point y. Obviously, $|y| < \frac{1}{\operatorname{Lip}(f^{-1})}$. We infer from [**78**, pages 74–75] that y is a stable value of the homeomorphism g. That is, for each positive number ε there is a positive number γ such that if $G\colon \overline{B(0,1)} \to \mathbb{R}^n$ is a continuous map that is uniformly within γ distance of g then $y \in G\Big[\overline{B(0,1)}\Big]$. Let k_j be such that g_{k_j} is uniformly within $\frac{\gamma}{3}$ of g and $|y_{k_j} - y| < \frac{\gamma}{3}$. By deforming g_{k_j} within the neighborhood $B\big(y, \frac{\gamma}{3}\big)$ in such a way as to move y_{k_j} to y, we can form a continuous map $G\colon \overline{B(0,1)} \to \mathbb{R}^n$ that is uniformly within γ of g and $y \notin G\Big[\overline{B(0,1)}\Big]$. That is, y is not a stable point of g, a contradiction. Thereby, the lemma is proved.

Let us return to the proof of Buczolich's theorem. With $0 < \eta < 1$, we have

$$\begin{aligned}
&\frac{\lambda_n\big(B\big(0, \frac{\eta r}{\operatorname{Lip}(f^{-1})}\big) \cap \tilde{f}\big[B(0,r)\big]\big)}{\lambda_n\big(B\big(0, \frac{\eta r}{\operatorname{Lip}(f^{-1})}\big)\big)} \\
&\qquad \leq \frac{\lambda_n\big(B\big(0, \frac{\eta r}{\operatorname{Lip}(f^{-1})}\big) \cap T\big) + \lambda_n\big(\tilde{f}\big[B(0,r) \setminus S\big]\big)}{\lambda_n\big(B\big(0, \frac{\eta r}{\operatorname{Lip}(f^{-1})}\big)\big)} \\
&\qquad \leq \frac{\lambda_n\big(B\big(0, \frac{\eta r}{\operatorname{Lip}(f^{-1})}\big) \cap T\big)}{\lambda_n\big(B\big(0, \frac{\eta r}{\operatorname{Lip}(f^{-1})}\big)\big)} \\
&\qquad\qquad + \left(2\operatorname{Lip}(\tilde{f})\,\frac{\operatorname{Lip}(f^{-1})}{\eta}\right)^n \frac{\lambda_n\big(B(0,r) \setminus S\big)}{\lambda_n\big(B(0,r)\big)}.
\end{aligned}$$

On passing to the limit as r tends to 0, we have that the density of T at 0 is equal to 1 whenever 0 is a point of density of S. The theorem is completely proved.

A.3. Bounded variation

Familiarity with the definition and properties of the notion of functions of bounded variations is assumed. The notion of an absolutely continuous function is also assumed to be known. The finiteness of the total variation of a function f has many analytical implications. In particular, recall the decomposition of such functions as the difference of two monotone bounded functions, the existence of a derivative $f'(x)$ at λ-almost every point x, and the validity of the inequality

$$\int_a^b |f'(x)|\, d\lambda(x) \leq \mathrm{V}\big(f, [a,b]\big),$$

where the equality holds when and only when f is an absolutely continuous function.

A.3.1. CBV class. When a function of bounded variation is continuous, the above decomposition can be made by continuous monotone functions. It is quite obvious from the definitions that a Lipschitzian function f is an absolutely continuous function on every bounded interval $[a,b]$. The usual Cantor function is not an absolutely continuous function.

A.3.2. Banach indicatrix. The well-known Banach indicatrix theorem has many useful consequences. In order to state the theorem we must define the indicatrix function. Let $f\colon X \to Y$ be a function, y be a point in Y and A be a subset of X. Then $\mathrm{N}(f, A, y)$ is defined to be the extended real number

$$\mathrm{N}(f, A, y) = \begin{cases} \operatorname{card}\left(f^{-1}[y] \cap A\right) & \text{if } f^{-1}[y] \cap A \text{ is finite,} \\ +\infty & \text{if } f^{-1}[y] \cap A \text{ is infinite.} \end{cases}$$

THEOREM (BANACH INDICATRIX). *Let $f\colon [a,b] \to \mathbb{R}$ be a continuous function. If $[c,d] \subset [a,b]$, then $\mathrm{N}(f, [c,d], \cdot\,)$ is Borel measurable and*

$$\int_{-\infty}^{+\infty} \mathrm{N}(f, [c,d], y)\, dy = \mathrm{V}(f, [c,d]).$$

A proof can be found in [**30**].

We have as a consequence of the Banach indicatrix theorem the following proposition.

PROPOSITION A.11. *If $f\colon [a,b] \to \mathbb{R}$ is a continuous function with finite total variation and if E is a compact subset of $\mathbb{R}$ with $\lambda(E) = 0$, then for each positive number ε there is a positive number δ such that*

$$\sum_{i=1}^{n} \mathrm{V}(f, [a_i, b_i]) < \varepsilon$$

whenever $[a_i, b_i]$, $i = 1, \dots, n$, is a nonoverlapping collection of intervals such that $f\big[[a_i, b_i]\big]$ is contained in the δ-neighborhood of E for each i.

A.3.3. Lower semicontinuity. An important property of the total variation of a function is its lower semicontinuity under pointwise convergence.

THEOREM A.12. *If $f_k\colon [a,b] \to \mathbb{R}$, $k = 1, 2, \dots$, is a sequence of functions that converges pointwise to $f\colon [a,b] \to \mathbb{R}$, then*

$$\mathrm{V}(f, [a,b]) \leq \liminf_{k\to\infty} \mathrm{V}(f_k, [a,b]).$$

The proof is straightforward and will be left out.

There is another mode of convergence that is very useful in analysis, that of almost everywhere convergence with respect to Lebesgue measure. When the functions that are involved are Lebesgue measurable, the points of approximate continuity of the functions provides us with a way of computing an appropriate variation function called the essential variation of the function. For a measurable function $f\colon [a,b] \to \mathbb{R}$, let E denote its set of points of approximate continuity and define the **essential variation** to be

$$\operatorname{ess} \mathrm{V}(f,[a,b]) = \mathrm{V}_E(f,[a,b]) = \sup \sum_{i=1}^{k} |f(x_i) - f(x_{i-1})|,$$

where the supremum is taken over all finite collections of points x_i, $i = 1, \ldots, k$, in E with $x_{i-1} < x_i$. It is known that E is a set of full measure in $[a,b]$. Moreover, if E_0 is a subset of E that is also of full measure in $[a,b]$, then $\mathrm{V}_E(f,[a,b]) = \mathrm{V}_{E_0}(f,[a,b])$. Consequently, if f_k, $k = 1, 2, \ldots$, is a sequence of measurable functions that converges almost everywhere to a function f, then

$$\operatorname{ess} \mathrm{V}(f,[a,b]) \leq \liminf_{k\to\infty} \operatorname{ess} \mathrm{V}(f_k,[a,b]).$$

It is easily shown that the above definition of $\operatorname{ess} \mathrm{V}(f,[a,b])$ agrees with the one given on page 65 of Chapter 5. Indeed, this is a consequence of the above lower semicontinuity property.

If $\operatorname{ess} \mathrm{V}(f,[a,b]) < +\infty$, then

$$\lim_{t\to x+} \operatorname{ess} \mathrm{V}(f,[a,t])$$

exists for each x in $[a,b)$, whence there is a BV function $\widetilde{f}$ such that $f = \widetilde{f}$ λ-almost everywhere and $\widetilde{f}$ is continuous from the right. These functions $\widetilde{f}$ are examples of regulated BV functions, that is, BV functions f for which

$$\lim_{t\to x-} f(t) \leq f(x) \leq \lim_{t\to x+} f(t)$$

for every x. (We may choose $\widetilde{f}$ so that $\widetilde{f}(x) = \frac{1}{2}\big(\widetilde{f}(x+) + \widetilde{f}(x-)\big)$ for every x.)

The Banach indicatrix theorem has a natural generalization to regulated BV functions. To establish such a theorem, we must define an appropriate indicatrix function for measurable functions f. Let $C_{ap}(f)$ be the set of all points of approximate continuity of f. For each open interval (a,b), define $K(a,b)$ to be the closed convex hull of $f\big[(a,b) \cap C_{ap}(f)\big]$. And, for each x, define Y_x to be the closed interval (possibly unbounded)

$$Y_x = \bigcap \{\, K(a,b) : x \in (a,b) \,\}.$$

Obviously, $Y_x = \{\, f(x) \,\}$ whenever x is a point of continuity of f. We define the indicatrix function of f to be

$$\mathrm{N}(f,A,y) = \operatorname{card}\{\, x : x \in A \ \text{ and } \ y \in Y_x \,\}.$$

When the essential total variation of f is finite, there is a right-continuous function $\widetilde{f}$ for which $\operatorname{ess} \mathrm{V}(f,[a,b]) = \mathrm{V}(\widetilde{f},[a,b])$ and $f = \widetilde{f}$ almost everywhere. Also, the connected set Y_x is the interval that corresponds to the jump discontinuity of $\widetilde{f}$ whenever x is a point of discontinuity of $\widetilde{f}$. Clearly, $\mathrm{N}(f,[a,b],y) = \mathrm{N}(\widetilde{f},[a,b],y)$ for each y. The measurability of $\mathrm{N}(\widetilde{f},[a,b],y)$ as a function of y is easily established. Moreover, if g_n, $n = 1, 2, \ldots$, is a sequence of inscribed piecewise linear, continuous functions that converges almost everywhere to $\widetilde{f}$, then

$$\lim_{n\to\infty} \mathrm{N}(g_n,[a,b],y) = \mathrm{N}(\widetilde{f},[a,b],y) \quad \text{for almost every } y.$$

As

$$\lim_{n\to\infty} \mathrm{V}(g_n,[a,b]) = \mathrm{V}(\widetilde{f},[a,b]),$$

we have the following.

THEOREM A.13. *If f is a measurable function with finite essential total variation, then*

$$\operatorname{ess} \mathrm{V}(f, [a,b]) = \int_{-\infty}^{+\infty} \mathrm{N}(f, [a,b], y)\, dy.$$

Approximate Continuity

A.4. Density topology

It is assumed that the reader is familiar with the elementary facts about points of density of subsets E of $\mathbb{R}^n$. A point x is said to be a **point of density** of a set E if the density of E at x is equal to 1.

A measurable subset E of $\mathbb{R}^n$ is said to be **density-open** if each point of E is a point of density of E. That the collection of all such sets is a topology was observed by Haupt and Pauc [**76**] who noted that the intersection of two density-open sets is a density-open set and that the union of a collection of density-open sets is a density-open set. The only serious problem is to show that the union is measurable. This follows by use of the Vitali covering theorem. Thus, we call this topology the **density topology** on $\mathbb{R}^n$. The closure operator and the closed sets in this topology will be called the **density-closure** and the **density-closed** sets.

It is clear that the density topology is finer that the usual Euclidean topology on $\mathbb{R}^n$. A real-valued function on $\mathbb{R}^n$ that is continuous with respect to the density topology is said to be **approximately continuous**. There is a classical definition of approximate continuity. Of course, the two definitions are equivalent. Moreover, the density topology is completely regular (see [**146, 96**] and especially [**63**, Theorem 3]) and is not normal [**63**, Theorem 2].

A.4.1. Zero-sets and cozero-sets. A subset Z of a topological space X is called a **zero-set** in the topology if $Z = f^{-1}[0]$ for some continuous function $f\colon X \to [0,1]$, and a set is a **cozero-set** if its complement is a zero-set.

We infer from Zahorski's lemma [**146**, Lemma 11] (see also [**63**, Lemma 4]) that the cozero-sets of approximately continuous real-valued functions are precisely those density-open sets that are F_σ sets in the usual topology. (The cozero-set of a real-valued function is the set of points at which the function takes nonzero values; the zero-set is the set of points at which the value is 0.) Clearly one can replace $\mathbb{R}$ with the space of extended real numbers endowed with the order topology to yield approximately continuous, extended real-valued functions. Here the order structure of the extended real number system replaces the arithmetic properties of the space $\mathbb{R}$. Hence, proofs often will be different in the case of extended real-valued functions.

A.4.2. Antiderivatives. Two useful facts about every bounded, approximately continuous function $f\colon \mathbb{R} \to \mathbb{R}$ are that f has an antiderivative and that each x in $\mathbb{R}$ is a Lebesgue point of f. But these facts need not hold for approximately continuous, extended real-valued functions. Hence more care is required for extended real-valued functions.

A.4.3. Connected sets. It is known that $\mathbb{R}^n$ is connected in the usual Euclidean topology. Surprisingly, it is also connected in the density topology (see Goffman and Waterman [**66**]). This will follow from a couple of easily proved lemmas.

LEMMA A.14. *Let I be a closed interval in $\mathbb{R}^n$ and let ε be a positive number. If A and B are measurable sets of positive measures so that $I = A \cup B$, then there is a closed interval $\widetilde{I}$ contained in the interior of I such that* $\operatorname{diam} \widetilde{I} < \varepsilon$ *and*

$$\lambda_n(\widetilde{I} \cap A) \geq \tfrac{1}{2}\lambda_n(\widetilde{I}) \quad \textit{and} \quad \lambda_n(\widetilde{I} \cap B) \geq \tfrac{1}{2}\lambda_n(\widetilde{I}).$$

PROOF. The proof is trivial when $\lambda_n(A \cap B) > 0$. So we shall assume that $\lambda_n(A \cap B) = 0$. As the sets A and B both have positive measures, there is a closed interval I_0 contained in the interior of I with $\lambda_n(I_0 \cap A) \geq 0$ and $\lambda_n(I_0 \cap B) \geq 0$. Let $\varepsilon_0 = \operatorname{dist}(I_0, \mathbb{R}^n \setminus I)$. Then there is a positive number δ with $\delta < \min\{\varepsilon_0, \varepsilon\}$ such that any cube $I(x)$ with diameter equal to δ and centered at x satisfies

(1) $I(x) \subset I$ whenever $x \in I_0$,
(2) $\lambda_n(I(x) \cap A) > \frac{1}{2}\lambda_n(I(x))$ for some x in I_0,
(3) $\lambda_n(I(x) \cap A) < \frac{1}{2}\lambda_n(I(x))$ for some x in I_0.

As the function

$$\varphi(x) = \frac{\lambda_n(I(x) \cap A)}{\lambda_n(I(x))}, \qquad x \in I_0,$$

is continuous, we infer from the conditions (2) and (3) that there is a point c in I_0 such that $\varphi(c) = \frac{1}{2}$. Since $\lambda_n(A \cap B) = 0$, we have $\lambda_n(I(c) \cap B) = \frac{1}{2}\lambda_n(I(c))$. Thereby the lemma is proved.

The following is a consequence of the nested interval theorem.

LEMMA A.15. *Let A and B be measurable subsets of a closed interval I of $\mathbb{R}^n$ such that both A and B have positive measures, their union is I, and $\lambda_n(A \cap B) = 0$. Then there is a point c in the interior of I that is neither a point of density of A nor a point of density of B.*

THEOREM A.16 (GOFFMAN-WATERMAN). *If U is a Euclidean connected, open subset of $\mathbb{R}^n$, then U as well as the density-closure of U are connected in the density topology.*

PROOF. Let I be a open interval in $\mathbb{R}^n$. Suppose that I is not connected in the density topology. Let V and W be density topology open sets such that $V \cap I \neq \emptyset$, $W \cap I \neq \emptyset$, $I \subset V \cup W$, and $V \cap W \cap I = \emptyset$. We infer from the last lemma that there is a point c of I such that c is neither a density point of $V \cap I$ nor a density point of $W \cap I$. That is, $c \in I \setminus (V \cup W) = \emptyset$, a contradiction. Hence I is connected in the density topology.

Now, if U is connected in the Euclidean topology, then each pair of points of U is contained in a chain I_k, $k = 1, 2, \ldots, m$, of open intervals contained in U with $I_k \cap I_{k+1} \neq \emptyset$. Hence U is connected in the density topology, whence so is the density-closure of U.

The connected subsets of $\mathbb{R}^1$ are the same for the usual topology and for the density topology. But, for $n > 1$, there are open connected sets in the usual topology of $\mathbb{R}^n$ whose usual closure is not connected in the density topology.

A.5. Approximately continuous maps into metric spaces

We present here two results on approximately continuous mappings from [**66**].

THEOREM A.17. *If $f: X \to Y$ is an approximately continuous mapping from a nonempty, density-open subset X of $\mathbb{R}^n$ onto a metric space Y, then Y is a separable space.*

PROOF. Suppose that the metric space $Y = f[X]$ is not separable. Then there is an uncountable subset M of Y and a positive number δ such that any 2 distinct members of M are no closer than 2δ in distance. The collection of neighborhoods

$$\{ B(y, \delta) : y \in M \}$$

is a disjointed one. As X is a density-open set and f is continuous in the density topology, the collection

$$\{ f^{-1}[B(y, \delta)] : y \in M \}$$

is an uncountable collection of mutually disjoint, measurable sets of positive measure. But no such collection can exist in $\mathbb{R}^n$; a contradiction has been established.

Observe that an uncountable set X of measure zero, when endowed the subspace topology generated by the density topology, is discrete (that is, points are both open and closed) and thereby discretely metrizable. Consequently, in the above theorem, it is necessary that X be at least the density-closure of a density-open set in $\mathbb{R}^n$.

THEOREM A.18. *If $f: \mathbb{R}^n \to Y$ is an approximately continuous mapping into a metric space, then f is a function in the Borel class 1. In particular, $f^{-1}[U]$ is an F_σ set whenever U is an open subset of Y.*

PROOF. Let M be a countable dense subset of $f[\mathbb{R}^n]$. For each y in M, let $g_y: Y \to \mathbb{R}$ be given by

$$g_y(\eta) = \rho(\eta, y), \quad \eta \in Y,$$

where ρ is the metric of Y. The real-valued function that results from the composition $g_y \circ f$ is approximately continuous, and therefore is in the Baire class 1. So, $(g_y \circ f)^{-1}[W]$ is an F_σ set for each open set W of $\mathbb{R}$. Let U be an open set in Y. Then there is a countable collection of open sets W_k, $k = 1, 2, \ldots$, and a countable collection y_k, $k = 1, 2, \ldots$, in the set M so that

$$U = \bigcup_{k=1}^{\infty} {g_{y_k}}^{-1}[W_k].$$

We now have that $f^{-1}[U]$ is an F_σ set since

$$f^{-1}[U] = \bigcup_{k=1}^{\infty} (g_{y_k} \circ f)^{-1}[W_k].$$

Thus we have established that f is in the Borel class 1.

For arbitrary metric spaces Y, it is well-known that not every Borel class 1 map of $\mathbb{R}^n$ is a Baire class 1 map. Indeed, consider Y to be the closure in $\mathbb{R}^2$ of the familiar graph of $\sin \frac{1}{x}$, $x \in \mathbb{R}$. The one-to-one map

$$\varphi(x) = \begin{cases} (x, \sin \frac{1}{x}) & \text{for } x \neq 0, \\ (0, -1) & \text{for } x = 0, \end{cases}$$

is certainly a Borel class 1 map which is not a Baire class 1 map of $\mathbb{R}$ into Y. To see this, let φ_k, $k = 1, 2, \ldots$, be a sequence of continuous maps into Y that converges pointwise and that satisfies $\varphi_k(-1) \to \varphi(-1)$ as $k \to \infty$. By way of a contradiction, we shall show that $\varphi_k(1)$ fails to converge to $\varphi(1)$. Suppose that $\varphi_k(1)$ converges to $\varphi(1)$. Then the first coordinate of $\varphi_k(1)$ must converge to 1, which is the first coordinate of $\varphi(1)$. Let k be such that

$$|\varphi_k(-1) - \varphi(-1)| < \tfrac{1}{4} \quad \text{and} \quad |\varphi_k(1) - \varphi(1)| < \tfrac{1}{4}.$$

Then $\varphi_k\big[[-1,1]\big]$ is an arcwise connected subset of Y whose projection into the x-axis of $\mathbb{R}^2$ must be a connected set that contains 0 in its interior. But the space Y does not contain any arcs that join $\varphi(-\frac{3}{4})$ to $\varphi(\frac{3}{4})$ and a contradiction has been reached. Hence, $\varphi_k(1)$ does not converge to $\varphi(1)$. Thereby, φ has been shown not to be in the Baire class 1 for functions of $\mathbb{R}$ into Y.

In passing, we remark that there is a homeomorphism $h\colon \mathbb{R} \to \mathbb{R}$ such that

$$g(x) = \begin{cases} \sin \frac{1}{h(x)} & \text{for } x \neq 0, \\ -1 & \text{for } x = 0, \end{cases}$$

is an approximately continuous, real-valued function. Consequently,

$$(\varphi \circ h)[\mathbb{R}] = \varphi[\mathbb{R}] = Y.$$

As $\varphi \circ h\colon \mathbb{R} \to \mathbb{R}^2$ is approximately continuous, we have that $\varphi \circ h\colon \mathbb{R} \to Y$ is also approximately continuous. Hence this approximately continuous function is in the Borel class 1 but not in the Baire class 1.

We infer from Theorem 1 of Rogers [**112**] that the following two conditions on a metric space Y will assure that each Borel class 1 map of $\mathbb{R}^n$ into Y will also be a Baire class 1 map.

Extension Property. For each closed subset F of $\mathbb{R}^n$ and each continuous map f of F into Y, there is a continuous extension $\widetilde{f}$ of f mapping $\mathbb{R}^n$ into Y.

Second Local Extension Property. For each point y of Y and for each neighborhood U of y, there is a neighborhood W of y such that for each closed subset F of $\mathbb{R}^n$ and for each continuous function f of F into W, there is a continuous extension $\widetilde{f}$ of f mapping $\mathbb{R}^n$ into U.

We take this opportunity to mention the error of the omission of the above two conditions in the statement of the hypothesis of Theorem 2 of [**66**].

Hausdorff Measure and Packing

As Hausdorff measures and packings will play important roles in the book, a brief discussion of them will be given here. Related to Hausdorff measures and packings are various indices called the Hausdorff dimensions and the rarefaction indices. These will be defined here also.

A.6. Hausdorff dimension

A.6.1. Hausdorff measures. We shall concentrate on Hausdorff measures on $\mathbb{R}^n$. Let us begin with the definition.

Definition A.19. Let α be a nonnegative real number. For each positive number δ and each subset E of $\mathbb{R}^n$, a **δ-cover of E** is a countable collection $\{\, C_j : j = 1, 2, \dots \}$ of subsets of $\mathbb{R}^n$ such that

$$E \subset \bigcup_{j=1}^{\infty} C_j \quad \text{and} \quad \operatorname{diam} C_j < \delta \quad \text{for each} \quad j.$$

Define

$$\mathsf{H}_\alpha^\delta(E) = \inf \sum_{j=1}^{\infty} \frac{\pi^{\frac{\alpha}{2}}}{\Gamma(\frac{\alpha}{2}+1)} \left(\frac{\operatorname{diam} C_j}{2}\right)^\alpha,$$

where the infimum is taken over all δ-covers of E. Then define

$$\mathsf{H}_\alpha(E) = \lim_{\delta \to 0} \mathsf{H}_\alpha^\delta(E).$$

The function H_α is called the **outer Hausdorff α-measure** on $\mathbb{R}^n$.

It is well-known that H_α is a metric outer measure and that H_n is the Lebesgue measure on $\mathbb{R}^n$. Also, H_0 is the counting measure on $\mathbb{R}^n$, and $\mathsf{H}_\alpha(\mathbb{R}^n) = 0$ for $\alpha > n$.

The following proposition is easily proved.

PROPOSITION A.20. *If $E \subset \mathbb{R}^n$ and $f\colon E \to \mathbb{R}^k$ is Lipschitzian, then*

$$\mathsf{H}_\alpha^\delta\big(f[E]\big) \le \big(\operatorname{Lip}(f)\big)^\alpha\, \mathsf{H}_\alpha^\delta(E) \quad \textit{for} \quad \delta > 0,$$

and

$$\mathsf{H}_\alpha\big(f[E]\big) \le \big(\operatorname{Lip}(f)\big)^\alpha\, \mathsf{H}_\alpha(E).$$

A.6.2. Dimension. The following facts are well-known.

(1) If $\mathsf{H}_\alpha(E) < +\infty$ for a positive number α, then $\mathsf{H}_\beta(E) = +\infty$ for $0 \le \beta < \alpha$.
(2) If $\mathsf{H}_\alpha(E) < +\infty$ for some α, then $\mathsf{H}_\beta(E) = 0$ for $\alpha < \beta$.

Hence for each subset E of $\mathbb{R}^n$ there is a unique number, called the **Hausdorff dimension of** E, given by

$$\dim_{\mathsf{H}} E = \inf\{\,\alpha : \mathsf{H}_\alpha(E) < +\infty\,\}.$$

Clearly,

$$\dim_{\mathsf{H}} E \le n \quad \text{for} \quad E \subset \mathbb{R}^n.$$

It is known that, for each number α with $0 \le \alpha \le n$, there is a nonempty closed set F with $\dim_{\mathsf{H}} F = \alpha$.

A.6.3. Hausdorff dimension 0. It is clear from the above discussion that there are nonempty perfect sets P with $\dim_{\mathsf{H}} P = \alpha$ for $0 < \alpha \le n$. What is not so obvious is the existence of a nonempty perfect set P with $\dim_{\mathsf{H}} P = 0$. We infer from a theorem of Besicovitch that such sets exist. The reader is referred to the book [**40**, pages 67–70] for a discussion of this theorem and for related references.

A.7. Hausdorff packing

A.7.1. Packing. Unlike Hausdorff measures, the development of Hausdorff packing is a multi-staged process. Despite its complicated formulation, it has many applications. Our application will be in the setting of $\mathbb{R}^1$ where open balls are quite nice and manageable. Indeed, our application uses only the first stage of the process in which only the most naive form of packing occurs. Hence our discussion here will deal only with what we shall call the naive theory.

Let E be a bounded subset of $\mathbb{R}$ and δ be a positive number. By a **naive δ-packing of** E we mean a countable collection $\{\, I_n : n = 1, 2, \dots \}$ (possibly finite) of disjoint open intervals I_n such that $\operatorname{diam} I_n < \delta$ and $\overline{I_n} \cap \overline{E} \ne \emptyset$ for every n. Note that a δ-packing need not be a covering.

Let α be a positive number and define

$$\mathsf{npH}^{\delta}_{\alpha}(E) = \sup \sum_{n=1}^{\infty} (\operatorname{diam} I_n)^{\alpha},$$

where the supremum is taken over all δ-packings of E. Then define

$$\mathsf{npH}_{\alpha}(E) = \lim_{\delta \to 0} \mathsf{npH}^{\delta}_{\alpha}(E).$$

Due to the requirement that $\overline{I_n} \cap \overline{E} \neq \emptyset$ for each I_n in the δ-packing of E, it is clear that

$$\mathsf{npH}^{\delta}_{\alpha}(E) = \mathsf{npH}^{\delta}_{\alpha}(\overline{E}) \quad \text{for} \quad \delta > 0,$$

and

$$\mathsf{npH}_{\alpha}(E) = \mathsf{npH}_{\alpha}(\overline{E}).$$

We have the following two elementary facts.

PROPOSITION A.21. *Suppose that α satisfies $0 < \alpha < 1$ and E is a bounded subset of $\mathbb{R}$ with $\lambda_1(\overline{E}) > 0$. Then*

$$\mathsf{npH}_{\alpha}(E) = +\infty.$$

PROOF. We infer from the Vitali covering theorem that there exists a δ-packing $\{I_n : n = 1, 2, \dots\}$ of $\overline{E}$ with

$$\lambda_1(\overline{E}) < 2 \sum_{n=1}^{\infty} \operatorname{diam} I_n.$$

Hence

$$\lambda_1(\overline{E}) < 2\,\delta^{1-\alpha}\, \mathsf{npH}^{\delta}_{\alpha}(\overline{E}).$$

Now the proposition follows easily.

We will use the following notation in the next proposition. Let E be a nonempty, compact set in $\mathbb{R}$ and let $\{J_i : i = 1, 2, \dots\}$ be an enumeration of the bounded, nondegenerate components of $\mathbb{R} \setminus E$, where $\lambda(J_i) \geq \lambda(J_{i+1})$. Then, for $0 < \alpha \leq 1$, we define $\Gamma_{\alpha}(E)$ to be

$$\Gamma_{\alpha}(E) = \sum_{i=1}^{\infty} (\lambda_1(J_i))^{\alpha}.$$

PROPOSITION A.22. *Suppose that α satisfies $0 < \alpha < 1$ and E is a bounded, infinite subset of $\mathbb{R}$ with $\lambda_1(\overline{E}) = 0$.*

Then the following statements are equivalent.

(1) $\Gamma_{\alpha}(E) < +\infty$.
(2) $\mathsf{npH}^{\delta}_{\alpha}(E) < +\infty$ *for some positive number* δ.
(3) $\mathsf{npH}^{\delta}_{\alpha}(E) < +\infty$ *for all positive numbers* δ.
(4) $\mathsf{npH}_{\alpha}(E) = 0$.

PROOF. Assume that statement (1) holds and let $\delta = \operatorname{diam} E$. Denote by $[a, b]$ the smallest closed interval that contains E. For each interval I_n in any δ-packing

$\{ I_n : n = 1, 2, \dots \}$ of E with $I_n \subset [a, b]$ we have

$$(\operatorname{diam} I_n)^\alpha \le \sum_{i=1}^\infty (\operatorname{diam}(I_n \cap J_i))^\alpha$$

(where $\operatorname{diam}\emptyset = 0$) because $0 < \alpha < 1$ and $\lambda_1(\overline{E}) = 0$. Furthermore, each interval J_i can meet at most two intervals from the δ-packing. Consequently,

$$\mathsf{npH}_\alpha^\delta(E) \le 2 \sum_{i=1}^\infty (\lambda_1(J_i))^\alpha + 2\,\delta^\alpha. \tag{A.1}$$

Hence statement (2) holds.

Conversely, suppose that statement (2) holds. Let k be such that $\operatorname{diam} J_i < \delta$ for $i > k$. Then

$$\sum_{i=1}^\infty \lambda_1(J_i)^\alpha < \mathsf{npH}_\alpha^\delta(E) + \sum_{i=1}^k (\operatorname{diam} J_i)^\alpha < +\infty,$$

and statement (1) holds.

Clearly, statement (3) implies (2), and (4) implies (2).

We shall show that (1) implies (4). Let $\varepsilon > 0$ and let $N(\varepsilon)$ be the largest integer i such that $\lambda_1(J_i) > \varepsilon$. Employing a calculation analogous to that for the inequality (A.1), we have, for any δ-packing $\{ I_n : n = 1, 2, \dots \}$ of E with $\delta < \varepsilon$,

$$\frac{1}{2} \sum_{n=1}^\infty (\operatorname{diam} I_n)^\alpha \le \sum_{i=N(\varepsilon)}^\infty (\operatorname{diam} J_i)^\alpha + \big(N(\varepsilon) + 1\big)\,\varepsilon^\alpha$$

and, for $\eta > \varepsilon$,

$$\le \sum_{i=N(\varepsilon)}^\infty (\operatorname{diam} J_i)^\alpha + \sum_{i=N(\eta)}^\infty (\operatorname{diam} J_i)^\alpha + \big(N(\eta) + 2\big)\,\varepsilon^\alpha.$$

By letting ε tend to 0 first and then letting η tend to 0, we conclude that statement (4) follows.

Finally, let us show that (1) implies (3). For a positive number δ, let $\varepsilon = 2\delta$. Define the integer $N(\varepsilon)$ as in the proof of the previous implication. Then we will have, for any δ-packing $\{ I_n : n = 1, 2, \dots \}$ of E,

$$\frac{1}{2} \sum_{n=1}^\infty (\operatorname{diam} I_n)^\alpha \le \sum_{i=N(\varepsilon)}^\infty (\operatorname{diam} J_i)^\alpha + \big(N(\varepsilon) + 1\big)\,\varepsilon^\alpha \le \Gamma_\alpha(E) + \big(N(\varepsilon) + 1\big)\,\varepsilon^\alpha,$$

and $\mathsf{npH}_\alpha^\delta(E) < +\infty$ has been established.

The proposition is now proved.

The following are easily calculated.

$$\mathsf{npH}_\alpha([0,1]) = +\infty \qquad \text{for} \quad 0 < \alpha < 1,$$
$$\mathsf{npH}_1([0,1]) = 1,$$

and

$$\mathsf{npH}_\alpha(\{0\}) = 0 \qquad \text{for} \quad \alpha > 0.$$

The above discussion on packing has been extracted from the paper [**125**] by Taylor and Tricot. As we have said earlier, our discussion concerns the most naive approach to packings. Other serious packing questions require more restrictive packings which are discussed in that paper. Moreover, we have not touched on the packing measure. Only the premeasure has been discussed since it is the premeasure that is important for our purposes. The reader is referred to the the paper by Taylor and Tricot for more on the packing measure.

A.7.2. Packing dimensions. The packing dimensions are defined in a manner analogous to that of Hausdorff dimension given above. Only the most naive situation will be considered here. We can now see that there is a **rarefaction index** $\Lambda(E)$ for each bounded subset E of $\mathbb{R}$ which is defined to be the unique number

$$\Lambda(E) = \inf\{\,\alpha : \mathsf{npH}_\alpha(E) = 0\,\}.$$

A.7.3. A construction. The existence of a special nondecreasing function was needed in Chapter 3. The construction of this function is given here.

For compact subsets E of $[0,1]$ and numbers α and δ with $\Lambda(E) < \alpha < 1$ and $0 < \delta$, we shall construct the required continuous nondecreasing functions $F_\alpha^\delta\colon [0,1] \to \mathbb{R}$. Since $\Lambda(E) < \alpha < 1$, we have that E is nowhere dense and $\mathsf{npH}_\alpha(E) = 0$. So, $\lambda(E) = 0$. We first define F_α^δ on E. The definition will use a packing that is even more naive than δ-packings. Let $\delta > 0$ and $x \in E$. Then $F_\alpha^\delta(x)$ is the real number defined as follows: For each x in E,

$$F_\alpha^\delta(x) = \sup \sum_{i=1}^{\infty} (\operatorname{diam} I_i)^\alpha,$$

where the supremum is taken over all δ-packings of $E \cap [0,x]$ by intervals I_i such that $\partial I_i \subset E$.

This function F_α^δ that has been defined on E is real-valued, nondecreasing and satisfies

$$0 \le F_\alpha^\delta(x') - F_\alpha^\delta(x) \le \mathsf{npH}_\alpha^\delta(E \cap [x,x'])$$

whenever $0 < x < x' < 1$ and $x, x' \in E$. By considering limits from the left and limits from the right, one can prove that F_α^δ is continuous on E as well. Hence one can extend it to $F_\alpha^\delta\colon [0,1] \to \mathbb{R}$ in a nondecreasing continuous manner. Observe that

$$|x - x'|^\alpha \le F_\alpha^\delta(x') - F_\alpha^\delta(x) \le \mathsf{npH}_\alpha^\delta(E \cap [x,x'])$$

whenever $0 < x' - x < \delta$ and $x, x' \in E$. Note that in the construction we used only the property that $\mathsf{npH}_\alpha(E) = 0$ with $0 < \alpha < 1$. We state the result of this construction as a proposition.

PROPOSITION A.23. *Let E be a compact subset of $[0,1]$ and α be such that $\mathsf{npH}_\alpha(E) = 0$ with $0 < \alpha < 1$. Then for each positive number δ there exists a nondecreasing continuous function $F_\alpha^\delta\colon [0,1] \to [0, \mathsf{npH}_\alpha^\delta(E)]$ such that, whenever $x \in E$,*

$$F_\alpha^\delta(x) = \sup \sum_{i=1}^{\infty} (\operatorname{diam} I_i)^\alpha$$

where the supremum is taken over all δ-packings of $E \cap [0,x]$ by intervals I_i such that $\partial I_i \subset E$. And,

$$|x - x'|^\alpha \le F_\alpha^\delta(x') - F_\alpha^\delta(x) \le \mathsf{npH}_\alpha^\delta(E \cap [x,x']),$$

whenever $0 < x' - x < \delta$ *and* $x, x' \in E$.

Nonparametric Length and Area

The notion of the total variation of a real-valued function f is intimately connected with the notion of the lengths of polygonal curves that are inscribed in the nonparametric curve defined by f. This section of the appendix will present brief discussions concerning (1) the pitfalls that result when one uses inscribed polyhedral surfaces to define nonparametric area, (2) the idea due to Lebesgue of the lower semicontinuous definition of area as a way to avoid these pitfalls, and finally (3) the distribution derivative of a function of bounded variation.

A.8. Nonparametric length

The nonparametric length of a function f on the interval $[a, b]$ has a natural definition by means of inscribed polygonal curves. A polygonal curve is a piecewise linear, continuous function g. That is, $g\colon [a, b] \to \mathbb{R}$ is defined by a finite partition (denoted by $\mathcal{P}$) $a = x_0 < x_1 < \cdots < x_n = b$ of $[a, b]$ such that g is linear on each subinterval $[x_{i-1}, x_i]$. The length of g is given by

$$\mathrm{E}_1(g, [a, b]) = \sum_{i=1}^{n} \sqrt{(\Delta x_i)^2 + (\Delta g_i)^2}.$$

A polygonal curve is inscribed in f if $g(x_i) = f(x_i)$ for $i = 0, 1, \ldots, n$. The nonparametric length of f on $[a, b]$ is the extended real number

$$\ell(f, [a, b]) = \sup \mathrm{E}_1(g, [a, b]),$$

where the supremum is taken over all inscribed polygonal curves of f.

For a continuous f, the length is also a limit of the inscribed polygonal curves that converge pointwise to f on $[a, b]$. Indeed, let $\mathcal{P}_g$ be the partition that defines the polygonal curve g inscribed in f. With $\delta(\mathcal{P}_g)$ being the usual mesh of the partition $\mathcal{P}_g$, we have

$$\lim_{\delta(\mathcal{P}_g) \to 0} \mathrm{E}_1(g, [a, b]) = \ell(f, [a, b])$$

whenever f is continuous.

A.9. Schwarz's example

We shall describe here a very simple nonparametric surface for which inscribed polyhedral surfaces lead to unpleasant area behaviors. The example that is described here is just a nonparametric variant of the example due to Schwarz from 1890. His example is a parametric right circular cylindrical surface.

The function is given by

$$z = f(x, y) = \sqrt{1 - x^2}, \quad (x, y) \in [-1, 1] \times [0, 1].$$

Let us first describe inscribed polyhedral surfaces g_n whose elementary areas converge to the correct answer for the area of the nonparametric surface given by the function f. Using polar coordinates, we locate $2n + 1$ points on the top half of the circle $x^2 + z^2 = 1$ in the following manner.

$$(x_j, z_j) = (\cos \tfrac{j\pi}{4n}, \sin \tfrac{j\pi}{4n}), \quad j = 0, 1, 2, \ldots, 4n.$$

Now we shall triangulate the rectangle $[-1,1]\times[0,1]$ by using the zigzag pattern defined by the points P_j in the (x,y)-plane given by

$$P_j = (x_j, y_j) = (\cos\tfrac{j\pi}{4n}, \tfrac{1-(-1)^j}{2}), \quad j = 0,1,2,\ldots,4n.$$

The inscribed piecewise linear, continuous function is determined by values of f at the vertices of the triangulation. The vertices of this triangulation are the points P_j and the two points $Q_0 = (-1,0)$ and $Q_1 = (1,0)$. Clearly, the values of f at these vertices are $f(P_j) = \sin\frac{j\pi}{4n}$, and $f(Q_0) = f(Q_1) = 0$. It is easily seen that the sequence g_n, $n = 1,2,\ldots$, converges uniformly to f and that

$$\lim_{n\to\infty} \mathrm{E}_2\big(g_n, [-1,1]\times[0,1]\big) = \pi.$$

Let us form triangulations of $[-1,1]\times[0,1]$ that will yield inscribe polyhedral surfaces whose areas converge to $+\infty$. We begin by dividing the interval $[0,1]$ into $2m$ equal parts. The first part is $[0,\frac{1}{2m}]$. In this part we put the above function g_n by shrinking in a linear manner the interval $[0,1]$ into $[0,\frac{1}{2m}]$, call this function $\widetilde{g}_{n,m}$. The triangles of the triangulation of $[-1,1]\times[0,1]$ consist of those which are congruent to the triangles $\triangle P_0P_1P_2$ or $\triangle P_0P_1Q_0$. Hence we have

$$\begin{aligned}\mathrm{E}_2\big(\widetilde{g}_{n,m}, [-1,1]\times[0,\tfrac{1}{2m}]\big) &= \big((\tfrac{1}{2m})^2 + (\sin\tfrac{\pi}{4n})^2\big)^{\frac12}\, \mathrm{E}_2\big(g_n, [-1,1]\times[0,1]\big)\\ &\quad - \Big(\big((\tfrac{1}{2m})^2 + (\sin\tfrac{\pi}{4n})^2\big)^{\frac12} - \tfrac{1}{2m}\Big)\sin\tfrac{\pi}{4n}.\end{aligned}$$

We extend $\widetilde{g}_{n,m}$ to the interval $[-1,1]\times[0,\frac{1}{m}]$ by using symmetry about the line $y = \frac{1}{2m}$. Then, in an accordion fashion, we extend $\widetilde{g}_{n,m}$ to $[-1,1]\times[0,1]$ by periodicity. We infer from the above equality that

$$\mathrm{E}_2\big(\widetilde{g}_{n,m}, [-1,1]\times[0,1]\big) \geq 2m\sin\tfrac{\pi}{4n}\,\mathrm{E}_2\big(g_n,[-1,1]\times[0,1]\big) - 2m\sin^2\tfrac{\pi}{4n}.$$

Then we have, with $m = n^2$, that

$$\mathrm{E}_2\big(\widetilde{g}_{n,n^2}, [-1,1]\times[0,1]\big) > 2n^2\sin\tfrac{\pi}{4n}\,\mathrm{E}_2\big(g_n,[-1,1]\times[0,1]\big) - 2(\tfrac{\pi}{4})^2.$$

Consequently,

$$\lim_{n\to\infty}\mathrm{E}_2\big(\widetilde{g}_{n,n^2}, [-1,1]\times[0,1]\big) = +\infty\,\pi - 2(\tfrac{\pi}{4})^2 = +\infty.$$

Clearly the sequence $\widetilde{g}_{n,n^2}$, $n = 1,2,\ldots$, converges uniformly to f. Even more, the supremum of the areas of polyhedral surfaces that are inscribed to f is also $+\infty$.

For a discussion of the original Schwarz example, the reader is referred to [**46**, page 150].

A.10. Lebesgue's lower semicontinuous area

Clearly, a naive parroting of the inscribed curves approach to length will not yield a definition of area. It was Lebesgue [**92**] who proposed a lower semicontinuous functional approach to area. He observed that the length and area functionals were lower semicontinuous when dealing with the class of piecewise linear, continuous functions or maps. More precisely, if g_n, $n = 1,2,\ldots$, is a sequence of piecewise linear, continuous functions that converges uniformly to a piecewise linear, continuous function g, then

$$\mathrm{E}_1\big(g,[a,b]\big) \leq \liminf_{n\to\infty}\mathrm{E}_1\big(g_n,[a,b]\big),$$

in the case of length, and

$$\mathrm{E}_2\big(g,[a,b]\times[c,d]\big) \leq \liminf_{n\to\infty}\mathrm{E}_2\big(g_n,[a,b]\times[c,d]\big),$$

in the case of area. According to Lebesgue, the natural choice of the definitions of length or area are the smallest possible numbers on the right-hand sides of the above inequalities when the piecewise linear requirement on g is replaced by general continuity.

As one can see, the idea employed here is that of extensions of lower semicontinuous functionals from a subset of a metric space S of functions to a lower semicontinuous functional defined on the closure of the subset S. This procedure was formalized by Fréchet [**44**] to an abstract setting as follows.

Let S be a metric space and let $F\colon S \to [0,+\infty]$ be a lower semicontinuous function that satisfies the property

A: For each f in S, there is a sequence f_n, $n = 1, 2, \dots$, of elements of $S \setminus \{f\}$ converging to f such that $\lim_{n\to\infty} F(f_n) = F(f)$.

Let T be the completion of S. That is, the elements ξ of T are equivalence classes of Cauchy sequences in S. The function F is extended by defining

$$F(\xi) = \inf \liminf_{n\to\infty} F(f_n),$$

where the infimum is taken over all Cauchy sequences in ξ. That $F\colon T \to [0,+\infty]$ is an extension and is lower semicontinuous are easily established. Moreover, the following property also holds.

B: For each ξ, there is a Cauchy sequence f_n, $n = 1, 2, \dots$, in the equivalence class ξ such that $F(\xi) = \lim_{n\to\infty} F(f_n)$.

We quote Fréchet's theorem.

THEOREM A.24 (FRÉCHET). *Every nonnegative, extended real-valued, lower semicontinuous function on a metric space S with property* A *can be extended to a unique lower semicontinuous function on the completion T of S with property* B.

Lebesgue used the mode of uniform convergence to define the metric space of continuous functions. In Chapter 5, the mode of almost everywhere convergence was used. This, of course, introduces discontinuous functions into the completion.

A.11. Distribution derivatives for one real variable

For locally integrable functions $f\colon \mathbb{R} \to \mathbb{R}$, one defines its distribution to be the functional

$$f(\varphi) = \int_{-\infty}^{+\infty} f(x)\varphi(x)\,dx,$$

where φ is an infinitely differentiable function with compact support. Its distribution derivative is defined by a formula that comes from integration by parts, namely the formula

$$\mathrm{D}f(\varphi) = -\int_{-\infty}^{+\infty} f(x)\,\mathrm{D}\,\varphi(x)\,dx,$$

where the D found in the integral on the right-hand side of the formula is the usual derivative operator. When this linear functional is continuous, the Riesz representation theorem tells us that the functional is represented by a measure, that is,

$$\mathrm{D}f(\varphi) = \int_{\mathbb{R}} \varphi(x)\,d\mu(x).$$

The motivation of this formula for $\mathrm{D}f$ comes from the integration by parts formula for the Riemann-Stieltjes integral. For functions f of bounded variation,

$$\int_a^b d(\varphi f)(x) = \int_a^b \varphi(x)\, df(x) + \int_a^b f(x)\, \mathrm{D}\,\varphi(x)\, dx.$$

When the support of φ is contained in $[a,b]$, we have $\int_a^b d(\varphi f)(x) = 0$, whence

$$-\int_a^b f(x)\, \mathrm{D}\,\varphi(x)\, dx = \int_a^b \varphi(x)\, df(x).$$

Hence

$$\mathrm{D}f(\varphi) = \int_a^b \varphi(x)\, df(x).$$

So the finite signed measure μ that gives the representation of the distribution derivative $\mathrm{D}f$ satisfies $d\mu = df$, where f is assumed to be continuous from the right.

For functions of more than one independent variable, the BV functions in the above discussion must be replaced with functions of bounded variation in the sense of Cesari. For this, we refer the reader to [**39**].

Bibliography

1. M. Avdispahić, Master's thesis, Sarajevo, 1982.
2. ———, *On the classes* ΛBV *and* V[ν], Proc. Amer. Math. Soc. **95** (1985), 230–234.
3. ———, *Concepts of generalized bounded variation and the theory of Fourier series*, Internat. J. Math. Sci. **9** (1986), 223–244.
4. A. Baernstein, *On the Fourier series of functions of bounded Φ-variation*, Studia Math. **42** (1972), 243–248.
5. A. Baernstein and D. Waterman, *Functions whose Fourier series converge uniformly for every change of variable*, Indiana Univ. Math. J. **22** (1972), 569–576.
6. S. Banach, *Sur les lignes rectifiables et les surfaces dout l'aire est fini*, Fund. Math. **7** (1925), 225–236.
7. ———, *Théorème sur les ensembles de première catégoriè*, Fund. Math. **16** (1930), 395–398.
8. N. K. Bary, *A treatise on trigonometric series*, Pergamon, Macmillan, Oxford-New York, 1964.
9. R. D. Berman, L. Brown, and W. S. Cohn, *Moduli of continuity and generalized BCH sets*, Rocky Mountain J. Math. **17** (1987), 315–338.
10. A. S. Besicovitch, *On the definition and value of the area of a surface*, Quart. J. Math. Oxford Ser. **16** (1945), 86–102.
11. W. A. Blankinship, *Generalization of a construction of Antoine*, Ann. of Math. **53** (1951), 276–297.
12. H. Blumberg, *New properties of all real functions*, Trans. Amer. Math. Soc. **24** (1922), 113–128.
13. ———, *Measurable boundaries of an arbitrary function*, Acta Math. **65** (1935), 263–282.
14. H. Bohr, *Über einen Satz von J. Pál*, Acta Sci. Math. (Szeged) **7** (1935), 129–135.
15. J. C. Bradford and C. Goffman, *Metric spaces in which Blumberg's theorem holds*, Proc. Amer. Math. Soc. **11** (1960), 667–670.
16. J. C. Breckenridge and T. Nishiura, *Two examples in surface area theory*, Michigan Math. J. **19** (1972), 157–160.
17. M. Brown, *A proof of the generalized Schoenflies theorem*, Bull. Amer. Math. Soc. **66** (1960), 74–76.
18. A. M. Bruckner, *Density-preserving homeomorphisms and the theorem of Maximoff*, Quart. J. Math. Oxford Ser. (2) **21** (1970), 337–347.
19. ———, *Differentiability a.e. and approximate differentiability a.e.*, Proc. Amer. Math. Soc. **66** (1977), 294–298.
20. ———, *Differentiation of real functions*, Lecture Notes in Math., vol. 659, Springer-Verlag, Berlin, 1978.
21. ———, *Differentiation of real functions*, CRM Monogr. Ser., vol. 5, Amer. Math. Soc., Providence, 1994.
22. A. M. Bruckner, R. O. Davies, and C. Goffman, *Transformations into Baire 1 functions*, Proc. Amer. Math. Soc. **67** (1977), 62–66.
23. A. M. Bruckner and C. Goffman, *Differentiability through change of variables*, Proc. Amer. Math. Soc. **61** (1976), 235–241.
24. A. M. Bruckner and J. L. Leonard, *On differentiable functions having an everywhere dense set of intervals of constancy*, Canad. Math. Bull. **8** (1965), 73–76.
25. Z. Buczolich, *Density points and bi-Lipschitz functions in* $\mathbb{R}^m$, Proc. Amer. Math. Soc. **116** (1992), 53–59.

26. L. Cesari, *Sulle funzioni a variazione limita*, Ann. Scuola Norm. Sup. Pisa (2) **5** (1936), 299–313.
27. ______, *Sulla funzione di due variabli a variazone limitata secondo Tonelli e sulla convergenza della relative serie doppie di Fourier*, Rend. Sem. Mat. Univ. Roma (4) **1** (1937), 277–294.
28. ______, *Surface area*, Ann. of Math. Stud. No. 35, Princeton University Press, Princeton, 1956.
29. ______, *Rectifiable curves and the Weierstrass integral*, Amer. Math. Monthly **65** (1958), 485–500.
30. ______, *Variation, multiplicity, and semicontinuity*, Amer. Math. Monthly **65** (1958), 317–332.
31. ______, *Recent results in surface area theory*, Amer. Math. Monthly **66** (1959), 173–192.
32. Z. A. Chanturiya, *The modulus of variation of a function and its application in the theory of Fourier series*, Dokl. Akad. Nauk SSSR **214** (1974), 63–66 (Russian), translation: Soviet Math. Dokl. **15** (1974), 67–71.
33. ______, *Absolute convergence of Fourier series*, Mat. Zametki **18** (1975), 185–192 (Russian), translation: Math. Notes **18** (1975), 695–700.
34. K. Ciesielski, *Density and $\mathcal{I}$-density continuous homeomorphisms*, Real Anal. Exchange **18** (1992–93), 367–384.
35. K. Ciesielski and L. Larson, *The space of density continuous functions*, Acta Math. Hungar. **58** (1991), 289–296.
36. P. J. Cohen, *On a conjecture of Littlewood and idempotent measures*, Amer. J. Math. **82** (1960), 191–212.
37. H. Davenport, *On a theorem of P. J. Cohen*, Matematika **7** (1960), 93–97.
38. R. Engelking, *General topology*, PWN–Polish Scientific Publishers, Warsaw, 1977.
39. L. C. Evans and R. F. Gariepy, *Measure theory and fine properties of functions*, CRC Press, Boca Raton, 1992.
40. K. J. Falconer, *The geometry of fractal sets*, Cambridge University Press, Cambridge, 1985.
41. H. Federer, *Geometric measure theory*, Springer-Verlag, New York, 1969.
42. R. Fleissner and J. Foran, *A note on the Garsia-Sawyer class*, Real Anal. Exchange **6** (1980–81), 245–246.
43. W. H. Fleming, *Functions whose partial derivatives are measures*, Illinois J. Math. **4** (1960), 452–478.
44. M. Fréchet, *Sur le prolongement des fonctionelles semi-continues et sur l'aire des surfaces courbes*, Fund. Math. **7** (1925), 210–224.
45. A. M. Garsia and S. Sawyer, *On some classes of continuous functions with convergent Fourier series*, J. of Math. Mech. **13** (1964), 589–601.
46. B. R. Gelbaum and J. M. H. Olmsted, *Counterexamples in analysis*, Holden-Day Inc., San Francisco, 1965.
47. L. Gillman and M. Jerison, *Rings of continuous functions*, Van Nostrand, Princeton, NJ, 1960.
48. A. Gleyzal, *Interval-functions*, Duke Math. J. **8** (1941), 223–230.
49. C. Goffman, *The approximation of arbitrary biunique transformations*, Duke Math. J. **10** (1943), 1–4.
50. ______, *Proof of a theorem of Saks and Sierpiński*, Bull. Amer. Math. Soc. **54** (1948), 950–52.
51. ______, *Lower semi-continuity and area functionals, I. The nonparametric case*, Rend. Circ. Mat. Palermo **2** (1953), 203–235.
52. ______, *On a theorem of H. Blumberg*, Mich. Math. J. **2** (1953), 21–22.
53. ______, *One-one measurable transformation*, Acta Math. **89** (1953), 261–278.
54. ______, *Real functions*, Rinehart, New York, 1953.
55. ______, *Lower semi-continuity and area functionals, II. The Banach area*, Amer. J. Math. **76** (1954), 679–688.
56. ______, *Non-parametric surfaces given by linearly continuous functions*, Acta Math. **103** (1960), 269–291.
57. ______, *A characterization of linearly continuous functions whose partial derivatives are measures*, Acta Math. **117** (1967), 165–190.

58. ______, *Everywhere convergence of Fourier series*, Indiana Univ. Math. J. **20** (1970/1971), 107–112.
59. ______, *Everywhere differentiable functions and the density topology*, Proc. Amer. Math. Soc. **51** (1975), 250.
60. C. Goffman and F-C Liu, *Derivative measures*, Proc. Amer. Math. Soc. **78** (1980), 218–220.
61. C. Goffman, F. C. Liu, and D. Waterman, *A remark on the spaces* $V^p_{\Lambda,\alpha}$, Proc. Amer. Math. Soc. **82** (1981), 366–388.
62. ______, *A differentiable function for which localization for double Fourier series fails*, Real Anal. Exchange **8** (1982–1983), 222–226.
63. C. Goffman, C. J. Neugebauer, and T. Nishiura, *Density topology and approximate continuity*, Duke Math. J. **28** (1961), 497–505.
64. C. Goffman and C.J. Neugebauer, *On approximate derivatives*, Proc. Amer. Math. Soc. **11** (1960), 962–966.
65. C. Goffman and G. Pedrick, *A proof of the homeomorphism of Lebesgue-Stieltjes measure with Lebesgue measure*, Proc. Amer. Math. Soc. **52** (1975), 196–198, MR 51, #13170.
66. C. Goffman and D. Waterman, *On upper and lower limits in measure*, Fund. Math. **48** (1959/60), 127–133.
67. ______, *Functions whose Fourier series converge for every change of variable*, Proc. Amer. Math. Soc. **19** (1968), 80–86.
68. ______, *Some aspects of Fourier series*, Amer. Math. Monthly **77** (1970), 119–133.
69. ______, *A characterization of the class of functions whose Fourier series converge for every change of variable*, J. London Math. Soc. **10** (1975), 69–74.
70. ______, *On localization for double Fourier series*, Proc. Nat. Acad. Sci. U.S.A. **75** (1978), 590–591.
71. ______, *The localization principle for double Fourier series*, Studia Math. **69** (1980), 41–57.
72. C. Goffman and R. E. Zink, *On upper and lower limits in measure*, Fund. Math. **48** (1959/60), 105–111.
73. W. J. Gorman III, *The homeomorphic transformation of c-sets into d-sets*, Proc. Amer. Math. Soc. **17** (1966), 825–830.
74. ______, *Lebesgue equivalence to functions of the first Baire class*, Proc. Amer. Math. Soc. **17** (1966), 831–834.
75. J. Hadamard, *Essai sur l'étude des fonctions données par leur développment de Taylor*, J. de Math. **8** (1892), 101–186, also: Oeuvres de Jacques Hadamard, vol.1, 7-92, Editions C.N.R.S., Paris, 1968.
76. O. Haupt and Ch. Pauc, *La topologie de Denjoy envisagée comme vraie topologie*, C. R. Acad. Sci. Paris **234** (1952), 390–392.
77. R. E. Hughs, *Functions of BVC type*, Proc. Amer. Math. Soc. **12** (1961), 698–701.
78. W. Hurewicz and H. Wallman, *Dimension theory*, Princeton University Press, Princeton, 1948.
79. S. Igari, *On the localization property of multiple Fourier series*, J. Approx. Theory **1** (1968), 182–188.
80. J. R. Isbell, *Uniform space*, Math. Surveys Monogr., Number 12, American Mathematical Society, Providence, 1964.
81. W. Jurkat and D. Waterman, *Conjugate functions and the Bohr-Pál theorem*, Abstracts, Internat. Cong. Mathematicians, Berkeley, 1986.
82. ______, *Conjugate functions and the Bohr-Pál theorem*, Complex Variables Theory Appl. **12** (1989), 67–70.
83. J.-P. Kahane and Y. Katznelson, *Homéomorphismes du cercle et séries de Fourier absolument convergente*, C. R. Acad. Sci. Paris Sér. I. Math. **292** (1981), 271–273.
84. ______, *Quatre leçons sur les homéomorphismes du cercle et les séries de Fourier*, Topics in Modern Harmonic Analysis, Proc. Sem. Torino-Milano, May-June 1982, vol. II (Rome), 1983, pp. 955–990.
85. ______, *Séries de Fourier des fonctions bornées*, Studies in Pure Math. in Memory of Pál Turán (Budapest), 1983, (Publ. Orsay 1978), pp. 395–410.
86. Y. Katznelson, *An introduction to harmonic analysis*, Wiley, New York, 1968.
87. Y. Katznelson and K. Stromberg, *Everywhere differentiable, nowhere monotone functions*, Amer. Math. Monthly **81** (1974), 349–354.
88. J. L. Kelley, *General topology*, D. van Nostrand, New York, 1955.

89. S. Kempisty, *Sur les fonctions quasicontinues*, Fund. Math. **19** (1932), 184–197.
90. K. Krickeberg, *Distributionen, Funktionen beschränkter variation und Lebesguescher Inhalt nichtparametrischer Flächen*, Ann. Mat. Pura Appl. (4) **44** (1957), 105–133.
91. M. Laczkovich and D. Preiss, *α-Variation and transformation into C^n functions*, Indiana Univ. Math. J. **34** (1985), 405–424.
92. H. Lebesgue, *Intégrale, longueur, aire*, Ann. Mat. Pura Appl. **7** (1902), 231–359.
93. R. Lesńiewicz and W. Orlicz, *On generalized variations (II)*, Studia Math. **45** (1973), 71–109.
94. R. Levy, *A totally ordered Baire space for which Blumberg's theorem fails*, Proc. Amer. Math. Soc. **41** (1973), 304.
95. J. Lukeš, J. Malý, and L. Zajíček, *Fine topology methods in real analysis and potential theory*, Lecture Notes in Math., vol. 1189, Springer-Verlag, Berlin, 1986.
96. I. Maximoff, *On density points and approximately continuous functions*, Tôhoku Math. J. **47** (1940), 237–250.
97. ———, *Sur la transformation continue de quelque fonctions en dérivées exactes*, Bull. Soc. Phys. Math. Kazan **12** (1940), no. 3, 57–81.
98. K. Morita and J. Nagata, *Topics in general topology*, North-Holland Mathematical Library, vol. 41, North Holland, Amsterdam, 1989.
99. J. Musielak and W. Orlicz, *On generalized variations (I)*, Studia Math. **18** (1959), 11–41.
100. C. J. Neugebauer, *Blumberg sets and quasi-continuity*, Math. Z. **79** (1962), 451–455.
101. ———, *Darboux function of Baire class one and derivatives*, Proc. Amer. Math. Soc. **13** (1962), 838–843.
102. A. M. Olevskii, *Change of variable and absolute convergence of Fourier series*, Dokl. Akad. Nauk SSSR **256** (1981), 284–288 (Russian), translation: Soviet Math. Dokl. **23** (1981), 76-79.
103. ———, *Modifications of functions and Fourier series*, Uspekhi Mat. Nauk **40** (1985), 157–193 (Russian), translation: Russian Math. Surveys **40** (1985), 181-224.
104. G. T. Oniani, *Topological characterization of a set of continuous functions whose conjugate functions are continuous and have bounded variation*, Trudy Tbiliss. Nat. Inst. Razmadze Akad. Nauk Gruzin. SSR **86** (1987), 110–113 (Russian), translation: Integral Operators and Boundary Properties of Functions. Fourier Series, 155–161, Nova Science, New York, 1992.
105. K. I. Oskolkov, *Generalized variation, the Banach indicatrix and the uniform convergence of Fourier series*, Mat. Zametki **12** (1972), 313–324 (Russian), translation: Math. Notes **12** (1972), 619–625.
106. J. C. Oxtoby, *Measure and category*, second ed., Grad. Texts in Math., Springer-Verlag, New York, 1980.
107. J. C. Oxtoby and S. M. Ulam, *Measure-preserving homeomorphisms and metrical transitivity*, Ann. of Math. (2) **42** (1941), 874–920.
108. J. Pál, *Sur les transformations de fonctions qui font converger leurs séries de Fourier*, C. R. Acad. Sci. Paris **158** (1914), 101–103.
109. P. Pierce and D. Waterman, *Regulated functions whose Fourier series converge for every change of variable*, (to appear).
110. D. Preiss, *Maximoff's theorem*, Real Anal. Exchange **5** (1979–80), 92–104.
111. F. Prus-Wiśniowski, *General properties of functions of bounded λ-variation*, Ph.D. thesis, Syracuse University, 1995.
112. C. A. Rogers, *Functions of the first Baire class*, J. London Math. Soc. (2) **37** (1988), 535–544.
113. A. A. Saakyan, *Integral moduli of smoothness and Fourier coefficients of compositions of functions*, Mat. Sb. **110** (1979), 597–608 (Russian), translation: Math. USSR-Sb. **38** (1981), 549–561.
114. ———, *On properties of Fourier coefficients of compositions of functions*, Dokl. Akad. Nauk SSSR **248** (1979), 302–305 (Russian), translation: Soviet Math. Dokl. **20** (1979), 1018–1022.
115. ———, *The Bohr theorem for multiple trigonometric series*, Mat. Zametki **46** (1989), 94–103 (Russian), translation: Math. Notes **46** (1989), 639–646.
116. S. Saks and W. Sierpiński, *Sur une propriété générale de fonctions*, Fund. Math. **11** (1928), 105–112.
117. R. Salem, *Sur un test générale pour la convergence uniforme des séries de Fourier*, C. R. Acad. Sci. Paris **207** (1938), 662–664.
118. ———, *Essais sur les séries trigonométriques*, Actualités Sci. Ind. No. 862, Hermann, Paris, 1940, Chapter VI.
119. ———, *On a theorem of Bohr and Pál*, Bull. Amer. Math. Soc. **50** (1944), 579–580.

120. ———, *Oeuvres Mathématiques*, Hermann, Paris, 1967.
121. M. Schramm and D. Waterman, *On the magnitude of the Fourier coefficients*, Proc. Amer. Math. Soc. **85** (1982), 407–410.
122. J. Serrin, *On the differentiability of functions of several variables*, Arch. Rational Mech. Anal. **7** (1961), 359–372.
123. W. Sierpiński and E. Szpilrajn (= Marczewski), *Remarque sur le problème de la measure*, Fund. Math. **26** (1936), 256–261.
124. F. D. Tall, *The density topology*, Pacific J. Math. **62** (1976), 275–284.
125. S. J. Taylor and C. Tricot, *Packing measure, and its evaluation for a Brownian path*, Trans. Amer. Math. Soc. **288** (1985), 679–699.
126. B. S. Thomson, *Real functions*, Lecture Notes in Math., vol. 1170, Springer-Verlag, Berlin, 1985.
127. L. Tonelli, *Serie trigonometriche*, Zanichelli, Bologna, 1928, Cap.9.
128. J. von Neumann, *Collected works, Vol II: Operators, ergodic theory and almost periodic functions in a group*, Pergamon, New York, 1961, p. 558.
129. S. L. Wang, *Properties of the functions of* Λ*-bounded variation*, Sci. Sinica Ser. A. **25** (1982), 149–160.
130. D. Waterman, *On convergence of Fourier series of functions of generalized variation*, Studia Math. **44** (1972), 107–117.
131. ———, *On functions of bounded deviation*, Acta Sci. Math. (Szeged) **36** (1974), 259–263.
132. ———, *On* Λ*-bounded variation*, Studia Math. **57** (1976), 33–45.
133. ———, *Functions whose Fourier series converge uniformly for every change of variable. II*, Indiana J. Math. **25** (1983), 257–264.
134. ———, *On the preservation of the order of magnitude of Fourier coefficients under every change of variable*, Analysis **6** (1986), 255–264.
135. ———, *On functions of bounded deviation II*, J. Math. Anal. Appl. **131** (1988), 113–117.
136. ———, *A generalization of the Salem test*, Proc. Amer. Math. Soc. **105** (1989), 129–133.
137. ———, *Uniform estimates of a trigonometric integral*, Colloq. Math. **60/61** (1990), 681–685.
138. ———, *An integral mean value theorem for regulated functions*, Real Anal. Exchange **21** (1995-96), 817–820.
139. C. Weil, *On nowhere monotone functions*, Proc. Amer. Math. Soc. **56** (1976), 388–389.
140. H. E. White, Jr., *Topological spaces in which Blumberg's theorem holds*, Proc. Amer. Math. Soc. **44** (1974), 454–462.
141. ———, *Topological spaces that are* α*-favorable for a player with perfect information*, TOPO 72–general topology and its applications (Proc. Second Pittsburgh Internat. Conf., Carnegie-Mellon Univ. and Univ. Pittsburgh, Pittsburgh, PA, 1972; dedicated to the memory of Johannes H. de Groot), Lecture Notes in Math., vol. 378, Springer-Verlag, Berlin, 1974, pp. 551–556.
142. ———, *An example involving Baire spaces*, Proc. Amer. Math. Soc. **48** (1975), 228–230.
143. L. C. Young, *Sur une généralisation de la notion de variation de puissance p-ième bornée au sense de N. Wiener, et sur la convergence des séries de Fourier*, C. R. Acad. Sci. Paris **204** (1937), 470–472.
144. J. W. T. Youngs, *Curves and surfaces*, Amer. Math. Monthly **51** (1944), 1–11.
145. Z. Zahorski, *Über die Menge der Punkte in welchen die Ableitung unendlich ist*, Tôhoku Math. J. **48** (1941), 321–330.
146. ———, *Sur la première dérivée*, Trans. Amer. Math. Soc. **69** (1950), 1–54.
147. A. Zygmund, *Trigonometric series*, second ed., Camridge University Press, Cambridge, 1977, (Combined volumes I and II).

Index

Selected Titles in This Series

(Continued from the front of this publication)

19 **Gregory V. Chudnovsky,** Contributions to the theory of transcendental numbers, 1984
18 **Frank B. Knight,** Essentials of Brownian motion and diffusion, 1981
17 **Le Baron O. Ferguson,** Approximation by polynomials with integral coefficients, 1980
16 **O. Timothy O'Meara,** Symplectic groups, 1978
15 **J. Diestel and J. J. Uhl, Jr.,** Vector measures, 1977
14 **V. Guillemin and S. Sternberg,** Geometric asymptotics, 1977
13 **C. Pearcy, Editor,** Topics in operator theory, 1974
12 **J. R. Isbell,** Uniform spaces, 1964
11 **J. Cronin,** Fixed points and topological degree in nonlinear analysis, 1964
10 **R. Ayoub,** An introduction to the analytic theory of numbers, 1963
9 **Arthur Sard,** Linear approximation, 1963
8 **J. Lehner,** Discontinuous groups and automorphic functions, 1964
7.2 **A. H. Clifford and G. B. Preston,** The algebraic theory of semigroups, Volume II, 1961
7.1 **A. H. Clifford and G. B. Preston,** The algebraic theory of semigroups, Volume I, 1961
6 **C. C. Chevalley,** Introduction to the theory of algebraic functions of one variable, 1951
5 **S. Bergman,** The kernel function and conformal mapping, 1950
4 **O. F. G. Schilling,** The theory of valuations, 1950
3 **M. Marden,** Geometry of polynomials, 1949
2 **N. Jacobson,** The theory of rings, 1943
1 **J. A. Shohat and J. D. Tamarkin,** The problem of moments, 1943